Systems Thinking in Practice

Applications of the Event Analysis of Systemic Teamwork Method

Transportation Human Factors: Aerospace, Aviation, Maritime, Rail, and Road Series

Series Editor

Professor Neville A Stanton

University of Southampton, UK

Automobile Automation
Distributed Cognition on the Road
Victoria A. Banks, Neville A. Stanton

Eco-Driving
From Strategies to Interfaces
Rich C. McIlroy, Neville A. Stanton

Systems Thinking in Practice
Applications of the Event Analysis of Systemic Teamwork Method
Neville A. Stanton, Paul M. Salmon, Guy H. Walker

For more information about this series, please visit: https://www.crcpress.com/Transportation-Human-Factors/book-series/CRCTRNHUMFACAER

Systems Thinking in Practice

Applications of the Event Analysis of Systemic Teamwork Method

Neville A. Stanton
Paul M. Salmon
Guy H. Walker

CRC Press
Taylor & Francis Group
Boca Raton London New York

CRC Press is an imprint of the
Taylor & Francis Group, an **informa** business

CRC Press
Taylor & Francis Group
6000 Broken Sound Parkway NW, Suite 300
Boca Raton, FL 33487-2742

Printed on acid-free paper

International Standard Book Number-13: 978-1-138-09787-2 (Hardback)
International Standard Book Number-13: 978-1-315-10468-3 (eBook)

Library of Congress Cataloging-in-Publication Data

Names: Stanton, Neville A. (Neville Anthony), 1960- author. | Salmon, Paul M., author. | Walker, Guy, author.
Title: Systems thinking in practice : applications of the Event Analysis of Systemic Teamwork method / Neville A. Stanton, Paul Salmon, and Guy Walker.
Description: First edition. | Boca Raton, FL : CRC Press/Taylor & Francis Group, 2018. | Series: Transportation human factors: Aerospace, aviation, maritime, rail, and road series | Includes bibliographical references and index.
Identifiers: LCCN 2018026593| ISBN 9781138097872 (hardback : alk. paper) | ISBN 9781315104683 (ebook)
Subjects: LCSH: Systems engineering. | Teams in the workplace.
Classification: LCC TA168 .S68 2018 | DDC 658.4/022011--dc23
LC record available at https://lccn.loc.gov/2018026593

Visit the Taylor & Francis Web site at
http://www.taylorandfrancis.com

and the CRC Press Web site at
http://www.crcpress.com

Contents

SECTION I Overview of EAST

SECTION II Applications of EAST

Preface

This book has arisen from the desire to share our insights into the practical application of the Event Analysis of Systemic Teamwork (EAST) framework. Over the past decade, we have applied EAST in many domains and have been impressed by the insights that the products of the analysis afford. EAST offers a systemic and systematic approach to the analysis, design and evaluation of sociotechnical systems. As a formative method, we have used it to design new concepts of operations, new teams and new ways of working. As a summative method, we have used it to gain new insights into existing systems and current ways of working.

The EAST method was initially developed for a programme of research into command and control funded by the UK Ministry of Defence called the Human Factors Integration Defence Technology Centre. This research began in 2003 and ended in 2012.

The total funding was for £30 million over 10 years, and one of our projects was to provide Human Factors advice for new networked architectures for command and control. The idea was that potentially all military systems across all of the forces (from the infantry soldier to the moving platforms [on land, in the air and at sea] to the joint operations headquarters) could be connected like the World Wide Web is for civilian activities.

This kind of connectivity meant we had to look at command and control in a completely new way. Rather than develop completely new methods, we developed EAST by integrating existing methods, as is explained in this book. Prof Chris Baber and I decided to use network methods to investigate these command and control networks. This began with investigations into both civilian and military examples of command and control, some of which are presented in this book. The approach has developed into the EAST method as we know it today. Over the years, we have developed and refined the approach. What we particularly like about the network methods is that they are both scalable and systemic in nature. We can analyse the networks both qualitatively and quantitatively (using network statistics, as demonstrated in this book).

We imagine people will approach this book with different purposes. For those new to the EAST approach, we advise reading Chapter 1 and then finding a chapter with your domain of interest. For those familiar with EAST, go straight to the chapters that are of interest. For researchers and those keen on extending the EAST approach, the final section of chapters on developments and future directions will be of most interest. Each of the chapters is intended to be read as a stand-alone article, so there is some inevitable repetition on the overview of EAST (although different emphases and approaches to the method are taken). Experts in EAST can skip over those sections. For those who use EAST and find it useful in their work, we say, welcome to the tribe of happy EASTers. That EASTer tribe already comprises the authors in this book, and we are aware of other groups using EAST around the world. All of the chapters were led by one or more of the lead authors of the book with the exception of Chapters 9 and 10, which were both led by Victoria Banks. We are grateful for the contributions and insights from our co-authors and the progress that has been made with the method since its original conception.

The EAST development journey is not at the end yet. There have been recent developments that have extending the use of EAST. EAST, as a systems method, has fared well, as systems approaches are very much in favour in the Ergonomics and Human Factors world. We have had some successes linking EAST to other systems methods, such as Systems Theoretic Accident Model and Process |(STAMP) and Cognitive Work Analysis (CWA). We have also used EAST in formative ways to predict system network resilience. The journey does not end here, however, and we are sure that we and others will continue to develop and extend the approach and apply it to even more domains of application.

Neville A. Stanton
Professor of Human Factors Engineering
University of Southampton

RECOMMENDED FURTHER READING ON EVENT ANALYSIS OF SYSTEMIC TEAMWORK

Salmon, P. M., Read, G. J. M., Goode, N., Grant, E., Dallat, C., Carden, T., Naweed, A., Walker, G. H. and Stanton, N. A. (2018). STAMP goes EAST: Integrating ergonomics methods for the analysis of railway level crossing safety management. *Safety Science*: doi.org/10.1016/j.ssci.2018.02.014.

Stanton, N. A. and Roberts, A. P. J. (2018). Examining task, social and information networks in submarine command and control. *IEEE Transactions on Human–Machine Systems*, doi: 10.1109/THMS.2017.2720659. (in press).

Banks, V. A. and Stanton, N. A. (2018). Analysis of driver roles: Modelling the changing role of the driver in automated driving systems using EAST. *Theoretical Issues in Ergonomics Science*, (in press).

Roberts, A. P. J, Stanton, N. A. and Fay, D. T. (2018). Go Deeper, Go Deeper: Understanding submarine command and control during the completion of dived tracking operations. *Applied Ergonomics*, 69, 162–175.

Banks, V. A., Stanton, N. A., Burnett, G. and Hermawati, S. (2018). Distributed Cognition on the road: Using EAST to explore future road transportation systems. *Applied Ergonomics*, 68, 258–266.

Roberts, A. P. J, Stanton, N. A. and Fay, D. T. (2017). Land Ahoy! – Understanding submarine command and control during the completion of inshore operations. *Human Factors*, 59(8), 1263–1288.

Stanton, N. A., Roberts, A. P. J. and Fay, D. T. (2017). Up periscope: Understanding submarine command and control teamwork during a simulated return to periscope depth. *Cognition, Technology and Work*, 19(2–3), 399–417.

Stanton, N. A. and Harvey, C. (2017). Beyond human error taxonomies in assessment of risk in sociotechnical systems: A new paradigm with the EAST 'broken-links' approach. *Ergonomics*, 60(2), 221–233.

Stanton, N. A., Salmon, P. M., Walker, G. H., Salas, E. and Hancock, P. A. (2017). State-of-science: Situation awareness in individuals, teams and systems. *Ergonomics*, 60(4), 449–466.

Plant, K. L and Stanton, N. A. (2016). Distributed cognition in Search and Rescue: Loosely coupled tasks and tightly coupled roles. *Ergonomics*, 59(10), 1353–1376.

Sorensen, L. J. and Stanton, N. A. (2016). Keeping it together: The role of transactional situation awareness in team performance. *International Journal of Industrial Ergonomics*, 53, 267–273.

Salmon, P. M., Lenné, M. G., Walker, G. H., Stanton, N. A. and Filtness, A. (2014). Using the Event Analysis of Systemic Teamwork (EAST) to explore conflicts between different road user groups when making right hand turns at urban intersections, *Ergonomics*, 57(11), 1628–1642.

Salmon, P., Lenné, M., Walker, G. H., Stanton, N. A. and Filtness, A. (2014). Exploring schema-driven differences in situation awareness across road users: An on-road study of driver, cyclist and motorcyclist situation awareness. *Ergonomics*, 57(2), 191–209.

Stanton, N. A. (2014). Representing distributed cognition in complex systems: How a submarine returns to periscope depth. *Ergonomics*, 57(3), 403–418.

Walker, G. H., Stanton, N. A., Baber, C., Wells, L., Gibson, H., Salmon, P. M. and Jenkins, D. P. (2010). From ethnography to the EAST method: A tractable approach for representing distributed cognition in air traffic control. *Ergonomics*, 53(2), 184–197.

Walker, G. H., Stanton, N. A., Salmon, P. M., Jenkins, D., Stewart, R. and Wells, L. (2009). Using an integrated methods approach to analyse the emergent properties of military command and control. *Applied Ergonomics*, 40(4) 636–647.

Stanton, N. A., Salmon, P. M., Walker, G. H., and Jenkins, D. P. (2009). Genotype and phenotype schema and their role in distributed situation awareness in collaborative systems. *Theoretical Issues in Ergonomics Science*, 10(1), 43–68.

Stanton, N. A., Salmon, P. M., Walker, G. H., and Jenkins, D. P. (2009). Genotype and phenotype schemata as models of situation awareness in dynamic command and control teams. *International Journal of Industrial Ergonomics*, 39(3), 480–489.

Salmon, P. M., Stanton, N. A., Walker, G. H., Jenkins, D. P., Baber, C., and McMaster, R. (2008). Representing situation awareness in collaborative systems: A case study in the energy distribution domain. *Ergonomics*, 51(3), 367–384.

Stewart, R., Stanton, N. A., Harris, D., Baber, C., Salmon, P., Mock, M., Tatlock, K., Wells, L. and Kay, A. (2008). Distributed situation awareness in an Airborne Warning and Control System: Application of novel ergonomics methodology. *Cognition, Technology and Work*, 10(3), 221–229.

Salmon, P. M., Stanton, N. A., Walker, G. H., Jenkins, D. P., Baber, C., and McMaster, R. (2008). Representing situation awareness in collaborative systems: A case study in the energy distribution domain. *Ergonomics*, 51(3), 367–384.

Walker, G. H., Gibson, H., Stanton, N. A., Baber, C., Salmon, P. and Green, D. (2006). Event Analysis of Systemic Teamwork (EAST): A novel integration of ergonomics methods to analyse C4i activity. *Ergonomics*, 49(12–13), 1345–1369.

Houghton, R. J., Baber, C., McMaster, R., Stanton, N. A., Salmon, P., Stewart, R. and Walker, G. (2006). Command and control in emergency services operations: A social network analysis. *Ergonomics*, 49(12–13), 1204–1225.

Stanton, N. A., Stewart, R., Harris, D., Houghton, R. J., Baber, C., McMaster, R., Salmon, P., Hoyle. G., Walker, G., Young. M. S. et al. (2006). Distributed situation awareness in dynamic systems: Theoretical development and application of an ergonomics methodology. *Ergonomics*, 49(12–13), 1288–1311.

THERE ARE ALSO THESE BOOKS:

Stanton, N. A., Salmon, P. M., Walker, G. H., Baber, C. and Jenkins, D. (2005). *Human Factors Methods: A Practical Guide for Engineering and Design* (first edition). Ashgate: Aldershot.

Stanton, N. A., Baber, C. and Harris, D. (2008). *Modelling Command and Control: Event Analysis of Systemic Teamwork*. Ashgate: Aldershot.

Stanton, N. A., Salmon, P. M., Rafferty, L. A., Walker, G. H., Baber, C. and Jenkins, D. (2013). *Human Factors Methods: A Practical Guide for Engineering and Design* (second edition). Ashgate: Aldershot.

Authors

Professor Neville Stanton, PhD, DSc, is a chartered psychologist, chartered ergonomist and chartered engineer. He holds the Chair in Human Factors Engineering in the Faculty of Engineering and the Environment at the University of Southampton in the United Kingdom. He earned degrees in Psychology, Applied Psychology and Human Factors and has worked at the Universities of Aston, Brunel, Cornell and MIT. His research interests include modelling, predicting, analysing and evaluating human performance in systems as well as designing the interfaces and interaction between humans and technology. Professor Stanton has worked on the design of automobiles, aircraft, ships and control rooms over the past 30 years on a variety of automation projects. He has published 40 books and over 300 journal papers on Ergonomics and Human Factors. In 1998, he was awarded the Institution of Electrical Engineers Divisional Premium Award for research into System Safety. The Institute of Ergonomics and Human Factors awarded him the Otto Edholm Medal in 2001, the President's Medal in 2008 and the Sir Frederic Bartlett Medal in 2012 for his contributions to basic and applied ergonomics research. The Royal Aeronautical Society awarded him and his colleagues the Hodgson Prize and Bronze Medal in 2006 for research on design-induced, flight-deck error published in The Aeronautical Journal. The University of Southampton awarded him a Doctor of Science in 2014 for his sustained contribution to the development and validation of Human Factors methods.

Professor Paul Salmon holds a Chair in Human Factors and is creator and director of the Centre for Human Factors and Sociotechnical Systems at the University of the Sunshine Coast. He currently holds a prestigious Australian Research Council Future Fellowship and has almost 15 years' experience in applied Human Factors research in a number of areas, including defence, transportation safety, sports and outdoor recreation and disaster management. Professor Salmon currently leads major research programmes in the areas of road and rail safety, identity theft and cybersecurity and led outdoor recreation accidents. He has co-authored 14 books, over 180 peer-reviewed journal articles and numerous conference articles and book chapters. He has received various accolades for his contributions to research and practice, including the Australian Human Factors and Ergonomics Societies 2016 Cumming Memorial medal, the UK Ergonomics Society's Presidents Medal, the Royal Aeronautical Society's Hodgson Prize for best research and paper and the University of the Sunshine Coast's Vice Chancellor and President's Medal for Research Excellence. Professor Salmon's current research interests relate to extending Human Factors and Sociotechnical Systems theory and methods to support the optimisation of systems in many areas. Specific areas of focus include accident prediction and analysis, systems thinking in transportation safety, the development of systemic accident countermeasures, human factors in elite sports and cybersecurity.

Professor Guy Walker works within the Institute for Infrastructure and Environment at Heriot-Watt University in Edinburgh. He lectures on Human Factors and is the author/co-author of over 100 peer-reviewed journal articles and 13 books. He and his co-authors have been awarded the Institute for Ergonomics and Human Factors President's Medal for the practical application of Ergonomics theory and the Peter Vulcan Prize for best research paper by the 2013 Australasian Road Safety Research Conference. In 2011, he also won Heriot-Watt University's Graduate's Prize for inspirational teaching. Prof Walker earned a BSc Honours degree in Psychology from the University of Southampton and a PhD in Human Factors from Brunel University. His research interests are wide ranging, spanning driver behaviour and the role of feedback in vehicles, using Human Factors methods to analyse black-box data recordings and the application of sociotechnical systems theory to the design and evaluation of civil engineering systems through to safety, risk and reliability. His research has featured in the popular media, from national newspapers, TV and radio through to an appearance on the Discovery Channel.

Board Members and Affiliations

Richard McMaster
Senior Human Factors Consultant
Abbott Risk Consulting (ARC) Ltd
London, UK

Anjum Naweed
Appleton Institute for Behavioural Science
Central Queensland University
Queensland, Australia

Gemma J. M. Read
Centre for Human Factors and Sociotechnical Systems
University of the Sunshine Coast
Queensland, Australia

Paul M. Salmon
Centre for Human Factors and Sociotechnical Systems
University of the Sunshine Coast
Queensland, Australia

Neville A. Stanton
Human Factors Engineering
Transportation Research Group
University of Southampton
Southampton, UK

Alison Starr
The National Composites Centre
Bristol and Bath Science Park
Bristol, UK

Nicholas J. Stevens
Centre for Human Factors and Sociotechnical Systems
University of the Sunshine Coast
Queensland, Australia

Rebecca Stewart
Lockheed Martin
Langstone Technology Park
Havant, UK

Guy H. Walker
School of the Built Environment
Heriot-Watt University
Edinburgh, UK

Linda Wells
BAE Systems
Somerset, UK

Section I

Overview of EAST

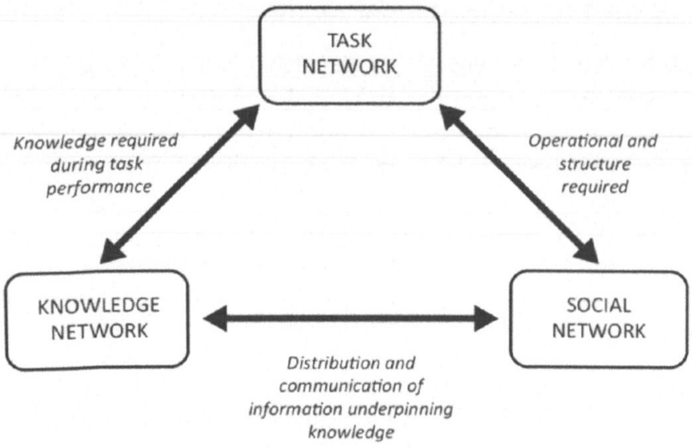

FIGURE 1.1 Network of networks approach.

DOMAIN OF APPLICATION

EAST is a generic approach that was developed originally for the analysis of teamwork in sociotechnical systems, but it has since been used to provide analyses at the micro (Salmon et al. 2014a), meso (Stanton 2014) and macro (Stanton and Harvey 2017) levels of sociotechnical systems. As such, it can be used in any domain in which social and technical elements are working together in pursuit of a common goal. The case study presented in this chapter, used to demonstrate EAST, is based on an application within the area of land use planning and in urban design (Stevens et al., 2018).

APPLICATION IN LAND USE PLANNING AND URBAN DESIGN (LUP & UD)

LUP & UD is most often a product of multi-disciplinary approaches to complex environments and sociotechnical systems (Stevens et al. 2018). Our projects and processes require different resources from a variety of participants over a range of time frames. Whilst we may not always define these approaches as teamwork, the EAST approach has the potential to offer critical insights into more effective and efficient cooperative processes and project performance.

PROCEDURE AND ADVICE

Step 1: Define Analysis Aims

First, the aims of the analysis should be clearly defined so that appropriate scenarios are used and relevant data are collected. In addition, not all components of the EAST framework may be required, so it is important to clearly define the aims at this point to ensure that the appropriate EAST methods are applied.

Step 2: Define the Task/System Under Analysis

Next, the task (or tasks) or scenario (or scenarios) under analysis should be clearly defined. This is dependent upon the aims of the analysis and may include a range of tasks or one task in particular. It is normally standard practice to develop a Hierarchical Task Analysis (HTA) (see Stanton 2006) for the task under analysis if sufficient data and subject matter expert (SME) access are available. This is useful later on in the analysis and is also enlightening, allowing the analyst to gain an understanding of the task before the observation and analysis begins.

Step 3: Data Collection

Once the aims of the analysis are clearly defined, the next step involves collecting targeted data about the system and its behaviour. The specific data collected are dependent on the analysis aims and the resources available; however, data collection for EAST typically involves observations, concurrent verbal protocols, structured or semi-structured interviews (e.g. the Critical Decision Method [CDM]; Klein et al., 1989), walkthrough analysis and documentation review (e.g. incident reports, standard operating procedures).

The observation step is often the most important part of the EAST procedure. Typically, a number of analysts are used to observe the system or scenario under analysis. All activities involved in the scenario under analysis should be recorded along an incident timeline, including a description of the activity undertaken, the agents involved, any communications made between agents and the technology involved. Additional notes should be made where required, including the purpose of the activity observed; any tools, documents or instructions used to support activity; the outcomes of activities; any errors made; and also any information that the agent involved feels is relevant. In addition, it is useful to video record the task and record verbal transcripts of all communications, if possible.

Once the task under analysis is complete, each 'key' agent (e.g. scenario commander, agents performing critical tasks) involved should be subjected to a CDM interview. This involves dividing the scenario into key incident phases and then interviewing the actor involved in each phase using a set of pre-defined CDM probes (e.g. O'Hare et al. 2000; see also Chapter 4 for more information on the CDM).

Step 4: Transcribe Data

Once all of the data are collected, it should be transcribed in order to make it compatible with the EAST analysis phase. An event transcript should then be constructed. This should describe the scenario over a timeline, including descriptions of activity, the actors involved, any communications made and the technology used. In order to ensure the validity of the data, the scenario transcript should be reviewed by one of the SMEs involved.

Step 5: Construct Task Network

The first analysis step involves constructing a task network. Prior to this, the initial HTA should be reviewed and refined based on the data collected during step 3.

The data transcription process allows the analyst to gain a deeper and more accurate understanding of the scenario under investigation. It also allows any discrepancies between the initial HTA scenario description and the actual activity observed to be resolved. Typically, activities in complex sociotechnical systems do not run entirely according to protocol, and certain tasks may have been performed during the scenario that were not described in the initial HTA description. The analyst should compare the scenario transcript to the initial HTA and add any changes as required.

Constructing the task network involves identifying high-level tasks and the relationships between them and creating a network to represent this. Some general rules around the construction of EAST networks are presented in Table 1.1.

TABLE 1.1
Analysis Rules Regarding the Relationships Between Nodes Within EAST Networks

Network	Nodes	Relationships	Examples
Task network	Represent high-level tasks that are required during the scenario under analysis. High-level tasks are typically extracted from the sub-ordinate goals level of the HTA	Represent instances where the conduct of one high-level grouping of tasks (i.e. task network node) influences, is undertaken in combination with or is dependent on another group of tasks	The nodes 'Identify legal constraints' and 'Identify site and zoning' are linked because the zoning cannot be established until the site has been legally identified
Social network	Represent human, technological, or organisational agents who undertake one or more of the tasks involved in the scenario under analysis (as identified in the HTA and task network)	Represent instances where agents within the social network interact with one another during the scenario under analysis	The nodes 'Urban planner' and 'community' are linked as the planner needs to communicate with and understand the local community if an informed analysis of the site is to be established
Information network	Represent grouped categories of information that is required by agents when undertaking scenario under analysis (as identified in the task and social network)	Represent instances where information influences other information or is used in combination with other information in the network during the scenario under analysis	The nodes 'views' and 'topography' are linked as the establishment of views requires appropriate topography

Step 6: Conduct Social Network Analysis

A Social Network Analysis (SNA) (Driskell and Mullen 2004) is used to analyse the relationships (e.g. communications, transactions) between the agents involved in the scenario under analysis. This involves first creating a social network matrix showing the relationships between agents followed by a social network diagram which provides a visual representation of the social network. Typically, the direction (i.e. from actor A to actor B) frequency, type and content of associations are recorded. It is normally useful to conduct a series of SNAs representing different phases of the task under analysis (using the task phases defined during the CDM part of the analysis).

Step 7: Construct Information Networks

The final step of the EAST analysis involves constructing information networks (see Chapter 7 for a full description) for each scenario phase identified during the CDM interviews. Following construction, information usage should be defined for each actor involved via shading of the information elements within the propositional networks.

Step 8: Construct Composite Networks

Composite networks are used to explore the relationships between tasks, agents and information (Stanton 2014). As such, composite networks are constructed by combining the different networks. For example, a *task by agents* network can be constructed by combining the task and social network to show which tasks are undertaken by which agents. This involves assigning a colour to the different agents within the social network and shading each node within the task network to show which agent performs that particular task. Useful composite networks to construct include

- Task by agents network (combined task and social network);
- Information by agents network (combined information and social network);
- Task and associated information network (combined task and information network);
- Information by agents and tasks network (combined task, social and information network).

Once the EAST networks are complete, it is pertinent to validate the outputs using appropriate SMEs and recordings of the scenario under analysis. Any problems identified should be corrected at this point.

Step 9: Analyse Networks

An important component of EAST analyses involved using network metrics to analyse the task, social and information networks. This enables analysis of the structure of the networks and identification of key nodes (e.g. tasks, agents, information) within the networks. Three popular network analysis metrics have previously been used to interrogate EAST networks:

1. *Network Density (overall network)*: Network density represents the level of interconnectivity of the network in terms of relations between nodes. Density is expressed as a value between 0 and 1, with 0 representing a network with no connections between nodes and 1 representing a network in which every node is connected to every other node (Walker et al. 2011). Higher density values are indicative of a well-connected network in which tasks, agents, information and controls are tightly coupled.
2. *Sociometric Status (individual nodes)*: Sociometric status provides a measure of how 'busy' a node is relative to the total number of nodes within the network under analysis (Houghton et al. 2006). In the present analysis, nodes with sociometric status values greater than the mean sociometric status value plus one standard deviation are taken to be the 'key' (i.e. most connected) nodes within each network. These nodes represent either key tasks, agents, pieces of information or controls. For example, in the case of the social network, the node with the highest sociometric status is the agent that is the most interrelated with other agents based on communication.
3. *Centrality (individual nodes)*: Centrality is used to examine the standing of a node within a network based on its geodesic distance from all other nodes in the network (Houghton et al. 2006). Central nodes represent those that are closer to the other nodes in the network as, for example, information passed from one to another node in the network would travel through less nodes. Houghton et al. (2006) point out that well-connected nodes can still achieve low centrality values as they may be on the periphery of the network. For example, in the case of the social network, nodes with higher centrality status values are those that are closest to all other agents in the network as they have direct rather than indirect links with them.

ADVANTAGES

- The analysis produced is extremely comprehensive and activities are analysed from various perspectives.
- The analysis is both qualitative (networks) and quantitative in nature (network analysis metrics).
- Composite networks enable analysts to explore the relationships between tasks, agents and information.
- The use of network analysis metrics enables analysts to identify key tasks, agents and information.
- The framework can be used both retrospectively and predictively to forecast system behaviour (e.g. Stanton and Harvey 2017).
- The framework approach allows methods to be chosen based on analysis requirements.

- EAST has been applied in a wide range of different domains for various purposes. The approach is generic and can be used to evaluate activities in any domain.
- Various Human Factor (HF) concepts can be examined, including distributed situation awareness, distributed cognition, decision making, teamwork and communications.
- It uses structured and valid HF methods and has a sound theoretical underpinning.

DISADVANTAGES

- When undertaken in full, the EAST framework is a very time-consuming approach.
- The use of various methods ensures that the framework incurs a high training time.
- In order to conduct an EAST analysis properly, a high level of access to the domain, task and SMEs is required.
- Some parts of the analysis can become overly time-consuming and laborious to complete.
- Some of the outputs can be large, unwieldy and difficult to present in reports, papers and presentations.
- Reliability and validity have not yet been formally tested.

RELATED METHODS

EAST uses HTA, social network analysis and information networks. Various methods can be used to collect the data required to construct task, social and information networks, including concurrent verbal protocol analysis, the CDM (Klein et al. 1986), observation, documentation review and content and thematic analysis.

APPROXIMATE TRAINING AND APPLICATION TIMES

Due to the number of different methods involved, the training time associated with the EAST framework is high. Similarly, application time is typically high, although this is dependent upon the task under analysis and the scope of the analysis.

RELIABILITY AND VALIDITY

Due the number of different methods involved, the reliability and validity of the EAST methods are difficulty to assess. Indeed, they have not yet been formally tested.

FLOWCHART

A flowchart showing the main phases of EAST is shown below, separated into the three phases of data collection, data analysis and representation methods (Figure 1.2).

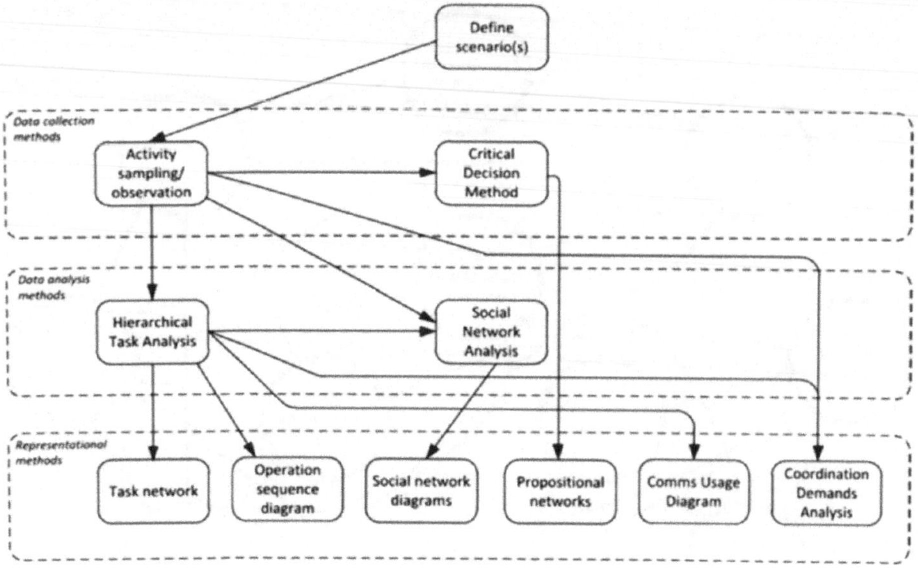

FIGURE 1.2 Flowchart showing phases and associated methods for EAST.

TOOLS NEEDED

Normally, video and audio recording devices are used to record the activities under analysis. A drawing software package such as Microsoft Visio is also typically used to reproduce the networks. The HTA tool or other task analysis tools can be used to support development of the task network. The AGNA social network analysis software tool is typically used to quantitatively analyse the networks, and the Leximaner thematic analysis tool can be used to construct information networks directly from verbal or communications transcripts (e.g. Salmon et al. 2014b).

EXAMPLE

In the following example, EAST was used to examine a generic land use and urban planning site analysis process (Stevens et al., 2018). Task, social and information networks were constructed to describe the key tasks, agents and information used during site analysis. Initially, a task network was constructed based on a HTA of a generic site analysis process.

As shown in Figure 1.3, 11 key interrelated tasks were identified. The task network is a dense one with many interdependencies between tasks, suggesting that the tasks required are tightly coupled. In particular, the tasks of analysing the neighbourhood and determining circulation patterns are the most connected within the task network, suggesting that they are central to the overall site analysis process.

A social network diagram was constructed based on identifying which agents are required to communicate with each other during the site analysis process (see

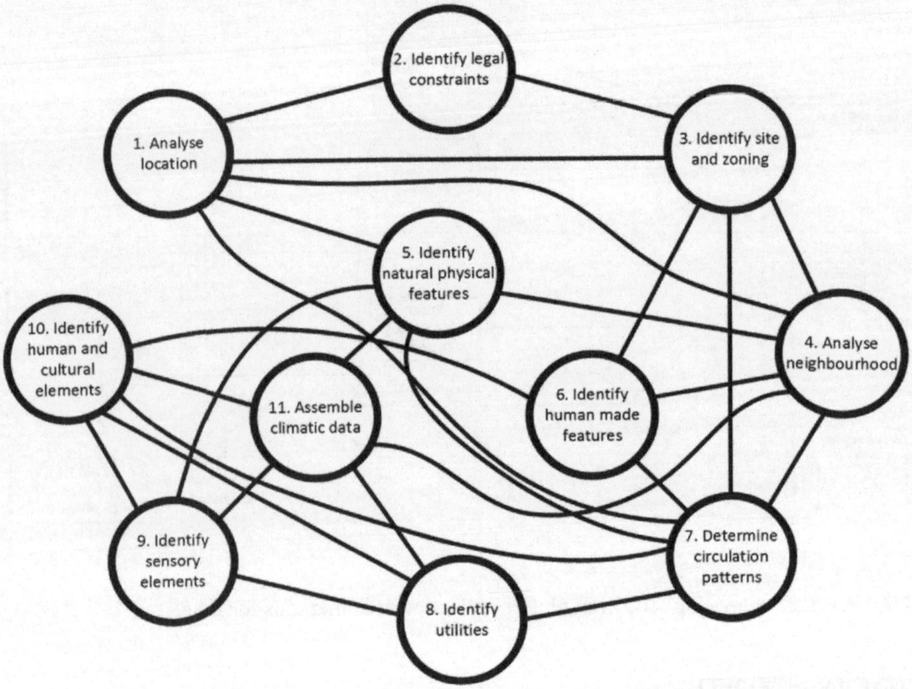

FIGURE 1.3 Task network of the generic site analysis.

Figure 1.4). The social network demonstrates that there are 19 distinct agents involved in the site analysis process. This reliance on multiple actors to assemble the required information for the site analyses is central the work within LUP & UD disciplines. In contrast to the well-connected task network, however, there are few connections between the agents within the social network, suggesting that the network of agents involved in site analysis are loosely coupled.

The social network suggests that, while many actors are necessary for the assembly of the required information, they are most often working independently, while a central organisational agent coordinates their responses. This central agent is the urban planner and designer, with the social network diagram revealing that the most connected agents within the site analysis process are urban planners and designers, with connections to all other agents involved in the process. Indeed, urban planners and designers are the only agent in the social network to have connections with more than four other agents. This indicates that urban planners and designers are the key agents within the site analysis process, with the local council being the next most connected. The structure of the social network diagram suggests that there may some simplistic interventions that could improve the site analysis process. Logical interventions would be to attempt to increase the connectivity of the network through incorporating a requirement for further communication between agents and to reduce the load placed on urban planners and designers.

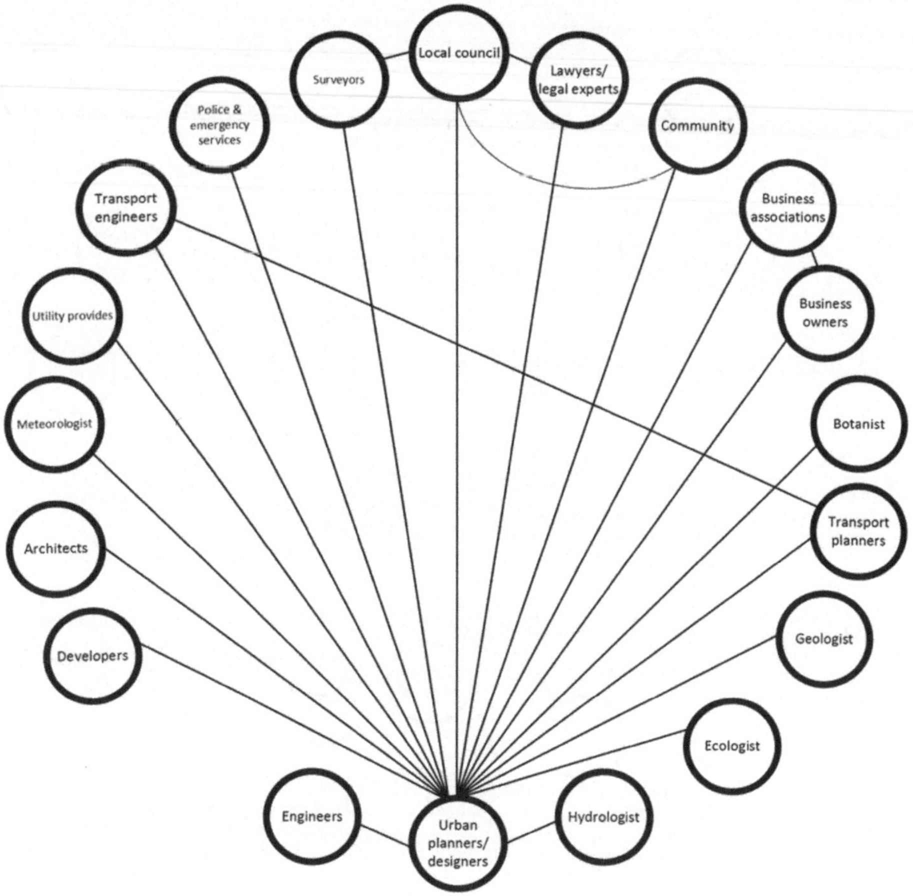

FIGURE 1.4 Social network of the agents in the site analysis process.

An information network showing the information required during the site analysis process was constructed based on the HTA (see Figure 1.5). According to the information network, multiple sources of information are required to complete the site analysis process, ranging from information on locations, topography, views, drainage, climate and utilities to traffic, urban form, sensory elements and commercial, retail, residential and community functions.

Examining the connectedness of different nodes within the network suggests that there are various critical pieces of information required, including locations, commercial functions, neighbourhood context, road hierarchy, residential functions, retail functions, pathways and travel times.

To demonstrate the composite network function of EAST, the task and social networks were combined to show which agents are involved in the 11 key site analysis tasks. The task by agent network is presented in Figure 1.6.

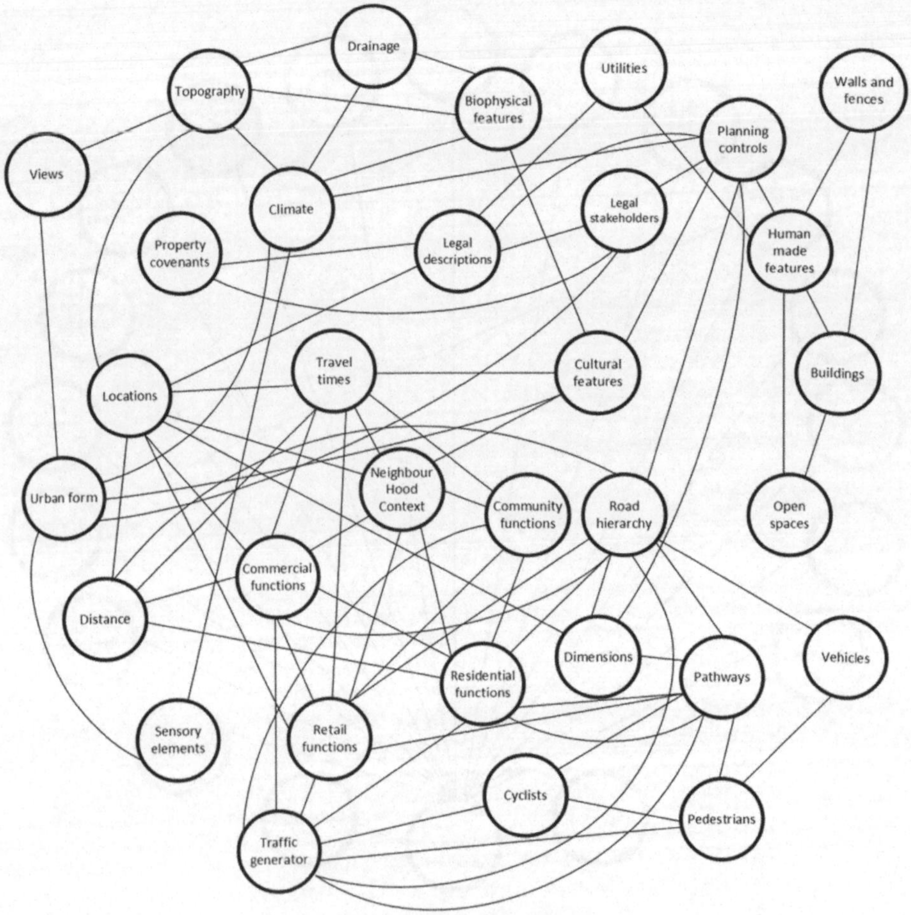

FIGURE 1.5 Information network for the site analysis task.

The composite task and social network shows the allocation of tasks across different agents. Again, the critical role of the urban planners and designers is highlighted, as they are the only agent that is constant across all 11 site analysis tasks. In addition, the network demonstrates the key requirement for multi-stakeholder input across the tasks. For example, tasks such as 'Identify human and cultural elements' and 'Identify natural physical features' have eight and seven agents involved, respectively. When this is considered with the social network, which shows that there are few agents communicating with one another through the site analysis process, it suggests that the process could be made more efficient through the addition of mechanisms designed to enhance communications and interactions between all of the stakeholders involved. The remainder of this book will show applications of EAST to a variety of domains as well as some developments and extensions to the approach.

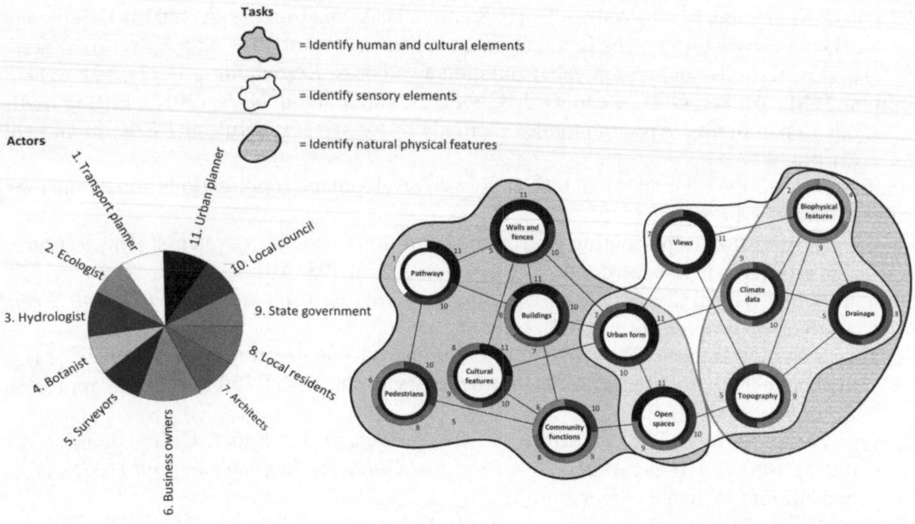

FIGURE 1.6 Combined task, social and information network showing site analysis tasks by agent.

REFERENCES

Driskell, J. E. and Mullen, B. (2004). Social network analysis. In: N. A. Stanton, A. Hedge E. Salas, H. Hendrick and K. Brookhuis (Eds.), (2005). *Handbook of Human Factors and Ergonomics Methods* (pp. 58.1–58.6). Taylor & Francis: London.

Hollnagel, E. (2012). *FRAM: The Functional Resonance Analysis Method: Modelling Complex Socio-Technical Systems*. Ashgate: Aldershot.

Houghton, R. J., Baber, C., Cowton, M., Walker, G. H. and Stanton, N. A. (2008). WESTT (Workload, Error, Situational Awareness, Time and Teamwork): An analytical prototyping system for command and control. *Cognition, Technology and Work*, 10 (3), 199–207.

Houghton, R. J., Baber, C., McMaster, R., Stanton, N. A., Salmon, P., Stewart, R. and Walker, G. (2006). Command and control in emergency services operations: A social network analysis. *Ergonomics*, 49 (12–13), 1204–1225.

Karsh, B. T., Waterson, P. and Holden, R. J. (2014). Crossing levels in systems ergonomics: A framework to support 'mesoergonomic' inquiry. *Applied Ergonomics*, 45 (1), 45–54.

Klein, G. A., Calderwood, R., and MacGregor, D. (1989). Critical decision method for eliciting knowledge. *IEEE Transactions on Systems, Man, and Cybernetics*, 19 (3), 462–472.

Kleiner, B.M. (2006). Macroergonomics: Analysis and design of work systems. *Applied Ergonomics*, 37 (1), 81–89.

Leveson, N. (2004). A new accident model for engineering safer systems. *Safety Science*, 42 (4), 237–270.

Read, G. J. M., Salmon, P. M. and Lenné, M. G. (2016). When paradigms collide at the road–rail interface: Evaluation of a sociotechnical systems theory design toolkit for cognitive work analysis. *Ergonomics*, 59 (9), 1135–1157.

Salmon, P. M., Lenné, M. G., Walker, G. H., Stanton, N. A. and Filtness, A. (2014a). Using the Event Analysis of Systemic Teamwork (EAST) to explore conflicts between different road user groups when making right hand turns at urban intersections. *Ergonomics*, 57 (11), 1628–1642.

Salmon, P. M., Lenné, M. G., Walker, G. H., Stanton, N. A. and Filtness, A. (2014b). Exploring schema-driven differences in situation awareness across road users: An on-road study of driver, cyclist and motorcyclist situation awareness. *Ergonomics*, 57 (2), 191–209.

Salmon, P. M., Walker, G. H., Read, G. J., Goode, N. and Stanton, N. A. (2017). Fitting methods to paradigms: Are ergonomics methods fit for systems thinking? *Ergonomics*, 60 (2) 194–205.

Stanton, N. A. (2006). Hierarchical task analysis: Developments, applications and extensions. *Applied Ergonomics*, 37, 55–79.

Stanton, N. A. (2014). Representing distributed cognition in complex systems: How a submarine returns to periscope depth. *Ergonomics*, 57 (3), 403–418.

Stanton, N. A., Baber, C. and Harris, D. (2008). *Modelling Command and Control: Event Analysis of Systemic Teamwork*. Ashgate: Aldershot.

Stanton, N. A. and Harvey, C. (2017). Beyond human error taxonomies in assessment of risk in sociotechnical systems: A new paradigm with the EAST 'broken-links' approach. *Ergonomics*, 60 (2), 221–233.

Stanton, N. A., Salmon, P. M., Rafferty, L. A., Walker, G. H., Baber, C. and Jenkins, D. (2013). *Human Factors Methods: A Practical Guide for Engineering and Design* (second edition). Ashgate: Aldershot.

Stanton, N. A., Salmon, P. M., Walker, G. H., Baber, C. and Jenkins, D. (2005). *Human Factors Methods: A Practical Guide for Engineering and Design* (first edition). Ashgate: Aldershot.

Stanton, N. A., Stewart, R., Harris, D., Houghton, R. J., Baber, C., McMaster, R., Salmon, P., et al. (2006). Distributed situation awareness in dynamic systems: Theoretical development and application of an ergonomics methodology. *Ergonomics*, 49 (12–13), 1288–1311.

Stevens, N., Salmon P. M., Walker, G. H. and Stanton N. A. (2018). *Human Factors in Land Use Planning and Urban Design: Methods, Practical Guidance, and Applications*. CRC Press: Boca Raton, FL.

Stewart, R., Stanton, N. A., Harris, D., Baber, C., Salmon, P., Mock, M., Tatlock, K., Wells, L. and Kay, A. (2008). Distributed situation awareness in an airborne warning and control system: Application of novel ergonomics methodology. *Cognition, Technology and Work*, 10 (3), 221–229.

Svendung, I. and Rasmussen, J. (2002). Graphic representation of accident scenarios: Mapping system structure and the causation of accidents. *Safety Science*, 40 (5), 397–417.

Vicente, K. (1999). *Cognitive Work Analysis: Towards Safe, Productive and Healthy Work*. Lawrence Erlbaum Associates: Mahwah, NJ.

Walker, G. H., Salmon, P. M., Bedinger, M. and Stanton, N. A. (2017). Quantum ergonomics: Shifting the paradigm of the systems agenda. *Ergonomics*, 60 (2), 157–166.

Walker, G. H., Stanton, N. A., Baber, C., Wells, L., Gibson, H., Salmon, P. M. and Jenkins, D. P. (2010). From ethnography to the EAST method: A tractable approach for representing distributed cognition in air traffic control. *Ergonomics*, 53 (2), 184–197.

Walker, G. H., Stanton, N. A. and Salmon, P. M. (2011). Cognitive compatibility of motorcyclists and car drivers. *Accident Analysis & Prevention*, 43 (3), 878–888.

Section II

Applications of EAST

2 EAST in Air Traffic Control

With Chris Baber, Linda Wells, Huw Gibson and Daniel P. Jenkins

INTRODUCTION

COMMAND AND CONTROL

In command and control scenarios, there is a common goal (comprising interacting sub-goals), there are multiple individuals who need to communicate and coordinate with each other in order to attain these goals and, increasingly, there are ever more complex ways of facilitating this process with technology. Command and control, at the most generic level, can be viewed as a form of management infrastructure for planning and organisation (Harris and White 1987). It involves the exercise of authority and direction by properly designated individual(s) over assigned resources, as well as planning, directing, coordinating and controlling how those resources are deployed (Builder et al. 1999). 'Command' can be viewed as the definition of overall system objectives or goals, whereas 'control' is the management of process and activities that lead to the achievement of these objectives (or compensate for changes in the environment that hinder their achievement). Many contemporary sociotechnical systems involve authority, planning, directing and coordinating and can be considered as forms of command and control. Air Traffic Control (ATC) is one example.

DISTRIBUTED COGNITION

From a distributed cognition perspective (e.g. Hutchins 1995a), the task of ATC can be viewed as a form of '"computation" to maintain separation between aircraft in a region of airspace' (Fields et al. 1998, p. 86). It is further argued that the computations do not reside solely in the heads of individual controllers; instead, they are distributed across the entire ATC system, comprising numerous controllers, teams and technical artefacts. The essence of distributed cognition is on 'how [these computations] transcend the boundaries of the individual actor' (Rogers 1997, p. 1; Hutchins 1995a; Hollan et al. 2000).

The language of representational states is used to describe the visible and external manifestations of various 'environmental contributions' to the total system (Rogers and Ellis 1994; Fields et al. 1998). Representational states subsume the full range of observable interactions between people and artefacts, as well as the resulting states (and state changes) that arise from the various 'computations to maintain separation between aircraft'. For example, an observable interaction might be a controller issuing an instruction to an aircraft. The resulting state might be a corresponding change in the aircraft's representation on the radar display.

In command and control situations, these computations and representational states interact. A change in representational state leads to further computations, further representational states and further computations. But whilst these simple low-level mechanisms can be multiplied in simple ways to form the total ATC system, the high-level function, the system's aggregate behaviour, can be highly complex and adaptive (Chalmers 1990). Phenomena 'wherein complex, interesting high-level function is produced as a result of combining simple low-level mechanisms in simple ways' are referred to as 'emergence' (Chalmers 1990, p. 2). A key emergent property of ATC is the so-called 'picture' or, in ergonomics parlance, *situational awareness* (SA).

Distributed Situation Awareness

The ability to sense changes in representational states, understand them and then perform some kind of computation based on them not only describes the essence of distributed cognition but also that of SA. At an individual level, SA is about the psychological processes, and information in working memory required, for developing 'the picture' (e.g. Endsley 1995; Bell and Lyon 2000). The controller's picture is not the same as the radar display; it does not arise solely from any one physical or human component but from the interaction of many such components. It arises both from the parts of the ATC task that have been overtly designed (i.e. radar displays and other prescribed forms of communication) and those parts of the task that have not been overtly designed (i.e. the clicking of new flight data strips, which informs controllers that a new aircraft is about to enter their sector). Because the picture, or SA, arises from these myriad individual components, yet its totality cannot be predicted solely from any one of them individually, it can be referred to as *emergent*.

Individualistic approaches to SA dominate ergonomics, but whilst they may be appropriate for tasks that are performed by individuals in isolation, few complex tasks are performed entirely independently of others (Perry 2003). In systems terms, SA is what helps entire sociotechnical systems such as ATC to be orientated towards and 'tightly coupled to the dynamics of the environment' (Moray 2004, p. 4). A distributed cognition perspective applied to command and control scenarios requires a shift from traditional notions of SA that focus on the individual (e.g. Endsley 1995) to those that focus on the system (e.g. Sandom 2001; Gorman et al. 2006; Salmon et al. 2008). At face value, the response to this might be the concept of team SA (e.g. Salas et al. 1995; Perla et al. 2000) but even here there are problems. 'The degree to which every team member possesses the SA required for his or her responsibilities' (Endsley 1995, p. 39) runs into difficulties when confronted with the twin concepts of 'overlapping SA' (i.e. portions of SA that are identically shared between people, normally represented as a Venn diagram) and 'compatible SA' (i.e. that which is 'not' overlapping between two or more people, but which fits together like a jigsaw; Salmon et al. 2008). The fundamental problem stems from a tacit assumption that the 'situation' can be defined as a single, objective, external reality and that the goal of the people operating within the situation is to respond to all features appropriately. This is problematic on three counts:

1. There are many aspects of command and control scenarios that require the individual to make judgements and interpretations (so the assumption of the 'objective reality' of a situation is not always valid).
2. There are multiple sub-goals and multiple views of the situation (so the idea of a single reality is not valid either).
3. Different agents within the system use different representational states to inform and support their work, so the notion that there can be a single view of the situation (as opposed to several interlocking views) is also not easily supported.

One way to resolve the mismatch between mainstream thinking in SA and distributed cognition of the sort encountered in complex systems, such as ATC, is to consider one of the relatively invariant properties of it: information. According to Bell and Lyon (2000): 'all aspects of momentary SA are eventually reducible to some form of [. . .] information in working memory' (p. 42). Information, in the SA sense, refers to what, in distributed cognition language, are called *representational states*. The question to ask is whether 'working memory' is the only place where such states can be represented. Distributed cognition would suggest not. It suggests that non-human artefacts can create, manage and share such states (to some extent at least), meaning that the technical aspects of a sociotechnical system will be contributing to the exchange of representational states too. The totality of this will be a form of systems level awareness that is not traceable to any one individual and is not consistent with a distributed cognition view of the world, nor resides exclusively in the minds of humans. Thus, not only is the individual-level ATC 'picture' emergent, so too is the systems level 'picture'.

Another consideration is that representational states can be promulgated around the system with very little in the way of overt communication. One of the great strengths of expert operators is their ability to chunk information, to abstract and pattern match, to develop a high level of awareness from relatively little information in the world (e.g. Chase and Simon 1973). The update of a representational state for one agent might lead to partial updating of that used by another. For example, as an aeroplane moves across a sector, its route is plotted on the ATC displays and its position updated dynamically; if there is no definable risk, then the updating happens automatically and without the need for intervention.

Beyond Ethnography

Whilst there is much to say in favour of blending some of the ideas connected with SA and distributed cognition, there are several problematic issues when it comes to actually applying either approach. It should be clear by now that purely individualistic approaches to analysis may not capture all the required information about complex command and control scenarios. It is also the case that the traditional experimental approach may lack predictive efficiency in the face of significant numbers of emergent behaviours. Distributed cognition, focusing as it does 'on the material and social means of the construction of action and meaning' (Hollan et al. 2000, p. 178), employs ethnography to understand how information is used to support decision making, how it is represented and how it is manifest in the physical and social world (e.g. Rogers and Ellis 1994; Hutchins and Klausen 1996; Hutchins and Holder 2000; Hutchins and Holder 2001). Ethnography is a form of naturalistic, qualitative description based on observation. True ethnography places a requirement to live as a member of the 'tribe' for an extended period (possibly weeks or months) and to undertake the tasks and rituals of the tribe. The understanding gained from this participatory observation normally forms the basis of a report written from a first-person perspective as a participant observer. In the field of distributed cognition, ethnographic methods are deployed. The problems with ethnography in terms of applying it in practice are as follows:

1. The outputs of ethnographic approaches (as they are used currently) remain couched at a qualitative, often highly discursive level of analysis (e.g. Hutchins 1995b; Hutchins and Klausen 1996; Hutchins and Palen 1997). Such outputs may not be easily reconciled with a predominantly engineering, or at least non-social science, audience (McMaster and Baber 2005).
2. Depending on the level of analysis, having the necessary requirement for an observer to be 'imbued' in the culture of the scenario can, in some cases, be incompatible with objectivity and validity in measurement (e.g. Hutchins 1995b).
3. Ethnographic analyses are not always easily amenable to generalisation, concerned as they are with specific and localised scenarios and phenomena. It is certainly difficult to imagine some analyses being amenable to repetition.

These issues limit the practical value of distributed cognition in command and control settings and serve as a barrier to employing what is potentially a useful and enlightened approach to the analysis of complex systems. A response to this problem, however, appears to lie in the fundamentals of the distributed cognition approach itself. By characterising complex systems using the language and metaphors of cognitive science, many of the phenomena of interest, such as communications, shared awareness and other determinates of decision superiority, are rendered physically and observably manifest. That is to say, they no longer reside just in the heads of individual actors; they become manifest in the way that information is represented, modified, communicated and shared. The purpose of this chapter is to deploy the Event Analysis for Systemic Teamwork (EAST) method as a way of capturing these phenomena, to show that there are ways to leverage the favourable theoretical perspective of distributed cognition that are also appropriate and relevant to systems designers.

THE AIR TRAFFIC CONTROL WORK SETTING

A case study from the ATC work domain is used to illustrate the complementary approach to 'doing' distributed cognition. The primary strategic objective of ATC is safety. This is achieved by providing instructions on height, speed and route to aircraft pilots so that individual aircraft maintain legally mandated physical separation criteria (typically 3 or 5 miles horizontally and 1000 feet vertically; Civil Aviation Authority 2004). A secondary tactical objective is to optimise the routing of aircraft so that they take off and land at prescribed times and follow the most expeditious routes to destinations. These strategic goals translate into the following operational level activities:

- Keeping in radio and/or radar contact with aircraft.
- Instructing aircraft in relation to speed, altitude and direction.
- Providing information to aircraft about weather conditions.
- Ensuring that minimum distances are maintained between aircraft.
- Handling unexpected events, emergencies and unscheduled traffic.

Underlying these activities are several key non-human artefacts. Like their human counterparts, they contain, represent and modify information and are part of a system that a distributed cognition perspective is able to model. Fields et al. (1998) provide a cogent summary of the key informational artefacts contained within the ATC work domain. These are as listed below.

Charts and Standard Routes

These can be seen as a prescription of how certain manoeuvres should be made. The information is shared between controllers and pilots as well as being stored on aircraft flight management systems (Fields et al. 1998).

Flight Data Strip

This can be seen as a representation of the projected state of an aircraft. It takes the form of a paper strip that contains information on aircraft height, speed, heading and call sign. This information is modified by the controller by hand when instructions are issued and confirmation of receipt is received (Fields et al. 1998).

Flight Data Strip Bay

Whereas the individual flight data strip is a projected state of an individual aircraft, the totality of data strips that refer to a region of airspace provides a 'schematic model' of air traffic progress (Fields et al. 1998, p. 87).

Radar Display

'The radar screen provides controllers with a snapshot of the current horizontal [. . .] locations of aircraft' annotated with individual aircraft call signs and height (Fields et al. 1998, p. 88).

Aircraft Call Signs

These are the unique alpha-numeric codes given to individual aircraft. Aircraft call signs serve 'as an important (indeed the only) means of coordinating the information represented in [. . .] various media', such as the flight strip and radar display (Fields et al. 1998, p. 88).

Communications

These represent, in a practical sense, the 'mediation of control'. Fields et al. (1998) argue that communications can also be seen as a 'system of representations' (p. 4). Of critical importance is that communications 'are also situated in the network of artefacts and information, and are [only] made comprehensible by reference to a larger context of shared representations' (p. 88).

DISTRIBUTED COGNITION METHODOLOGY: THE IMPORTANCE OF METHODS

Explicit methods lie at the heart of ergonomics as a discipline, enabling the practitioner to vary their approach between scientist (i.e. testing and developing theories of human performance using rigorous data collection and analysis techniques) and practitioner (evaluating the effects of change, developing best practice and, fundamentally, addressing real-world problems). Ergonomics methods are useful in the scientist–practitioner model because of the structure and potential for repeatability that they offer over and above ethnography alone.

Descriptive vs. Formative Methods

This study uses the EAST method in an attempt to reconcile distributed cognition with the methodological traditions of ergonomics. EAST is based on the integration of seven individual methods, which in turn is a reflection of the multi-faceted nature of the command and control. In other words, no one method can adequately describe all of the degrees of freedom inherent in such a complex sociotechnical system. That said, no such claim is made for the EAST method, but it can be argued that at least some of the major human dimensions of the problem space can be explored by taking a multi-method approach.

EAST is a descriptive method. It does not specify a formal architecture and what 'should' happen. Even though it uses normative methods such as task analysis and process modelling, these are populated with data on what is actually observed. Neither does EAST focus on constraints, boundaries and a problem space defined formatively by the scope of what 'could happen', such as cognitive work analysis (CWA). EAST focuses on what 'did' happen. CWA admits the possibility of non-linear and emergent behaviour, whereas EAST is designed to identify specific instances of it.

Despite their descriptive vs. formative differences, EAST and CWA are both representative of a shift in methodological thinking. They share two key aspects: both acknowledge that complex sociotechnical systems require more than one approach

(EAST comprises seven individual methodologies, CWA comprises five 'phases') and both acknowledge that these perspectives are as interlinked as the complex sociotechnical phenomenon under analysis (i.e. they are both systemic in nature). This is a core principle of sociotechnical design (Clegg 2000).

Method Integration

The following formal methodologies combine to form EAST: hierarchical task analysis (HTA) (Annett); coordination demand analysis; communications usage diagram; social network analysis (SNA) (Driskell and Mullen 2005); propositional networks; and an enhanced form of operation sequence diagram. A multiple method approach has a number of compelling advantages. Not only does the integration of existing methods bring reassurance in terms of a validation history, but it also enables the same data to be analysed from multiple perspectives. Also, with over 200 existing methodologies to choose from (Stanton et al. 2005) there seemed little pragmatic need to develop yet more. Of course, multiple interconnected methods require greater effort to analyse, but a companion to EAST, called *workload, error, situation awareness, tasks and time*, is designed to help. This is a software tool that greatly streamlines and simplifies the application of the method and it was used in the current analysis (Houghton et al. 2006).

EAST is structured as follows. The HTA provides input into the analysis of teamworking (CDA), communications usage (CUD) and the linkage (via communications) between agents (SNA). Data for the HTA are gathered from live observation of the scenario. The output of all these methods (HTA, CDA, CUD and SNA) is given a summary visual form by using an enhanced Operation Sequence Diagram (OSD). Interview data, in the form of the critical decision method (CDM) (Klein and Armstrong 2005), are used to create a network of linked 'information objects' or representational states. This representation is called a *propositional network* and is rather similar in concept to semantic networks. It is important to note at this point that the purpose of this chapter is not to introduce the methodological intricacies of the EAST method (the reader is referred to Chapter 1 for further detail) but to show the effect of method integration in terms of enabling a distributed cognition perspective upon ATC.

AIR TRAFFIC CONTROL SCENARIOS

Data were collected from the ATC work domain between 28 and 30 June 2004 at a major UK terminal and area control centre. Separate analyses were performed on four discrete scenarios that were observed to take place repeatedly. By dividing up the controller's task in this way, it was possible to capture something of the dynamics of the system. In practice, multiple aircraft are presenting themselves to the controller(s), prompting them to engage interchangeably in one of the following four behaviours:

- Scenario 1: To bring aircraft inbound from a major air route into a holding stack and then pass them onto an aerodrome controller (the holding scenario)

- Scenario 2: To deal with aircraft that have left the holding stack and are en route to airfield(s) but have yet to enter the final approach phase (the approach scenario)
- Scenario 3: To deal with over-flying aircraft in such a way as to avoid conflict with the holding stack and other en-route aircraft (the over-flight scenario)
- Scenario 4: To deal with departing aircraft in such a way as to avoid conflict with the holding stack (the departure scenario)

Applying the Method

The first stage of the EAST method involves a two-step process of observation and interview (Annett 2005). This is similar in some respects to the ethnographically based approaches currently used in distributed cognition research, but here there is not a particular requirement for the observer to be directly 'imbued' in the scenario. The focus instead is on unobtrusive observation. The bulk of the analysis derives from live audio feeds, which detail who is communicating to whom and about what. These data are supplemented by the observers' notes and by in-depth technical critique and insight provided by subject matter experts. Whilst observational techniques provide information on the observable artefacts of interaction, they produce limited data on the representational states internal to individuals. It is to the CDM (Klein and Armstrong 2005) that relevant insights can be provided. The CDM is a semi-structured interview technique that uses cognitive probes in order to elicit information on expert decision making. This was administered to each participant in relation to each scenario.

Representing Distributed Cognition

The data collection methods provide information on the activities performed within ATC, how those activities are facilitated and how non-human artefacts participate in the joint cognitive system by containing, representing and transforming information. Using more conventional ethnographic techniques, these facets of a scenario would be described, exhaustively, in a first-person written account. While this can capture the nuances of the situation as it was observed and is often illustrated with diagrams and photographs of the work setting, it is not easy to summarise the account in a manner that brings out common features. In this chapter, the EAST method is used to present multiple views of the situation in such a way as to complement an ethnographic account, if such an analysis were to be performed (which, in a lot of ergonomics studies, would be unlikely). The EAST method maps the tasks being performed (in a task network), the communications between agents (in a social network) and the representational states being sensed (in a propositional network). Figure 2.1 illustrates the relationships between these three views.

Task Networks

HTA is a means of describing a system in terms of a structured hierarchy of goals and sub-goals with feedback loops (Annett 2005). Its appropriateness in this instance

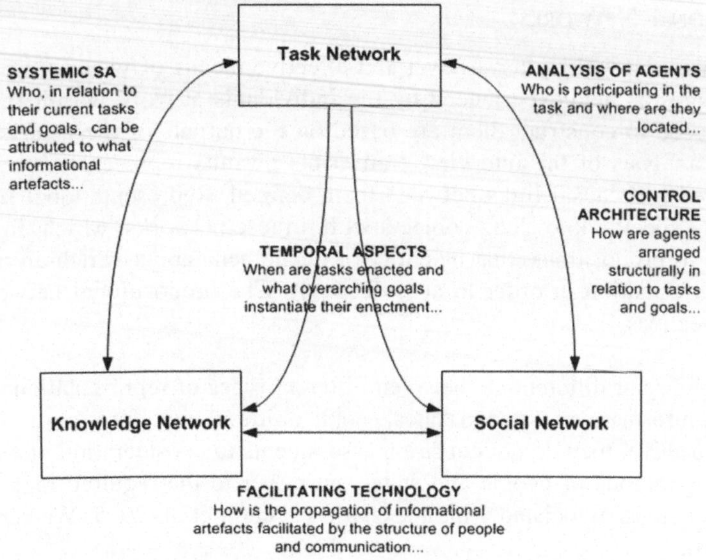

FIGURE 2.1 The three network-based approaches provide a detailed characterisation of a complex sociotechnical system.

can be put down to two key points. First, it is inherently flexible, and the approach can be used to describe any system. Second, it can be used for many ends, from person specification and training requirements to error prediction, team performance assessment and system design. Key to its present application is its ability to model the temporal relations between tasks and the external conditions upon which task activity is cued. The task network, specifically, is a way of representing the detailed task analysis in terms of the interaction of higher-level goals. This technique allows complex task analyses to be easily compared according to differences in overall task structure and type. The task analysis/network forms the foundation for developing insightful social and propositional networks.

Social Networks

SNA is particularly compatible with the distributed cognition perspective because it 'focuses on the relationships among actors embedded in their social context' (Driskell and Mullen 2005, p. 58). Furthermore, it can represent the technological mediation of communication and networks in which some of the nodes are non-human. A social network is a way to represent individuals or teams who are linked by communications to each other and to subject that network to mathematical analysis using Graph Theory (Driskell and Mullen 2005). Two mathematical indices are used in the present analysis, namely *centrality* (i.e. a numeric ranking allowing key agents in the network to be identified) and *density* (the interconnectivity of the network as a whole). The metrics reveal certain important characteristics of the networks to be revealed, in turn allowing comparisons between them.

Propositional Networks

Propositional networks offer a novel and effective means of visualising representational states as held/experienced by the individuals at work within the system. The data used to construct them are based on the outputs of the CDM, in which a content analysis of the interview transcripts permits representational states to be extracted and causal links between them defined. Representational states take the form of specific knowledge objects within these networks, which, in turn, are analogous to propositions, that is, entity or phenomena about which an individual requires information in order to act effectively. The propositional network offers four perspectives.

1. They do not differentiate between different types of representational state (e.g. information related to states, people or ideas); therefore, from a design perspective, they do not constrain assessments to consideration of existing configurations of people and states but rather to the required representational states associated with a scenario (Stanton et al. 2005; Walker et al. 2006).
2. The network shows the totality of information used in the scenario (within the constraints of the data collection techniques used) regardless of whether agents in the scenario are human or technical.
3. Shared SA can be accessed from the CDM, in which multiple agents can be attributed to common knowledge objects/states within the network.
4. The dynamic aspects of SA can also be captured by animating the propositional network. This is achieved by highlighting active and non-active knowledge objects/states occurring in different task scenarios.

These individual network-based outputs not only meet the need to go beyond reductionism (and focus on interconnections as well as parts) but they can also be linked to provide several different perspectives on the scenario.

APPLICATION TO AIR TRAFFIC CONTROL

Analysis of Agents in the Distributed Cognition System

The social network representation provides a particularly powerful example of distributed cognition, in particular, the idea of a joint cognitive system of collaborating human and non-human agents. The SNA of all the actors participating in the ATC task shows that there are 21 agents in the network, 14 human and 7 non-human (shown as shaded in Figure 2.2), joined by 22 communication links. Note that this network has been simplified somewhat. The number of aircraft that an individual controller would be expected to handle could be significantly larger than the four shown for illustration in Figure 2.2.

The metrics 'status' and 'centrality' are used to identify key agents in the scenario, and these are shown in Table 2.1.

Between them, they indicate the degree of connectedness an individual agent has and, consequently, the amount of influence that agent has on the performance

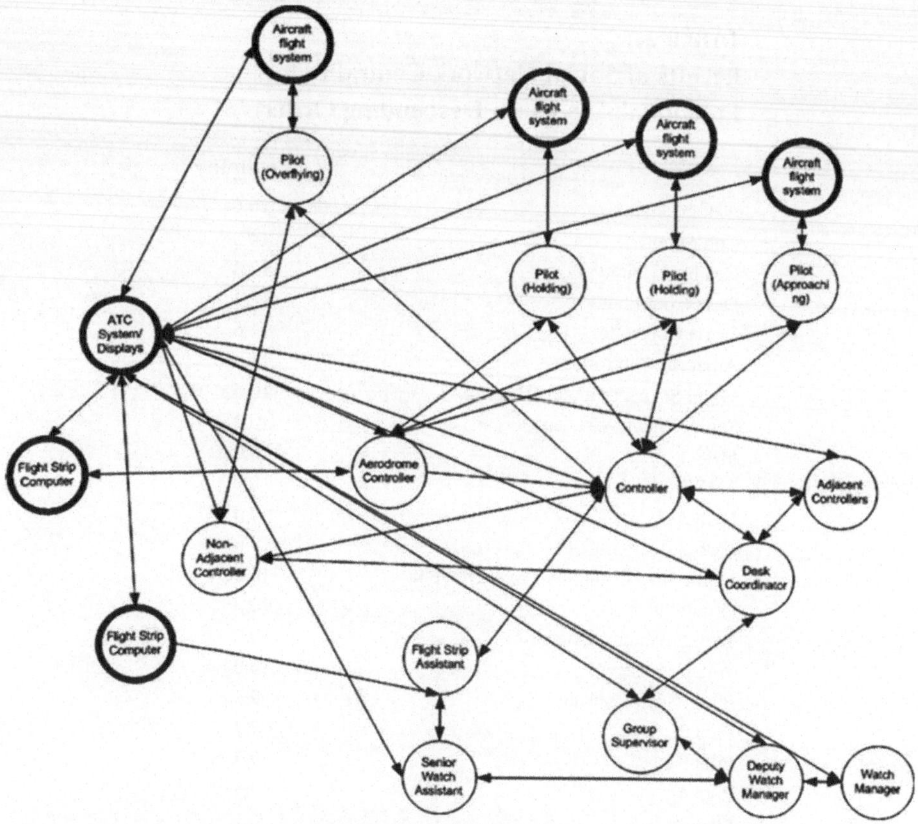

FIGURE 2.2 Social network diagram illustrating agents in the air traffic control (ATC) work domain linked by communication (emphasis denotes non-human agent).

of the network as a whole. Key agents can be defined as having high status and high connectedness. The analysis indicates that the controller (as one might expect) has high status and centrality in the network, but the metric also reveals that the most significant agent is in fact non-human; the ATC technical system (which is the technical infrastructure and assemblage of ground aids, communications systems, radar displays, etc.). A great deal of information is received, stored, displayed or otherwise transformed by this (non-human) agent. This is a similar finding to that of Houghton et al.'s (2006) analysis of police control rooms (in which the central logging computer was also the most central agent).

In addition to the distribution between collaborating human and non-human agents, cognition is also distributed in a geographical sense, between, for example, aircraft in flight within the sector, controller(s) based at the control centre and controllers based at airfield(s). Table 2.2 presents the number of times communication links cross these geographical boundaries.

Geographical dispersion is further illustrated in geographical regions. The type of communications media is physically constrained in cases where links between nodes

TABLE 2.1
Results of Social Network Centrality for Individual Agents (in Descending Order)

Agent	Centrality
ATC System	16.8
Controller	14.0
Heathrow Director	12.3
Desk Coordinator	12.3
Welin Controller	11.6
Adjacent Controller	11.3
Flight Strip Computer (Heathrow)	10.5
Senior Watch Assistant	10.2
FMS 1	10.0
FMS 2	10.0
FMS 3	10.0
FMS 4	10.0
Flight Strip Computer (West Drayton)	10.0
Group Supervisor	10.0
Deputy Watch Manager	10.0
Watch Manager	10.0
Flight Strip Assistant	9.5
Pilot 2	9.3
Pilot 3	9.3
Pilot 4	9.3
Pilot 1	9.1

cross from one shaded region to the other. For example, verbal communications are not physically possible between controllers on the ground and pilots in flight without some sort of facilitation by technical means (such as radio). There remain a host of other local conditions that may also require some form of technological mediation. For example, individual controllers have to remain at their workstations during their shift in order to communicate with controllers who are not immediately adjacent, the telephone is used and/or the agent acting in the role of desk coordinator may have to facilitate.

TABLE 2.2
Number of Communications Links That Exist between Geographically Disperse Locations

From/To	Number of Links
Aircraft to Airfield	3
Aircraft to Traffic Control Centre (TCC)	6
TCC to Airfield	3

Facilitating Technology

Figure 2.4 shows how ATC operations are facilitated by seven types of communication media.

Face-to-face communication comprises verbal communications. Face-to-face communication also contains a visual component. Controllers will point and demonstrate visually to aid in understanding instructions. Telephone and radio technology facilitates voice communication. Radio also facilitates the dissemination of data, as do data network facilities. Written communication is dominated by the flight data-strip aspect of the task, although machine interfaces (e.g. the radar display) also contain written information as well as visual representations. Table 2.3 provides a summary by crossing communications modality with communications media to provide a technology/modality matrix.

ATC is a highly evolved and proceduralised work domain, and the prominent role of implicit communications is also noted. The CDM interview gives access to some of these unobservable artefacts of interaction, one of which is a system of passing aircraft between sectors that does not rely on explicit verbal exchanges. The system works by putting aircraft into a particular position in the new sector at a pre-agreed height, speed and heading. As soon as the controller sees an aircraft in this position, he/she knows it is being passed onto them, and they can take control of it. A wider awareness is gained via open channel radio communications and instructions to aircraft overheard from other controllers. Several contextual factors, such as the high tempo of operations and the proceduralised nature of the task, enable this level of shared understanding and this form of communication to take place. From a distributed cognition perspective, it can be seen how a comparatively lengthy verbal exchange is often replaced with a far simpler visio-spatial task (facilitated by external technological artefacts such as the radar display and flight data strips; Hutchins and Klausen 1996). These are captured in the social networks. In other words, the presence of a link does not necessarily have to connote just an overt form of communication but also implicit ones.

Control Architecture

Although Figures 2.2 through 2.4 are visually complex, it is possible to discern certain features of the social network. It can be noted that there is a diverging hierarchy

TABLE 2.3
Communications Modality/Technology Matrix

	Technology					
Modality	**Phone**	**Radio**	**Network**	**Strips**	**MMI**	**In-Person**
Verbal	■	■				■
Written				■	■	■
Data		■	■			
Visual/Other				■	■	■

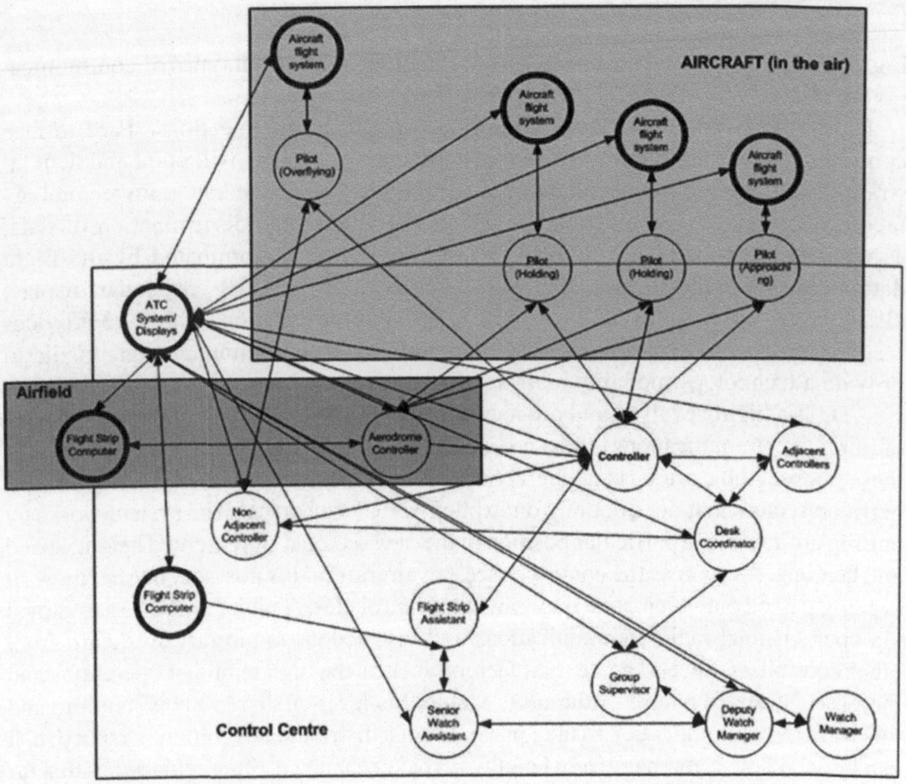

FIGURE 2.3 Geographical dispersion of human and non-human agents in the air traffic control (ATC) work domain.

from the Watch Manager level downwards. This hierarchy splits into the Group Supervisor level (and downwards towards individual controllers) and the Senior Watch Assistant level (and downwards towards Flight Strip Assistants). According to Dekker (2002), this so-called 'split architecture' 'provides some of the benefits of centralised planning with tactical adjustments to new information from subordinate units' (p. 5). At this level of 'air traffic management', the inherent complexity appears to justify a degree of hierarchical sub-division so that aspects of complexity can be spread across agents. The possible trade-off in situations where the state of the world is changing rapidly (a fast tempo), and where decisions need to be enacted quickly, is that any delay in the dissemination of information through a hierarchy becomes critical (Dekker 2002).

At the tactical or 'air traffic control' level, in cases where one agent requires some form of assistance, or to avoid the degradation of aircraft separation, the controller can interact sideways through the structure to an adjacent controller so that issues can be resolved and task load shared or re-distributed quickly. This pattern of communications is known as *peer to peer* or a *negotiation architecture*. In these instances, each agent is more or less independently responsible for a defined area (Dekker 2002) and is able to react promptly to rapid changes in the environment.

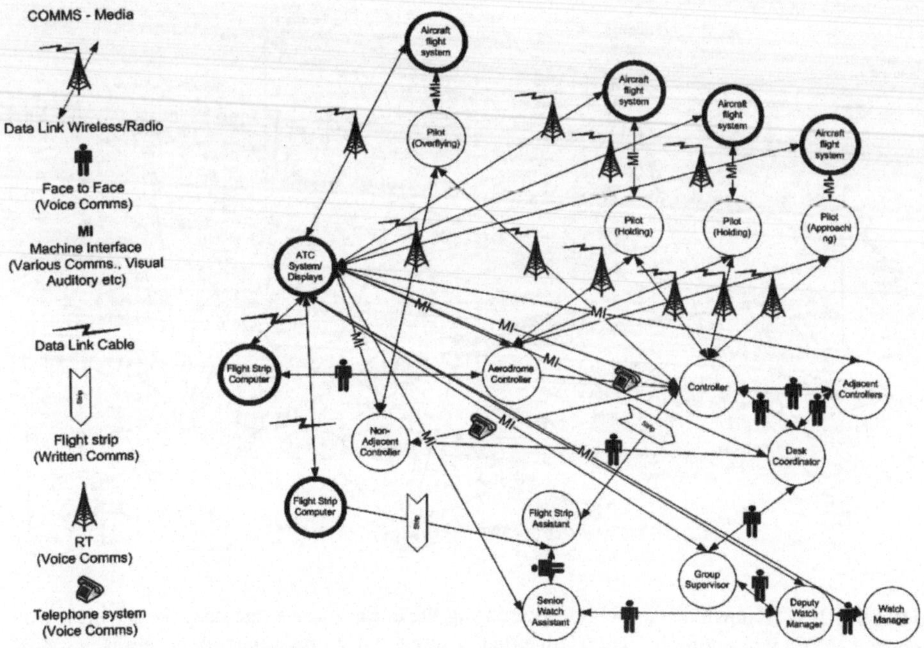

FIGURE 2.4 Social network overlain with the media that facilitates communication.

A further characteristic of the scenario (from a structural perspective) is the amount of information sharing. This relates back to the ATC system and its displays (which subsumes the entire radar, aircraft identification, warning, communication and ground systems in general) being the highest-ranking agent in terms of 'centrality'. Overlain across these two interleaved command architectures (split and negotiated) is the provision of high-quality information to facilitate decision making within them. This overall configuration, known as a *negotiation architecture with information sharing*, is an emerging paradigm in several alternate domains, where it is referred to as *network centric*. The benefits of this architecture are agility, fast response and the ability to quickly organise and re-organise. In the theoretical work by Dekker (2002), the negotiation architecture with information sharing proved to be the most effective in high-tempo tasks.

Systemic Situational Awareness

Systemic SA is modelled using propositional networks (Figure 2.5). From these networks, it is possible to divine certain structural characteristics of the knowledge base that underpins effective SA for the total ATC task.

Central nodes in the network are identified using a centrality metric based simply on divining nodes with five or greater links. Table 2.4 presents an analysis of core knowledge objects for each individual scenario based on this criterion. By representing and simplifying SA in this manner, the table shows that depending on the

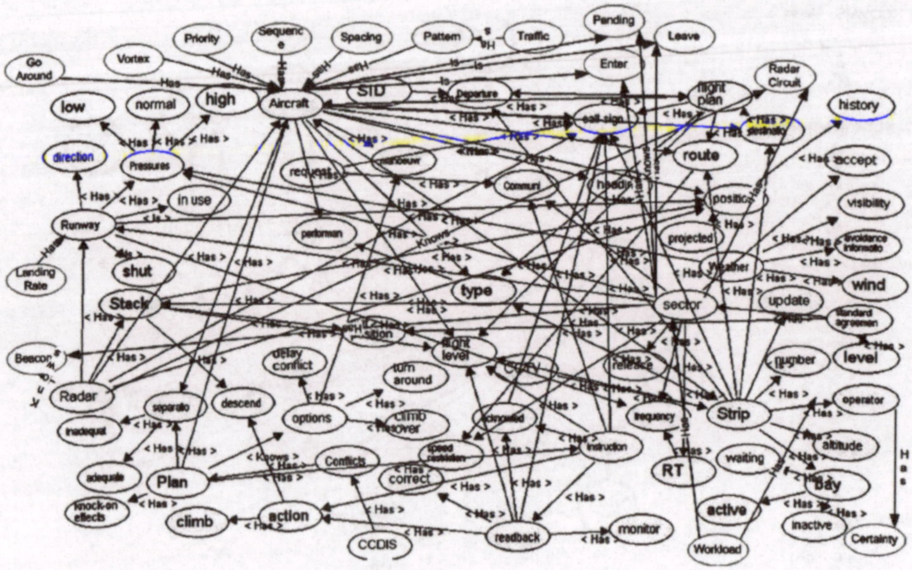

FIGURE 2.5 Knowledge network representing the entire knowledge base for the air traffic control (ATC) work domain. Knowledge that is active during task enactment represents situational awareness at the systems level.

TABLE 2.4
Summary of Key Knowledge Objects (KO) that are Active During Different Phases of Air Traffic Control Process

Knowledge Object	Departure	Over-Flight	Holding	Approach
Pressures	■	■		■
Runway	■			
Radar	■	■	■	
Stack	■			
Aircraft	■	■	■	■
Flight Level	■	■	■	■
Strip	■		■	
Position	■		■	
Read-back		■		
Instruction		■		■
Acknowledge		■		
Call Sign		■	■	■
Flight Plan		■	■	■
Plan			■	
Options		■	■	
Sector				

Note: The table demonstrates changes in systems level state of situational awareness.

task (which occurs at different points in time), not only does the type of information change, but also, by implication, so does the structure of it. This information can be used to explore and understand the information needs of controllers and, moreover, how one might go about supporting them.

Temporal Aspects of Command and Control in Air Traffic Control

The controller will typically switch backwards and forwards among the four scenarios detailed above. As they do so, not only do the type and structure of information change (as shown in Table 2.4) but so also do the characteristics of the task and social networks. To visualise the temporal aspects of the interrelations between the networks requires an alternate form of representation. Some form of animation seems a likely candidate for bringing this facet to life, but this and further refinements to the method, though eminently feasible, are within the purview of future work.

CONCLUSIONS

The purpose of this chapter has been to show how the EAST method puts ergonomic analyses in touch with the distributed cognition perspective, rendering the output much more tractable than comparable ethnographic techniques. Although this chapter is necessarily couched at a summary level of analysis, the key characteristics of the ATC work domain are identified in Table 2.5. Within the table is a checklist

TABLE 2.5
Key Characteristics of the Air Traffic Control Work Domain and the Network Approach in which Detailed Systems Level Insights Reside

	Specified in Detail by:		
Key Characteristic	Task Network	Social Network	Knowledge Network
The coordination of individuals and teams to achieve a common goal (comprising separate though interacting sub-goals, knowledge and situational awareness [SA])	■		■
The technologically mediated geographical dispersion of 'agents' (from controllers located in area control centres, or actual airfields, to specific aircraft in controlled airspace)		■	
A substantial technology infrastructure (providing high-quality, accurate, real-time information via seven different communications modalities), which supports the dispersion and representation of knowledge and a systems level state of SA		■	■
A high degree of information sharing (where most agents have access to parts of the technology infrastructure)		■	
A highly evolved and proceduralised mode of operation (involving implicit communication between and among separate actors in the scenario)	■		■

used to sign-post where the system designer will find detailed insights into the type, nature and structure of these distributed cognition artefacts.

This chapter has focused on a systems level description of the ATC scenario and serves to illustrate that this description can be achieved with the EAST method in live settings. The strength of this descriptive level of analysis can be summarised as follows:

- The methods avoid bias by focusing on objective and manifest phenomena.
- The methods are applicable to any domain, and the results gained are comparable across and within domains.
- The results are graphical and easily interpreted, yet amenable to further summarisation using tables and numerical indices.
- The summary level is underpinned by considerable detail that can be explored further in the context of system design.
- It is consistent with existing narrative approaches to distributed cognition analysis.

As well as providing a descriptive level of analysis, the real potential of the method lies in its ability to offer predictive insights, that is, to 'model' complex sociotechnical systems. In theory, it should be possible to subject the networks to known changes (in task, social or information structure) and to derive outputs as to the effect of these under various performance contexts. For example, an alteration in the command structure may influence the type of communications available to an agent, therefore affecting the type of information that is able to be communicated to other agents, leading to a deleterious affect on SA (or indeed vice versa). The modelling aspect of this work is at a nascent stage but shows promise for further development. What can be communicated about the current application of the method to live data is that the highly relevant theoretical perspective of distributed cognition is within the reach of systems designers and ergonomists.

ACKNOWLEDGEMENTS

This work from the Human Factors Integration Defence Technology Centre (HFI DTC) was part-funded by the Human Sciences Domain of the UK Ministry of Defence Scientific Research Programme.

REFERENCES

Annett, J. (2005). Hierarchical task analysis. In: N. A. Stanton, A. Hedge, K. Brookhuis, E. Salas and H. Hendrick (Eds.), *Handbook of Human Factors and Ergonomics Methods* (pp. 33.1–33.7). CRC: London.

Bell, H. H. and Lyon, D. R. (2000). Using observer ratings to assess situation awareness. In: M. R. Endsley (Ed.), *Situation Awareness Analysis and Measurement* (pp. 129–146). Lawrence Erlbaum Associates: Mahwah, NJ.

Builder, C. H., Bankes, S. C. and Nordin, R. (1999). *Command Concepts: A Theory Derived from the Practice of Command and Control*. Rand: Santa Monica, CA.

Chalmers, D. J. (1990). Thoughts on emergence. Available at: http://consc.net/notes/emergence.html (Accessed April 2005).

Chase, W. G. and Simon, H. A. (1973). Perception in chess. *Cognitive Psychology*, 4, 55–81.

Civil Aviation Authority (2004). *CAP 413: Radiotelephony Manual* (edition 15). Safety Regulation Group, Civil Aviation Authority: Gatwick, UK.

Clegg, C. W. (2000). Sociotechnical principles for system design. *Applied Ergonomics*, 31, 463–477.

Dekker, A. H. (2002). *C4ISR Architectures, Social Network Analysis and the FINC Methodology: An Experiment in Military Organisational Structure*. DSTO Electronics and Surveillance Research Laboratory: Edinburgh, Australia.

Driskell, J. E. and Mullen, B. (2005). Social network analysis. In: N. A. Stanton, A. Hedge, K. Brookhuis, E. Salas and H. Hendrick (Eds.), *Handbook of Human Factors and Ergonomics Methods* (pp. 58.1–58.6). CRC: London, UK.

Endsley, M. R. (1995). Toward a theory of situation awareness in dynamic systems. *Human Factors*, 37 (1), 32–64.

Fields, R., Wright, P., Marti, P. and Palmonari, M. (1998). Air traffic control as a distributed cognitive system: A case study of external representations. In: T. R. G. Green, L. Bannon, C. P. Warren and J. Buckley (Eds.), *ECCE9: Proceedings of the Ninth European Conference on Cognitive Ergonomics* (pp. 85–90). EACE: Limerick, Republic of Ireland.

Gorman, J. C., Cooke, N. J. and Winner, J. L. (2006). Measuring team situation awareness in decentralized command and control environments. *Ergonomics*, 49 (12–13), 1312–1325.

Harris, C. J. and White, I. (1987). *Advances in Command, Control & Communication Systems*. Peregrinus: London, UK.

Hollan, J., Hutchins, E. and Kirsh, D. (2000). Distributed cognition: Toward a new foundation for human–computer interaction research. *ACM Transactions on Computer Human Interaction*, 7, 174–196.

Houghton, R. J., Baber, C., McMaster, R., Stanton, N. A., Salmon, P. M., Stewart, R. and Walker, G. H. (2006). Command and control in emergency services operations: A social network analysis. *Ergonomics*, 49 (12–13), 1204–1225.

Hutchins, E. (1995a). How a cockpit remembers its speeds. *Cognitive Science*, 19, 265–288.

Hutchins, E. (1995b). *Cognition in the Wild*. MIT Press: Cambridge, MA.

Hutchins, E. and Holder, B. (2000). Conceptual models for understanding an encounter with a mountain wave. *Proceedings of HCI-Aero 2000*. Toulouse, France, September.

Hutchins, E. and Holder, B. (2001). What pilots learn about autoflight while flying on the line. *11th International Symposium on Aviation Psychology*, 6 March.

Hutchins, E. and Klausen, T. (1996). Distributed cognition in an airline cockpit. In: Y. Engestrom and D. Middleton (Eds), *Cognition and Communication at Work* (pp. 15–34). Cambridge University Press: Cambridge, UK.

Hutchins, E. L. and Palen, L. (1997). Constructing meaning from space, gesture, and speech. In: L. B. Resneck, R. Saljo, C. Pontecorvo and B. Burge (Eds.), *Discourse, Tools and Reasoning: Essays in Situated Cognition* (pp. 23–40). Springer-Verlag: Heidelberg, Germany.

Klein, G. and Armstrong, A. A. (2005). Critical decision method. In: N. A. Stanton, A. Hedge, K. Brookhuis, E. Salas and H. Hendrick (Eds.), *Handbook of Human Factors and Ergonomics Methods* (pp. 35.1–35.8). CRC: London, UK.

McMaster, R. and Baber, C. (2005). Integrating human factors into systems engineering through a distributed cognition notation. *The IEE and MoD HFI Symposium on People and Systems* (pp. 77–83). IEE: London, UK.

Moray, N. (2004). Où sont les neiges d'antan? In: D. A. Vincenzi, M. Mouloua and P. A. Hancock (Eds.), *Human Performance, Situation Awareness and Automation: Current Research and Trends* (pp. 1–31). Lawrence Erlbaum Associates: Mahwah, NJ.

Perla, P. P., Markowitz, M., Nofi, A., Weuve, C., Loughran, J. and Stahl, M. (2000). Gaming and shared situation awareness. Available at: http://www.thoughtlink.com/publicationsGamingSSA00Abstract.htm/[Accessed April 2005].

Perry, M. (2003). Distributed cognition. In: J. M. Carroll (Ed.), *HCI Models, Theories and Frameworks* (pp. 93–224). Morgan-Kaufmann: San Francisco, CA.

Rogers, Y. (1997). *A Brief Introduction to Distributed Cognition*. Interact Lab, University of Sussex: Brighton, UK.

Rogers, Y. and Ellis, J. (1994). Distributed cognition: An alternative framework for analysing and explaining collaborative working. *Journal of Information Technology*, 9 (2), 119–128.

Salas, E., Prince, C. and Baker, D. P. (1995). Situation awareness in team performance – implications for measurement and training. *Human Factors*, 37, 123–136.

Salmon, P. M., Stanton, N. A., Walker, G. H. and Jenkins, D. P. et al. (2008). What really is going on? Review of situation awareness models for individuals and teams. *Theoretical Issues in Ergonomics Science*, 9 (4), 297–323.

Sandom, C. (2001). Situation awareness. In: J. Noyes and M. Bransby (Eds.), *People in Control: Human Factors in Control Room Design* (pp. 51–68). Institution of Electrical Engineers: London, UK.

Stanton, N. A., Salmon, P., Walker, G. H., Baber, C. and Jenkins, D. (2005). *Human Factors Methods: A Practical Guide for Engineering and Design*. Ashgate: Aldershot.

Walker, G. H., Stanton, N. A., Gibson, H., Baber, C., Young, M. and Green, D. (2006). Analysing the role of communications technology in C4i scenarios: A distributed cognition approach. *Journal of Intelligent Systems Special Issue: Information and Communication Technology*, 15 (1–4), 299–328.

3 EAST in Military Command and Control

With Rebecca Stewart, Daniel P. Jenkins, Linda Wells and Chris Baber

INTRODUCTION

Event Analysis for Systemic Teamwork (EAST) is a macroergonomic method for extracting large-scale systems level data on the emergent properties of command and control scenarios (e.g. Kleiner 2006). The method is based on a number of observations: firstly, 'effective team performance is not an automatic occurrence' (Salas et al. 1995, p. 55); secondly, increased complexity in military operations gives them the appearance of 'different components and layers of subsystems with multiple non-linear interconnections that are difficult to recognise, manage and predict' (Marashi and Davis 2005; Johnson 2005, p. 1); and thirdly, the interaction of components and subsystems (teamworking + complexity) creates non-linear emergent properties at the level of the entire system. In other words, sociotechnical systems like this can be more (or indeed, much less) than the sum of their 'socio' and 'technical' parts. The challenge is to find ways to exploit complexity and non-linearity in order to 'obtain a disproportionate leverage from a given action' (Smith 2006, p. 40). Thus, focusing on the interrelations between command and control's component parts is perhaps as important as the parts in isolation. So, by shifting the unit of analysis from 'technical' to 'human' and shifting it again from 'individuals' to that of the 'system', and by deploying network-based methodologies as a form of non-linear modelling, the data that EAST provides ultimately speak to this goal.

DESCRIPTION OF COMMAND AND CONTROL SCENARIOS

ARMY LAND WARFARE AND THE COMBAT ESTIMATE

The focus of this chapter is on the application of the EAST method to military command and control and the specific case of army land warfare. Land warfare (and other services) rely on a highly evolved planning heuristic called the *Combat Estimate* (or 'the seven questions'), which forms the topic of current EAST analysis. The Combat Estimate describes the process by which plans are made, expected outcomes are defined and actions then have to be taken.

In broad terms, Questions 1 and 2 are concerned with the development of situational awareness concerning the spatial configuration of the battlespace and of mission objectives. The specific activities undertaken in Question 1 include Battlefield

Area Evaluation (BAE), which deals with the potential effects of the physical environment on military operations; and Threat Evaluation, which involves assessing the enemy's capabilities and tactics. Question 2 is concerned with Mission Analysis and the scrutiny of orders that have been received. Questions 4 to 7, in equally broad terms, can be subsumed under the heading 'Course of Action Development'. Figure 3.1 shows how the seven phases of the Combat Estimate relate to each other functionally and temporally. The diagram is a 'task network' based on the high-level goals of a comprehensive Hierarchical Task Analysis (HTA) of the scenario. The links between goals are specified by the HTA's top level 'Goal 0'.

Data Collection

Data for the EAST analysis were gathered by live observation of Command and Staff Training (CAST) exercises at the British Army's Land Warfare Centre in Warminster.

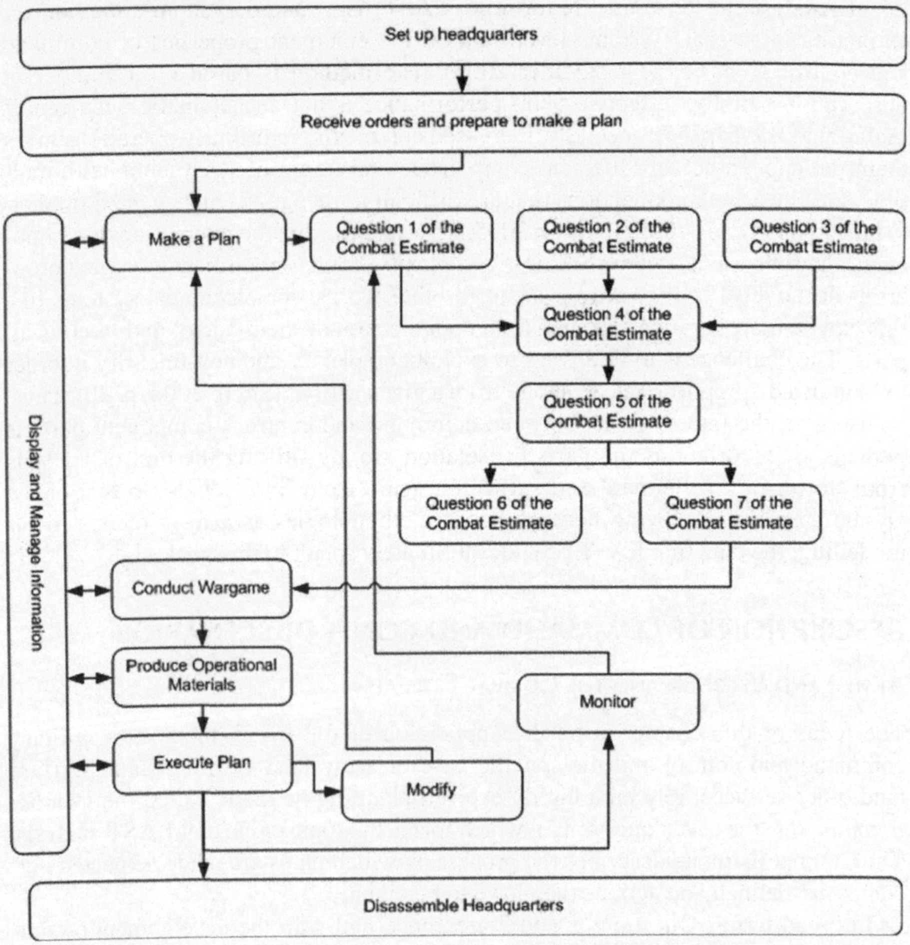

FIGURE 3.1 Task network for the observed military command and control scenario.

The exercises took place in a battlegroup command post set up on-site as it would be deployed in the field. The command post was set to work within a scenario which, in broad terms, required effects to be delivered to a large-scale enemy force passing in one direction through the battlegroup's area of operations. Enemy and friendly units were simulated by a team of remotely located operators who provided a form of augmented reality, supplying friendly radio traffic (simulating units that would ordinarily be located in the field) and updating digital data pertaining to enemy movements, reactions and counter-actions. A team of analysts and subject matter experts monitored and transcribed video and audio feeds from the battlegroup headquarters. Key personnel active in the scenario were further interviewed at key points in the scenario using the Critical Decision Method (Klein and Armstrong 2005).

DESCRIPTION OF THE EAST METHOD

The Importance of Methods

The importance of ergonomics methods cannot be overstated (e.g. Stanton et al. 2005a, 2005b; Wilson and Corlett 1995). Explicit methods lie at the heart of ergonomics as a discipline, enabling the practitioner to vary their approach between those of scientist (i.e. testing and developing theories of human performance using rigorous data collection and analysis techniques) and practitioner (evaluating the effects of change, developing best practice and, fundamentally, addressing real-world problems). Ergonomics methods are useful in the scientist–practitioner model because of the structure and potential for repeatability that they offer over and above informal methods.

Descriptive vs. Formative Methods

The EAST method is based on the integration of seven individual Ergonomics methods. What appears to be a sizeable modelling endeavour is a reflection of the multi-faceted nature of the command and control problem to which it is designed to apply. No one method can adequately describe all of the degrees of freedom inherent in such a complex sociotechnical system. No such claim is made for the EAST method; however, it can be argued that at least some of the major human dimensions of the problem space can be explored by taking a systems perspective on the diagnosis and description of 'what is going on' when command and control organisations are set to work.

EAST is a descriptive method. It does not specify a formal architecture and what 'should' happen. Even though it uses normative methods like task analysis and process modelling, they are populated based on what is actually observed. Neither does EAST focus on constraints, boundaries and a problem space defined formatively by what 'could happen' like Cognitive Work Analysis (CWA) (Vicente 1999): EAST focuses on what 'did' happen. CWA admits the possibility of non-linear and emergent behaviour: EAST is designed to identify specific instances of it.

Despite their descriptive versus formative differences, EAST and CWA are both representative of a shift in methodological thinking. They share two key aspects: both acknowledge that complex sociotechnical systems require more than one approach (EAST is comprised of seven individual methodologies, CWA is comprised of five 'phases'), and both acknowledge that these perspectives are as interlinked as the complex sociotechnical phenomenon under analysis (i.e. they are both systemic in nature). This is a core principle of sociotechnical design (Clegg 2000).

Method Integration

The following formal methodologies combine to form EAST: Hierarchical Task Analysis (HTA) (Annett 2005), Coordination Demand Analysis (CDA) Burke 2005), Communications Usage Diagram (CUD) Watts and Monk 2000), Social Network Analysis (SNA) Driskall and Mullen 2005), Propositional Networks (PN) (e.g. Ogden 1987) and an enhanced form of Operation Sequence Diagram (OSD) (Kirwan and Ainsworth 1992). A multiple method approach has a number of compelling advantages. Not only does the integration of existing methods bring reassurance in terms of a validation history but it also enables the same data to be analysed from multiple perspectives. With over 200 existing methodologies to choose from (Stanton et al. 2005a) there seemed little pragmatic need to develop yet more, hence the approach adopted, which was to integrate existing methods. The trade off, of course, is time. Multiple interconnected methods require greater effort to analyse, but in some sense, this is an artefact of the problem domain being analysed. Good news in this respect comes in the form of a companion to EAST called WESTT. This is a software tool that greatly streamlines and simplifies the application of the method and was used in the current analysis (Houghton et al. 2007).

The HTA provides input into the analysis of teamworking (CDA), communications usage (CUD) and the linkage (via communications) between agents (SNA). Data for the HTA are gathered from live observation and activity sampling as mentioned above. The output of all these methods (HTA, CDA, CUD and SNA) is given a summary visual form by using an enhanced Operation Sequence Diagram (OSD). Interview data, in the form of the Critical Decision Method (CDM: Klein and Armstrong 2005), are used to create the final part of EAST, a network of linked 'information objects' (or Propositional Networks; PN). This is a systemic, network-based approach to the concept of Situation Awareness (SA). The debates surrounding the concept of SA require more in-depth discussion in relation to EAST.

Situational Awareness

One of the key emergent properties from command and control scenarios, and one of the major determinants of decision superiority, is the concept of situation awareness (SA) Endsley 1995; Stanton et al. 2009; Salmon et al. 2008a). At an individual level, SA is about simply 'knowing what is going on' (Endsley 1995). At a systems level, SA enables decisions to be made in real time, and for sociotechnical systems like military command planning to be orientated towards and '*tightly coupled to the dynamics of the environment*' (Moray 2004, p. 4). A distributed cognition perspective applied to command and control scenarios requires a shift from traditional notions of SA that focus on the individual (e.g. Smith and Hancock 1995; Adams et al. 1995; Bedny and Meister 1999; Endsley 1995). Whilst these approaches may be appropriate for tasks that are performed by individuals in isolation, few complex command and control tasks are performed entirely independently of others. The idea of co-dependence in SA, of course, finds expression in several approaches to 'team SA' (e.g. Perla et al. 2000; Salas et al. 1995). Broadly speaking, these approaches to SA tacitly assume that the 'situation' can be defined as a single, objective, external reality, and that the goal of the people operating within the situation is to respond to all features appropriately. These approaches are problematic on three counts: first, there are many aspects of command and control scenarios that require the individual to make judgements and interpretations (so the assumption of the 'objective reality' of a situation is not always valid); second, there are multiple sub-goals and, therefore, multiple views of the situation (so the idea of a single reality is also not always valid either); and third, as mentioned above, different agents within the system use different pieces of information to inform and support their work, so the notion that there can be a single view of the situation (as opposed to several overlapping views) is not easily supported. Distributed cognition provides a way of coping with these conceptual issues and of providing a systems level view of SA (Stanton et al. 2009). This view rests on three key factors.

1. Firstly, a relatively invariant theoretical property of SA concepts is 'information', in so far as 'all aspects of momentary SA are eventually reducible to some form of … information in working memory' (Bell and Lyon 2000, p. 42). The question is whether 'working memory' is an individual phenomenon (as implied by Bell and Lyon 2000) or whether it is a 'system-level' representational state (as implied by the notions of Distributed Cognition); for the purposes of the current EAST analysis, the latter, more contentious, view is taken.
2. Secondly, the 'information' that underlies distributed SA is itself distributed across the entire system, including non-human artefacts that can create, manage and share representational states. This means that the system will be managing the exchange of aspects of representational states through the passage of information between agents.
3. Third, there is often implicit transaction of information rather than a conscious hand-over or exchange. Thus, the update of a representational state for one agent might lead to partial updating of that used by another, and this might not be the result of communication that is managed by human agents (or might not require detailed information processing by any of the agents). For example, as an enemy force element moves across a sector, its route is plotted on the various map displays and its position updated dynamically; if there is little immediate risk, then the updating happens without a corresponding need for intervention or overt 'awareness'.

The concept of SA, therefore, looks rather different when considered from a 'distributed' as opposed to an individual perspective. It is still about the dynamic orientation of a system to its operational context, but the units of analysis centre around information as opposed to individuals (and the psychological processes by which the state of SA is achieved). Information is held, exchanged, represented and transformed by human and non-human agents, its propagation is supported by whatever communication infrastructure is currently in place, managed by the underlying control structure, and evaluated by the underlying command activity.

Information objects, defined as entities or phenomena about which an 'agent' in the system requires information in order to act effectively, are extracted from the CDM interview transcripts using content analysis. Causal links between objects are established in order to create the propositional network. The totality of information residing at the systems level, when modelled as an interconnected web of information, can be viewed as 'systemic SA'. Different parts of the system use/share different items of information; different items of information, when active, relate to yet more information that is connected to it; and the whole configuration of usage and sharing in turn changes dynamically in response to the context.

Theoretical Basis

The combination of observation and interview, analysis and representational methods forms EAST. A more detailed description of how the specific method outputs are derived is provided in the next section (as well as in Walker et al. 2006a). A summary of EAST, and how its component methods relate to each other functionally, is shown below in Figure 3.2.

EAST is a human-centred approach but one whose component methods are all anchored to a common systems perspective. This is manifest in a number of underpinning ideas, principal among which is *distributed cognition*. Under this perspective, the 'computations' that comprise military command and control are not the exclusive province of individuals (Rogers 1997; Hollan et al. 2000; Hutchins 1995); instead, they are distributed across the entire command and control system, comprised of numerous individuals, teams and technical artefacts. The essence of distributed cognition is on 'how [these computations] transcend the boundaries of the individual actor', and to this end, the common language of representational states is used. Representational states are visible and external manifestations of various 'environmental contributions' to the total system (Fields et al. 1998; Rogers and Ellis 1994). They subsume the full range of observable interactions between people and artefacts, as well as the resulting states (and state changes) that arise. In essence, the focus of the EAST method is on describing artefacts that relate to

1. the changes that are made to these representational states
2. their influence and promulgation around a distributed network of human and non-human actors
3. how that influence, in turn, generates new changes to representational states.

FINDINGS

Coordination Demand Analysis (CDA)

According to Salas, Bowers and Cannon-Bowers (1995), 'the military is growing increasingly dependent on the ability of individuals to coalesce quickly into effective

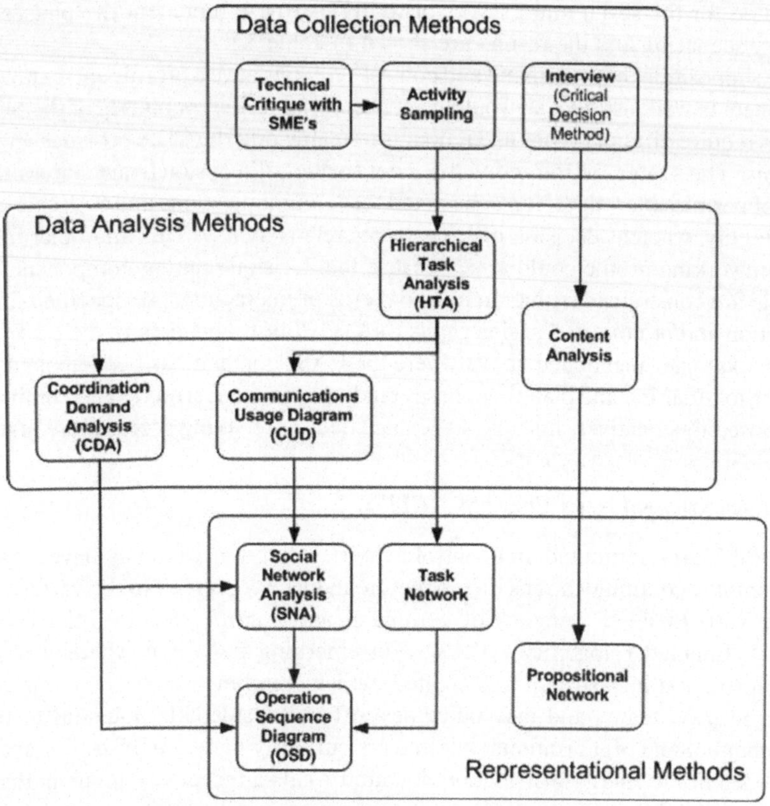

FIGURE 3.2 Structure of the EAST method.

teams' (p. 55). Therefore, it might be assumed that command and control scenarios will be dominated by teamwork activities, but this supposition can be checked in more detail. The Coordination Demand Analysis (CDA) procedure allows for the identification of teamwork skills needed for smooth coordination among team members. Individual tasks from the HTA were categorised into task or teamwork tracks (e.g. Salas et al. 1995; Morgan et al. 1986). The teamwork tasks were then scored against the CDA taxonomy of communication, situational awareness, decision making, mission analysis, leadership, adaptability and assertiveness (Burke 2005). Each CDA taxonomy item was scored from 1 to 3, where 1 is low coordination and 3 is high coordination. From these individual scores, a 'total coordination' figure can be derived, which is based on the mean of the component scores. In this scenario, the scoring was derived from a focus group comprised of the analyst team and subject matter experts. Each item was briefly discussed and its final score based on consensus among the group. Overall, the mean total coordination score for the military scenario is 1.2 (out of a maximum score of three). This score is broadly comparable to civilian command and control domains like air traffic control (e.g. Walker et al. 2006a), railway signalling and safety operations (e.g. Walker et al. 2006b) and bulk energy distribution (e.g. Salmon et al. 2008b). The mean coordination score was also calculated for the seven main stages of the HTA (these represent the phases in the military scenario), and the results are shown in Table 3.1.

The supposition that command and control activities have a prominent teamworking component is well justified, as would be expected, yet different phases of the planning process require different types and amounts of teamwork; the CDA provides a window onto this. The scores for the individual coordination dimensions vary across the full range of permissible values. It can be noted that communications and SA score consistently highly, whereas decision making scores relatively low. This and other facets of the teamworking profile could arise because the decision-making components of the scenario are constrained by hierarchical patterns of interaction, pre-specified forms of interaction and/or unitary decision rights (NATO 2006). This pattern differs from the civilian examples mentioned above where the EAST method has been applied. Here, the decision-making and planning phases tend to occur concurrently and continuously (as opposed to a relatively discrete stage), and decision making scores more highly.

Communications Usage Diagram (CUD)

Current military command and control activities do not necessarily have a particularly complex communications infrastructure in place compared to the various forms of 'Network Enabled' versions of command and control seen in other domains. Network Enabled Capability (NEC) is an emerging paradigm whereby a pervasive information infrastructure is created, which, combined with greater autonomy and a focus on teams and task outcomes, offers considerable potential to impact all the human and organisational parameters currently under analysis. For example, experience in the energy distribution domain reveals a recent development that sees 'peripatetic working' (focusing on autonomous and mobile teams) and the extensive use of tablet PCs and a dedicated information infrastructure (Salmon et al. 2008b). Indeed, EAST is motivated to a great extent by the growing demands of this emerging paradigm. The value in assessing an ostensibly non-NEC military scenario like the current one is to establish a form of baseline.

The advent of NEC technology like advanced communications and data networks has an immediate impact on communications; therefore, its proximal effects are likely to be seen first in this particular sub-method, the Communications Usage Diagram (CUD). In the current scenario, the military planning process involves verbal dialogues between planning personnel in close proximity (during meetings) and announcements in which the command staff will shout (e.g. timescales for upcoming meetings or deadlines). Radio communications are also an integral part of mediating voice communications and take the form of verbal dialogues using a standardised radio telephony method. Communications within the command centre are, therefore, conducted principally by voice. Having said that, the final group of communications media are the operational materials produced out of the planning process, along with the visual aids such as whiteboards and clear overlays that are used during it. All of these communication methods can be analysed using the CUD.

The CUD contains a description of the activity conducted at each geographical location, the communication between actors involved, the technology used for the communications, the advantages and disadvantages associated with that technology

TABLE 3.1
CDA Analysis Results According to Task Phase

Category	Prepare Plan	Display & Manage Information	Combat Estimate (Make Plan)	Translate Products of Q1-7 into Operational Graphics	Conduct Wargame	Execute Plan
Mean Communications	2.5	2	2.2		2	3
Mean SA	2	2	2.2		2	2
Mean DM		1.5	1.8		1	2
Mean MA	1	2	2.1		1	1
Mean Leadership	1.25	1.3	1.6		2	3
Mean Adaptability		2	2		3	2
Mean Assertiveness			1.8		1	2
Total Coordination	1.7	1.8	2.0	0	1.7	2.1

medium and a recommended technology, if there is one. The CUD method, therefore, is a structured way to represent communications within a scenario based on task flow (and thus the HTA). For the purposes of this chapter, the method output has been synthesised into a list of advantages and disadvantages (Table 3.2) followed by a critique of communications usage.

Advantages and Disadvantages of Existing Communications Media

Critique of Communications Usage

The survey of advantages and disadvantages forms part of a structured means to describe and critique existing communications technology. It should be added that the critique is not indicative of any actual or proposed recommendation; rather, it is a consideration of possible alternatives and issues based on the data collected.

Verbal Communications

A compelling advantage of verbal communications is the level of immediacy and redundancy it provides. Carvalho (2006) states that, 'operators use verbal exchanges to produce continuous, redundant and recursive interactions to successfully construct and maintain individual and mutual awareness' (p. 51). There remains, however, the possibility of psychological issues such as 'group-think' and bias in using this form of communication in this context, although there are techniques available to help overcome this (Janis 1982a, 1982b). It can be noted that verbal communications also dominate in civilian command and control scenarios.

Network Enabled Solutions

Gaining situational awareness of the battlespace, a key aspect of the Combat Estimate, requires a number of transformations to be undertaken in order to represent and to understand the state of the world. Voice, written and 2D imagery in the planning process all need to be visualise, and then acted upon in 4D space (3D plus time). Cognitive effort is required, therefore, to achieve adequate levels of SA from which to develop and resource courses of action. Thus, there is a lot of 'information in the head', which can be advantageous for SA. NEC approaches that embody positional data, 3D representations of the battlespace and live updating of it directly from the field embody considerably more 'information in the world'. The task of air traffic control, for example, would be virtually impossible to conduct at its current tempo without the sort of visualisation and live updating provided by the radar display, although even here, the representation is two dimensional. An issue that emerged from the CUD and CDA methods was that the combat estimate 'process' was often just as important as the 'outcome'. For example, a considerable degree of teamworking was devoted to developing SA. In addition, scrutinising the supporting communications methods using the CUD reveals advantages as well as prima-facia disadvantages of what could be considered simplistic planning apparatus.

Social Network Analysis (SNA)

Social Network Analysis (SNA) is a means to present and describe the underlying network structure of individuals or teams who are linked through communications

TABLE 3.2
Advantages and Disadvantages of Existing Communications Media

Media	Advantages	Disadvantages
In-Person Voice	Physical verification that correct individual (and planning role) is being referred to Favourable role of non-verbal communications in aiding shared understanding Possible favourable role of military rank in face-to-face communications Sharing of explanatory resources such as operational graphics/whiteboards etc.	Possible detrimental role of social status/military rank Possible contextually related distractions (e.g. noise and general confusion in the command centre) Possible ambiguity in physically pointing out and referring to shared resources (e.g. plans and whiteboards) Relatively static descriptions of a highly dynamic and spatially dispersed scenario
Radio Communications	Sound stable Possible for communications to be recorded for post-hoc analysis and training Hands free Time saving with common abbreviations and nomenclature Enhanced intelligibility with common abbreviations and nomenclature Read-back provides validation of shared understanding Open channel radio comms. aids shared SA among other units and members of planning staff Possibly favourable dilution of group-think/military rank artefacts on communication and comprehension errors	Intelligibility can be an issue with distortion/artefacts in radio comms. Language/accent ambiguities Unscheduled/ad-hoc presentation of comms. Any informality/abbreviation in comms. relies on assumption of shared meaning Relatively slow communications compared to other comms solutions Translation from verbal domain to visio-spatial domain required (and vice versa) Open channel radio comms. could permit simultaneous comms. on same frequency causing masking Read-back can be out of synchronisation with current activities if sender/recipient is slow to respond Unfavourable dilution of favourable aspects of military rank

(Continued)

TABLE 3.2 (CONTINUED)

Advantages and Disadvantages of Existing Communications Media

Media	Advantages	Disadvantages
Operational Graphics and Planning Aids	Paper-based materials can be substantially degraded without information loss Relatively easy to derive with little extra training required in having to use a pen and paper	Static representations of typically dynamic scenarios Training load relatively high in the use of methods to overcome the disadvantages of representing dynamic 3D phenomena as 2D paper-based representations Complexity and dynamism of scenario can cause administrative bottlenecks. Legibility of graphics and handwriting Whiteboards and overlays potentially cumbersome to handle Document tracking and administration potentially difficult

(Driskell and Mullen 2005). Social networks focus 'on the relationships among actors embedded in their social context' (Driskell and Mullen 2005, p. 58). NEC concepts embody a form of information commodification; thus, social networks can be used to represent the technological mediation of communication and networks where some of the nodes are non-human. The resulting network can then be subject to mathematical analysis using Graph Theory in order to simplify it and express its key features in a standardised form using a numeric index (Driskell and Mullen 2005). Two mathematical indices are used, namely *centrality* (i.e. a numeric ranking allowing key agents in the network to be identified) and *density* (the interconnectivity of the network as a whole). Both of these metrics can be understood in relation to other contextual factors to enable judgements to be made about what aspects of the network configuration constrain or enhance performance. The metrics, being emergent properties of the networks as well as a means to simplify them, permit easy comparison between alternate domains.

The social network for military command and control defines seven key actors, some of which have been grouped into sub-systems for simplicity (as illustrated in Figure 3.3). The actors or nodes include the Higher Command Formation, the Commander and Chief of Staff (COS) at the command headquarters, the 'Principal' Planning Staff (which subsumes individual roles such as G2 Intelligence and EngGeo), other command staff (responsible for more general tasks and information

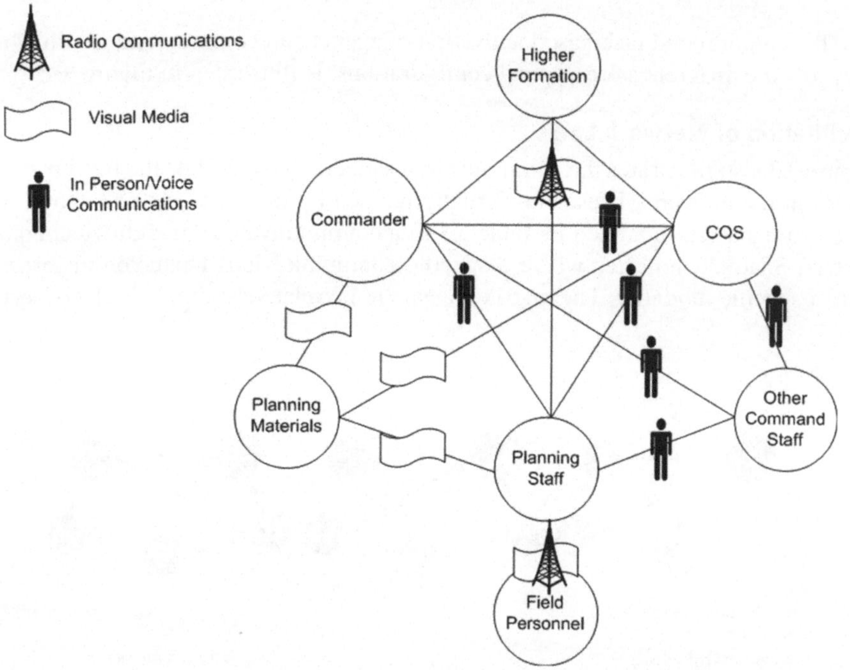

FIGURE 3.3 Social network for military command and control scenario (icons represent the communications media that facilitates the linkage).

management), personnel in the field and the collection of graphics and planning aids derived from the Combat Estimate process (an informational node that is non-human).

Activity Stereotypes

The social network is dynamic and adaptive, with different nodes and links becoming active under different activity stereotypes (revealed by detailed analysis of the HTA). The activity stereotypes are as follows:

- Briefing or providing direction: the Commander is directing communications and information outwards to subordinate staff in a prescribed and tightly coupled manner (particularly Questions 1 and 3 of the Combat Estimate planning technique).
- Reviewing: the planning staff communicate in a more collaborative, peer-to-peer manner, with mutual exchange of information and ad-hoc usage of planning materials and outputs (in particular Questions 2 and 5 of the Combat Estimate).
- Semi-autonomous working: members of the headquarters are working individually on assigned tasks and become relatively loosely coupled in terms of communication. The communication channels remain open but are used in an ad-hoc, un-prescribed manner (this occurs at various points in all phases of the Combat Estimate and the scenario more generally).

The temporal and task-based activation of agents and communications, in which they assume different stereotypical configurations, is illustrated in Figure 3.4.

Facilitation of Network Links

Figure 3.3 also illustrates the communications media that facilitate the links between nodes in the network. These are formally defined by the CUD method above. The results are also summarised in Table 3.3 as a communications/modality/technology matrix. Shading indicates where a specific communications technology is crossed with a specific modality. The matrix appears to be relatively simple in the observed

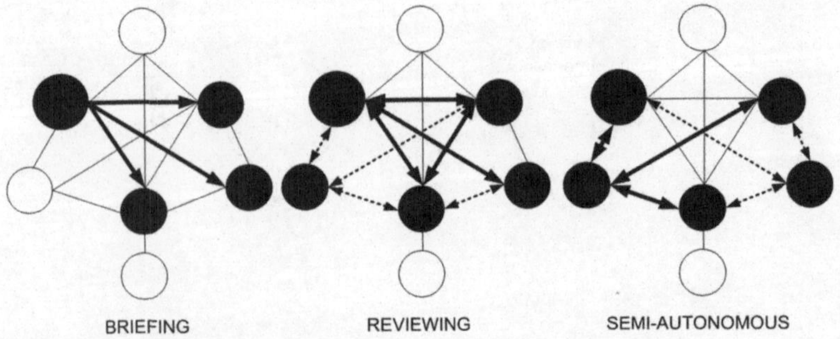

FIGURE 3.4 Social networks illustrating the characteristics of a number of network archetypes detected in the analysis (to be read in conjunction with Figure 3.3).

TABLE 3.3
Technology/Facilitation/Modality Matrix

	Technology/Facilitation		
Modality	**Radio**	**Planning Aids**	**In-Person Voice**
Verbal	■		■
Visual		■	■
Written		■	

Note: Shading Represents a Match Between Communications Technology and Communications Modality.

military scenarios. A heavy reliance on verbal information, which occurs in person and via radio, was noted. There is also heavy reliance on visual communication, as embodied by the various planning aids. In-person communications, it was noted, also incorporate visually mediated communication as staff point and gesture at maps and planning aids. Although simplistic compared to other civilian examples of NEC (like air traffic control) the underlying social and communications system is undeniably robust. Similar robustness might be regarded as a redundant and, indeed, an inefficient feature in civilian contexts. Clearly, there are opportunities, for example, to more rapidly acquire the state of SA through novel technology that does not necessarily rely on verbal communications and manual updating of maps. Such a system is realised in the civilian examples of air traffic control (Walker et al. 2010) and energy distribution (Salmon et al. 2008b), in which the resulting social networks show, in comparison, a much denser interconnection between actors using a more diverse array of technology to facilitate the kind of redundant, recursive interactions that are required in high-tempo tasks. Table 3.3 is based on Figure 3.3 (which in turn is based on the CUD method above) and shows the communication modality that is afforded by the various items of communications technology in the scenario. It is interesting to note that the dominant communications modality is well supported; it is, moreover, interesting to note that future digitisation of military command and control is likely to expand the visual and written modalities somewhat more than it will the verbal modality.

Calculation of Social Network Metrics

The 'most central agents' are revealed, by network mathematics, to be the principal planning staff, followed by the commander and chief of staff (Table 3.4). In NEC scenarios, it might be anticipated that the spread of centrality scores will be less pronounced as a result of more devolved decision rights and peer-to-peer interaction, and this is certainly evident in civilian examples. The network density figure of 0.31 is suggestive of a moderate level of connectivity within the network and is again comparable with civilian examples. The point here is that the total number of available communications links is more or less the same, but they are configured

TABLE 3.4
Network Metrics Illustrating Centrality (Key Agents in the Scenario) and Density (Network Connectivity) for the Social Network as a Whole

Agent	Agent Centrality	Network Density
Higher Formation	0.89	0.31
Commander	1.11	
COS	1.11	
Other Command Staff	0.67	
Principal Staff	1.33	
Field Personnel	0.22	
Planning Materials	0.67	

differently in NEC paradigms. These links define the structure of the network and also its function.

Table 3.5 presents the results of this analysis, showing how the properties of the network change to reflect the stereotypical ways in which it is configured. The change in network density for each activity stereotype is also indicative of a high degree of reconfigurability. This appears to be a relatively unique feature of military command and control. In civilian examples, the network density figures do not change as dramatically as the task progresses through its distinct phases. The reason is due to the complexity inherent in the operational context. Both civilian examples operate within a relatively stable and placid environment in contrast to the turbulent and complex military one.

Operation Sequence Diagrams (OSD)

The Operation Sequence Diagram (OSD) is the main descriptive summary representation within the EAST method. It is an activity-based representation showing who is performing what, when. Using colour coding, symbology and annotation, the OSD can represent most of the critical features of the preceding methods on one common representation. There is insufficient space to present the full OSD analysis of the current scenario, but an opportunity remains to present a small sub-set of the charts in order to reflect on the more distinctive overall features of the analysis so far.

The sample OSD in Figure 3.5 illustrates many of the facets dealt with in earlier methods. Teamworking and coordination is reflected not only by the explicit colour coding of operations (darker shading denotes higher levels of total coordination) but also in the varying patterns of connectivity between and among operations (represented by the symbols). The social network stereotypes are also represented by the configurations of links and operations. For example, in the earlier phases of the Combat Estimate (more Briefing activities), there is a focus on the hierarchical/

TABLE 3.5

Network Metrics Illustrating Centrality (Key Agents in the Scenario) and Density (Network Connectivity) for the Activity Stereotypes of Briefing, Reviewing and Semi-Autonomous Working

Agent	Centrality			Density		
	Briefing	Reviewing	Semi-Autonomous	Briefing	Reviewing	Semi-Autonomous
Higher Formation Orders				0.03	0.20	0.13
Commander	0.33	0.67	0.33			
COS	0.11	0.67	0.33			
Other Command Staff	0.11	0.33	0.33			
Principal Staff	0.11	0.67	0.33			
Field Personnel						
Planning Materials		0.33	0.67			

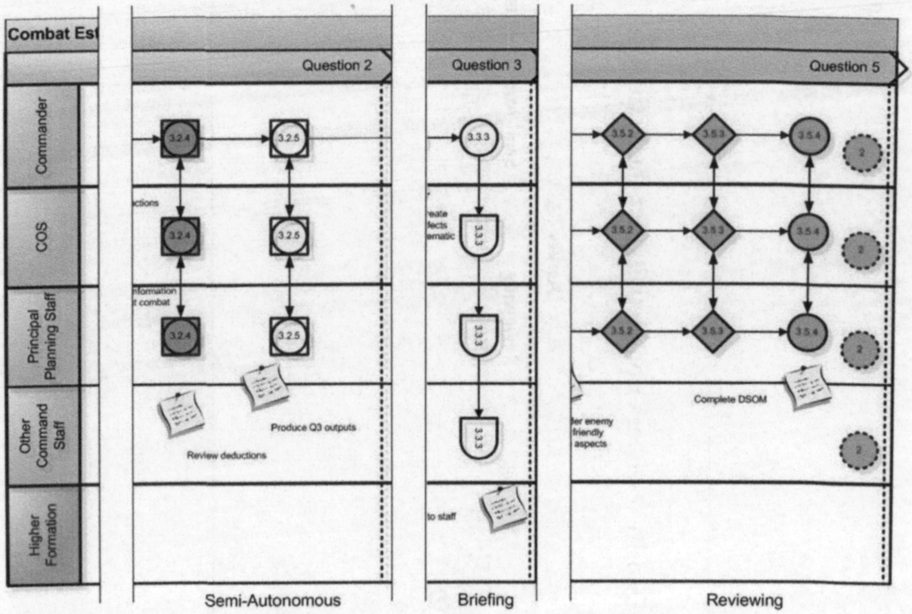

FIGURE 3.5 Sections through complete OSD chart for military scenario showing (left to right) semi-autonomous working, briefing and reviewing.

vertical flow of information, with the Commander producing it and other agents receiving it. In the second phase (more Reviewing activities), the operations (and agents) become closely interconnected, with decision-making components dominating (shown by the diamond symbol[s]). In the third and final phase (more Semi-Autonomous working), the pattern of operations once again assumes a vertical/hierarchical disposition. At the highest level, then, the command and control process seems to assume a pattern of information retrieval, closely knit decision-making processes and then dispersion and action.

Propositional Networks (PN)

From the CDM interview, it is possible to construct Propositional Networks (PN) (an example of which is shown in Figure 3.6) to show the information that is related to the scenario. The propositional network consists of a set of nodes that represent sources of information, agents and objects that are linked through specific causal paths (for example, the object [situation] 'has' the property of [updates] associated with it, etc.). As mentioned earlier, these objects are extracted from the CDM interview transcripts using content analysis. The deeper, more fundamental concept that this method refers to is SA, an important concept in decision making, agility and tempo. The advantage of the propositional network approach is that it represents a way of modelling the information that comprises the state of SA, from an individual as well as a systems perspective. In addition, because it is network based, it meshes with the social and task networks that form the basis for the rest of the EAST

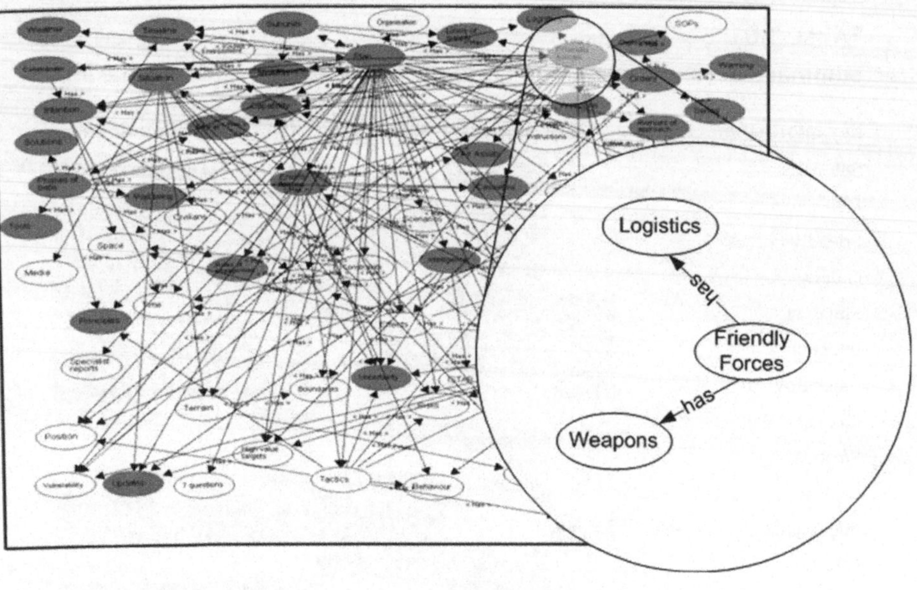

FIGURE 3.6 Illustration of propositional network representing a systems level view of information but also temporally activated information for a particular task phase (shaded objects). The overall network is presented for illustration; the 'zoomed' section shows some of the detail.

method. From the PN, it is possible to identify the structure and temporal nature of distributed SA (explained in full in Stanton et al. 2009) and the information underpinning decision making.

The summary table (Table 3.6) uses simple Graph Theory metrics (as used in the social networks above) to summarise the visually complex network(s) into a more tractable form. Based on an analysis of centrality, so-called 'core information objects' can be defined for each phase in the scenario (a CDM interview was carried out in relation to each phase, as was a separate PN). The table crosses each phase of the Combat Estimate planning technique with the list of core information objects. Shading denotes specifically what information objects are active in what phase. These core information objects also feed back up to the CUD method earlier. Their prescription enables an analysis of what information objects are shared between what agents, and, therefore, require some form of communications technology to mediate the sharing. In the CUD method, the appropriateness of this match forms one aspect of the basis by which communications technology is, and can be, critiqued (see Walker et al. 2006a for an in-depth treatment of how this specific approach can be realised).

Eighteen key information objects can be identified. As the Combat Estimate planning process progresses through its distinct phases, it can clearly be seen that the activation of these key objects changes. This is further indicative of changes in the type and structure of the propositional network and in the type of situational awareness possessed by individuals and the system as whole. This information can be

TABLE 3.6
Summary of Key Information Objects Active within Each Scenario

Key Information Objects	Q1	Q2 & 3	Q4	Q5	Q6	Implementing Plan
Sub-units		■			■	■
Plan		■	■		■	■
Friendly Forces		■	■	■	■	■
Orders		■		■	■	■
Situation	■	■		■	■	■
Intention	■	■	■		■	■
Capability	■	■			■	■
Phases of Battle					■	
Weapons		■			■	
Enemy		■	■	■	■	■
Intelligence	■		■	■	■	■
Effects		■	■	■		
Courses of Action		■			■	■
Uncertainty	■				■	
Terrain	■	■		■		
Position	■					
Tactics	■					
History	■			■		
Total Information Objects	9	12	6	8	13	10

Note: Shading indicates what information is specifically active during what stage of the combat estimate.

used to assess the extent to which the system is orientated towards the dynamics of its situation as well as to identify information requirements for different tasks and scenarios.

CONCLUSIONS

The principles of Human Factors Integration (HFI) encompass 'a balanced development of both the technical and human aspects of equipment provision. It provides a process that ensures the application of scientific knowledge about human characteristics through the specification, design and evaluation of systems' (MoD 2000, p. 6). The EAST method is couched firmly within this context. It is, at bottom, a way of capturing and describing the human view of complex sociotechnical systems – not what should or might happen when they are set to work but what actually does happen. Being based on the integration of seven existing ergonomics methodologies, the outputs are structured, systematic and standardised. The validation history that is associated with each of them brings the promise of repeatability and compatibility across situations and even domains.

The aim of the present analysis has been to demonstrate the applicability of the method within military contexts, which represent a particularly challenging and complex command and control environment. The analysis has been couched at a fairly coarse-grained level of analysis, anchored as it has been to the discrete phases of the Combat Estimate planning process. As much (or as little) resolution can be extracted depending on the questions being asked and analysis effort that actors are willing to expend. For the purposes of presenting EAST's human view of military command and control, and how those insights map across to other domains, the present level of analysis seems appropriate.

What does this application of the EAST method tell us? It tells us about the structural and temporal nature of the goals and tasks that have to be performed and how they are interlinked and interdependent. The structure of the task network (like any system) defines its function. In the case of the military, it is noted not just that there is a defined 'planning phase' but also that it is relatively sequential and recursive in nature. In the civilian domains that have been observed previously, planning, at least at strategic and operational levels, tends to occur concurrently with the task – 'on the fly', so to speak. That said, in civilian domains, the sociotechnical system's combined experience (of which it is an evolutionary expression) and its generally placid environment enables rule-based optimum means to ends to be learnt and rapid decision making to take place.

It can be noted that the function of the social networks also changes; they are reconfigurable (and readily reconfigured) to suit the task phase, often in ways that are not anticipated beforehand. This is, therefore, an emergent property of the interaction between the task and social networks. The links in the social network are informational, and the content of this information is represented by the PN, a representation that captures systems level information and, we argue, SA. SA is once again an emergent property of both the task and social networks. In particular, the way in which the social network is configured dictates the informational constraints placed on it by the facilitating communications technology. Overall, it is clear that cognition is distributed between socio and technical agents in the scenario. This, combined with the fact that the socio elements (the people) will adapt the technical elements to suit their needs and preferences, contributes to a degree of self-organisation and emergence that is apparent through analysis with EAST.

EAST's three main network representations, illustrated in Figure 3.7, are an analytical response to complexity and emergence but from a uniquely human perspective. A key point is that EAST acknowledges the sociotechnical principle of equifinality: that the same people and the same technology may reach the same goal by entirely different means and from entirely different initial conditions. This is not a conceptual inconvenience but actually one of the very promises of NEC. The inviolable fact that humans will adapt to the techno-organisational properties of a given system is what permits desirable emergent properties to 'emerge', like systems level 'shared awareness', tempo, agility and self-synchronisation. EAST seems to offer a way to capture and describe this in a way that deterministic (i.e. linear) methods may not.

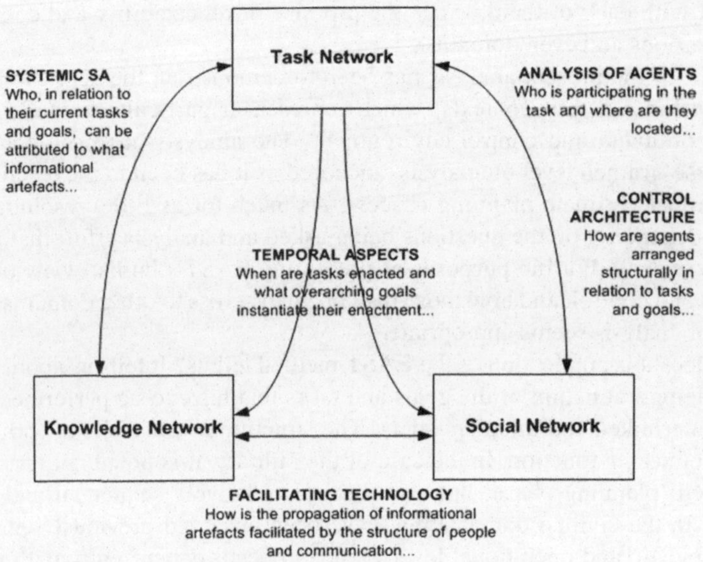

FIGURE 3.7 Summary of EAST's three network-based perspectives and the insights into emergent properties that they provide.

Driven off the higher-level network representations are a number of more explicit methods that relate more directly to requirements capture. In the CUD, for example, task and social networks combine to inform a structured assessment of communications technology. This analysis is further informed by the notions of distributed SA that are embodied in EAST. In other words, it becomes possible to assess what purpose a communication exchange is designed to serve, what information it is designed to convey and, moreover, what would be the best means of facilitating that exchange. In addition to this, the CDA method highlights certain important aspects of teamworking at different phases of the planning process. This sort of information is useful in the design of digital command planning systems. For example, there would be little point in providing 'decision support' for a planning phase that is actually revealed as requiring more 'mission analysis'. The benefits of the interlinked EAST method are that changes in any of the three underlying network representations (task, social and propositional networks) are reflected in each other and in the supporting methods like CUD and CDA. The disadvantage of the EAST method, at least at face value, is the amount of time it takes to perform such an analysis. The reader is referred to a companion software tool called WESTT that greatly simplifies and streamlines the application of EAST, automating many functions, notably the production of the various task, social and knowledge networks (Houghton et al. 2007).

In conclusion, EAST is offered as a means to describe the dynamic, emergent behaviour of complex sociotechnical systems from a human perspective. It appears consistent with wider trends in Ergonomics towards method integration and systems views, and it is eminently compatible with similar approaches like CWA. The aim of the present high-level analysis has been to illuminate some of these wider issues and the applicability of EAST to military contexts. More targeted forms of application will enable EAST to provide more targeted and direct answers to design and procurement questions in an age when the critical networks in NEC are not necessarily technological but human.

ACKNOWLEDGEMENTS

This work from the Human Factors Integration Defence Technology Centre (HFI DTC) was part-funded by the Human Sciences Domain of the UK Ministry of Defence Scientific Research Programme.

The assistance of LtCol Allan Ellis is gratefully acknowledged as is the assistance of the staff at the Land Warfare Centre and CAST in Warminster, in particular Majors Forster and Sharkey.

Information on the Combat Estimate draws heavily on material used during the experimental team's observation of the Combat Estimate training noted above. In particular, two presentations, the first entitled 'Command and Staff Trainer (South) 1 WFR MiniCAST 2/3 Aug 05' and the second 'Wargaming: Mastering Your Enemy' (based on the Army Field Manual Vol I (Combined Arms Operations) Part 2 (July 98) and 3(UK) Div Wargaming Aide Memoire), are used with permission.

REFERENCES

Adams, M. J., Tenney, Y. J. and Pew, R. W. (1995). Situation awareness and the cognitive management of complex systems. *Human Factors*, 37, 85–104.

Annett, J. (2005). Hierarchical task analysis. In: N. A. Stanton, A. Hedge, K. Brookhuis, E. Salas and H. Hendrick (Eds.), *Handbook of Human Factors and Ergonomics Methods* (pp. 33.1–33.7). CRC: London, UK.

Bar Yam, Y. (1997). *Dynamics of Complex Systems*. Perseus: Jackson, TN.

Bedny, G. and Meister, D. (1999). Theory of activity and situation awareness. *International Journal of Cognitive Ergonomics*, 3 (1), 63–72.

Bell, H. H. and Lyon, D. R. (2000). Using observer ratings to assess situation awareness. In: M. R. Endsley (Ed.), *Situation Awareness Analysis and Measurement* 129–146). Lawrence Erlbaum Associates: Mahwah, NJ.

Builder, C. H., Bankes, S. C. and Nordin, R. (1999). *Command Concepts: A Theory Derived from the Practice of Command and Control*. Rand: Santa Monica, CA.

Burke, S. C. (2005). Team task analysis. In: N. A. Stanton A. Hedge, K. Brookhuis, E. Salas and H. Hendrick (Eds.), *Handbook of Human Factors and Ergonomics Methods* (pp. 56.1–56.8). CRC: London, UK.

Carvalho, P. V. R. d. (2006). Ergonomic field studies in a nuclear power plant control room. *Progress in Nuclear Energy*, 48, 51–69.

Clegg, C. W. (2000). Sociotechnical principles for system design. *Applied Ergonomics*, 31, 463–477.

DoD (1999). *Realizing the Potential of C4I: Fundamental Challenges*. National Academy Press: Washington, DC.

Driskell, J. E. and Mullen, B. (2005). Social network analysis. In: N. A. Stanton, A. Hedge, K. Brookhuis, E. Salas and H. Hendrick (Eds.), *Handbook of Human Factors and Ergonomics Methods* (pp. 58.1–58.6). CRC: London, UK.

Endsley, M. R. (1995). Toward a theory of situation awareness in dynamic systems. *Human Factors*, 37 (1), 32–64.

Fields, R., Wright, P., Marti, P. and Palmonari, M. (1998). Air traffic control as a distributed cognitive system: A case study of external representations. In: T. R. G. Green, L. Bannon, C. P. Warren and J. Buckley (Eds.), *ECCE9: Proceedings of the Ninth European Conference on Cognitive Ergonomics* (pp. 85–90). EACE: Limerick, Ireland.

Harris, C. J. and White, I. (1987). *Advances in Command, Control & Communication Systems*. Peregrinus: London, UK.

Hollan, J., Hutchins, E. and Kirsh, D. (2000). Distributed cognition: Toward a new foundation for human–computer interaction research. *ACM Transactions on Human Computer Interaction*, 7, 174–196.

Hollnagel, E. (1993). *Human Reliability Analysis: Context and Control*. Academic Press: London, UK.

Houghton, R. J., Baber, C., Cowton, M., Walker, G. H and Stanton, N. (2007). WESTT (Workload, error, situational awareness, time and teamwork): An analytical prototyping system for command and control. *Cognition Technology and Work*, 10 (3), 199–207.

Hutchins, E. (1995). *Cognition in the Wild*. MIT Press: Cambridge, MA.

Janis, I. L. (1982a). *Groupthink*. Houghton Mifflin: Boston, MA.

Janis, I. L. (1982b). Counteracting the adverse effects of concurrence-seeking in policy planning groups: Theory and research perspectives. In: H. Brandstatter, J. H. Davis and G. Stocker-Kreichgauer (Eds.), *Group Decision Making*. Academic Press: London, UK.

Jones, C. (1993). *Toward a Science of Command, Control, and Communications*. American Institute of Aeronautics and Astronautics: Washington, D.C.

Johnson, C. W. (2005). *What Are Emergent Properties and How Do They Affect the Engineering of Complex Systems?* University of Glasgow, Department of Computing Science: Glasgow, UK.

Kirwan, B. and Ainsworth, L. K. (1992). *A Guide to Task Analysis*. Taylor and Francis: London, UK.

Klein, G. and Armstrong, A. A. (2005). Critical decision method. In: N. A. Stanton, A. Hedge, K. Brookhuis, E. Salas and H. Hendrick (Eds.), *Handbook of Human Factors and Ergonomics Methods* (pp. 35.1–35.8). CRC: London, UK.

Kleiner, B. M. (2006). Macroergonomics: Analysis and design of work systems. *Applied Ergonomics*, 37, 81–89.

Marashi, E. and Davis, J. P. (2005). A systems approach to resolving complex issues in a design process. *Workshop on Complexity in Design and Engineering*, 10–12 March, Glasgow, Scotland, pp. 140–149.

MoD (2000). *Human Factors Integration: An Introductory Guide*. HMSO: London, UK.

MoD (2005a). *Network Enabled Capability (JSP 777)*. HM Stationery Office: London, UK.

MoD (2005b). *The BG PSO Combat Estimate*. MoD: Warminster, UK.

Moray, N. (2004). Où sont les neiges d'antan? In: D. A. Vincenzi, M. Mouloua, and P. A. Hancock (Eds.), *Human Performance, Situation Awareness and Automation: Current Research and Trends*. Lawrence Erlbaum Associates: Mahwah, NJ.

Morgan, G. (1986). *Images of Organisation*. Sage: London, UK.

NATO (2006). SAS-050 conceptual model version 1.0. Available at: https://pdfs.semanticscholar.org/0937/20a588962d1a80a31360e869515b3f76de02.pdf.

Ogden, G. C. (1987). Concept, knowledge and thought. *Annual Review of Psychology*, 38, 203–227.

Perla, P., Markowitz, M., Nofi, A., Weuve, C., Loughran, J. and Stahl, M. (2000). Gaming and shared situation awareness. Available at: http://www.thoughtlink.com/publications/GamingSSA00Abstract.html.

Rasmussen, J., Petjerson, A. M. and Goodstein, L. P. (1994). *Cognitive Systems Engineering*. Wiley: New York, NY.

Rogers, Y. (1997). *A Brief Introduction to Distributed Cognition*. Interact Lab, University of Sussex: Brighton, UK.

Rogers, Y. and Ellis, J. (1994). Distributed cognition: An alternative framework for analysing and explaining collaborative working. *Journal of Information Technology*, 9 (2), 119–128.

Salas, E., Bowers, C. A. and Cannon-Bowers, J. A. (1995). Military team research: Ten years of progress. *Military Psychology*, 7, 55–75.

Salmon, P. M., Stanton, N. A., Walker, G. H., Baber, C., Jenkins, D. P., McMaster, R. and Young, M. S. (2008a). What really is going on? Review of situation awareness models for individuals and teams. *Theoretical Issues in Ergonomics Science*, 9 (4), 297–323.

Salmon, P. M., Stanton, N. A., Walker, G. H., Jenkins, D. P., Baber, C. and McMaster, R. M. (2008b). Representing situation awareness in collaborative systems: A case study in the energy distribution domain. *Ergonomics*, 51 (3), 367–384.

Smith, E. A. (2006). *Complexity, Networking, & Effects-Based Approaches to Operations*. CCRP Publication Series.

Smith, K. and Hancock, P. A. (1995). Situation awareness is adaptive, externally directed consciousness. *Human Factors*, 37 (1), 137–148.

Stanton, N. A., Salmon, P., Walker, G. H., Baber, C. and Jenkins, D. (2005a). *Human Factors Methods: A Practical Guide for Engineering and Design*. Ashgate: Aldershot, UK.

Stanton, N. A., Hedge, A., Brookhuis, K., Salas, E. and Hendrick, H. (2005b). *Handbook of Human Factors and Ergonomics Methods*. CRC: Boca Raton, FL.

Stanton, N. A., Salmon, P. M., Walker, G. H. and Jenkins, D. P. (2009). Genotype and phenotype schemata and their role in distributed situation awareness in collaborative systems. *Theoretical Issues in Ergonomics Science*, 10 (1), 43–68.

Vicente, K. J. (1999). *Cognitive Work Analysis: Toward Safe, Productive, and Healthy Computer-Based Work*. Lawrence Erlbaum Associates: Mahwah, NJ.

Walker, G. H., Gibson, H., Stanton, N. A., Baber, C., Salmon, P. M. and Green, D. (2006a). Event analysis of systemic teamwork (EAST): A novel integration of ergonomics methods to analyse C4i activity. *Ergonomics*, 49, 1345–1369.

Walker, G. H., Stanton, N. A., Gibson, H., Baber, C., Young, M. and Green, D. (2006b). Analysing the role of communications technology in C4i scenarios: A distributed cognition approach. *Journal of Intelligent Systems. Special Issue: Information and Communication Technology*, 15 (1–4), 299–328.

Walker, G. H., Stanton, N. A., Baber, C., Wells, L., Gibson, H., Salmon, P. M. and Jenkins, D. P. (2010). From ethnography to the EAST method: A tractable approach for representing distributed cognition in air traffic control. *Ergonomics: Special Issue on Aviation Safety*, 53 (4), 184–197.

Watts, L. A. and Monk, A. F. (2000). Reasoning about tasks, activities and technology to support collaboration. In: J. Annett and N. Stanton (Eds.), *Task Analysis* (pp. 55–78). Taylor and Francis: London, UK.

Wilson, J. R. and Corlett, E. N. (Eds.) (1995). *Evaluation of Human Work: Practical Ergonomics Methodology*. Taylor and Francis: London, UK.

4 EAST in Energy Distribution Operations

With Daniel P. Jenkins, Chris Baber and Richard McMaster

INTRODUCTION

The Event Analysis of Systemic Teamwork (EAST) analysis presented in this chapter focuses on command and control activities in the energy distribution domain. Command and control scenarios are characterised by multiple individuals and teams working together in pursuit of a common goal (comprising multiple interacting sub-goals). High levels of communication and coordination are required, and the onus is often placed on technologies to facilitate this. Various sub-constructs are also evident within command and control, including planning, directing, coordinating and control of resources, situation awareness and so on. When examining command and control, the descriptive constructs of interest can be distilled down to simply

- *why*: (the goals of the system, sub-system[s] and actor[s])
- *who*: (the actors performing the activity, including humans and technologies)
- *when*: (when activities take place and which actors are associated with them)
- *where*: (where activities and actors are physically located)
- *what*: (what activities are undertaken, what knowledge/decisions/processes/devices are used and what levels of workload are imposed)
- *how*: (how activities are performed and how actors communicate and collaborate to achieve goals).

More than likely, none of the Human Factors methods described in this book can independently cover all of these constructs. Using an integrated suite of methods, however, allows scenarios to be analysed exhaustively from the perspectives described, but more importantly, it allows the effects of constructs on other constructs to be considered, for example, how communications influence the way in which tasks are performed and in turn how the way in which the tasks are being performed influences the knowledge/decisions/processes being used. There are also further advantages associated with methods integration, because not only does the integration of existing methods bring reassurance in terms of a validation history, but it also enables the same data to be analysed from multiple perspectives. These multiple perspectives, as well as being inherent in the scenario that is being described and measured, also provide a form of internal validity. Assuming that the separate methods integrate on a theoretical level, then their application to the same data set offers a form of 'analysis triangulation'.

EAST (Stanton et al. 2005) provides one such framework of methods that allows collaborative performance to be comprehensively described and evaluated. Since its conception, the framework has been applied in many domains, including land and naval warfare (Stanton et al. 2006), aviation (Stewart et al. 2008), air traffic control (Walker et al. 2010), railway maintenance (Walker et al. 2006) and the emergency services (Houghton et al. 2006). EAST is underpinned by the notion that complex collaborative systems can be meaningfully understood through a network of networks approach (see Figure 4.1). Specifically, three networks are considered: task, social and knowledge networks. Task networks describe the goals and subsequent tasks being performed within the system. Social networks analyse the organisation of the system (i.e. communications structure) and the communications taking place between the actors working in the team. Finally, knowledge networks describe the information and knowledge (distributed situation awareness) that the different actors use and share during task performance.

The process begins with the conduct of an observational study (Chapter 2) of the scenario under analysis. Hierarchical Task Analysis (HTA) (Chapter 3) is then used to describe the goals, sub-goals and operations involved during the scenario. The resultant task network identifies the actors involved, what tasks are being performed, the temporal structure of tasks and the interrelations between tasks. The HTA output is then used to construct an Operation Sequence Diagram (OSD) (Chapter 3) of the task, detailing all activities and interactions between actors involved. The social network is embodied by Social Network Analysis (SNA) (Chapter 8), which considers the associations between agents during the scenario. It is important to note here that agents may be human or technological, and so SNA caters too for human–machine interactions. The Critical Decision Method (CDM) (Chapter 4) focuses on the decision-making processes used during task performance and the information and

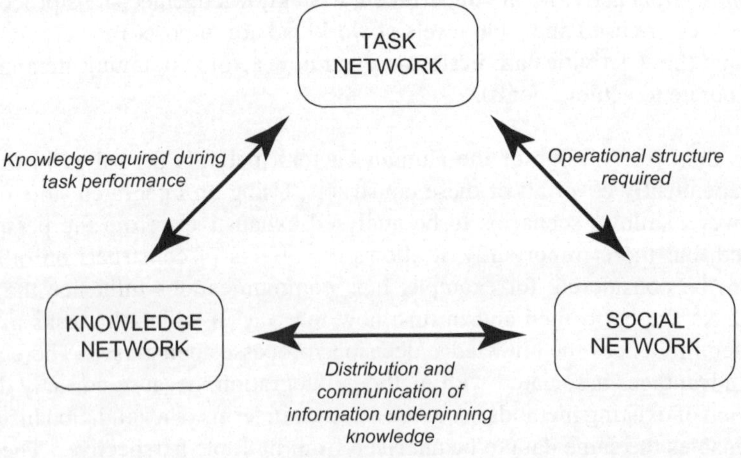

FIGURE 4.1 Network of networks approach.

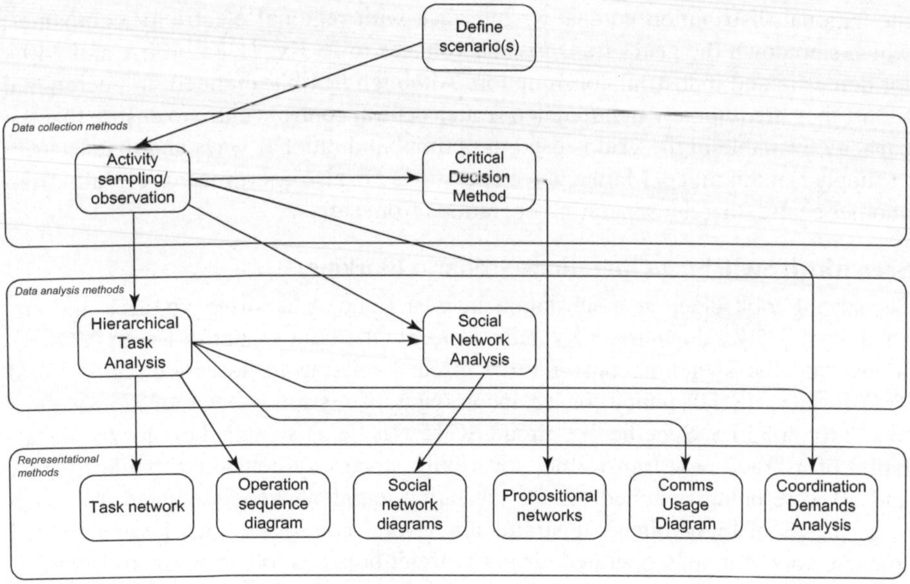

FIGURE 4.2 Internal structure of EAST framework.

knowledge underpinning decision making. Based on these data, the propositional network approach (Chapter 6) represents situation awareness during the scenario, from both the point of view of the overall system and that of the individual agents performing activities within the system. Thus, the type, structure and distribution of knowledge throughout the scenario is represented. The overlap between methods and the constructs they access is explained by the multiple perspectives provided on issues such as the 'Who' and the 'What'. For example, the HTA deals with 'what' tasks and goals, the CDM deals with 'what' decisions are required to achieve goals and the propositional networks deal with 'what' knowledge or situation awareness underpins the tasks being performed and decisions being made. Each is a different but complementary perspective on the same descriptive construct, and a different but complementary perspective on the same data derived from observation and interview, which is an example of analysis triangulation. The structure of the EAST framework is represented in Figure 4.2.

Example Case Study: A Human Factors Analysis of Civilian Command and Control

The analysis presented in this chapter concerns two scenarios undertaken on a major UK electrical distribution network. The distribution grid in question consists of 341 geographically dispersed substations in England and Wales, which are used to distribute electricity to consumers. Power stations (and feeds from continental Europe) energise the grid, which uses an interconnected network of 400,000 volt (400 Kv), 275 Kv (the super grid network) and 132 Kv overhead lines and towers or cables running in tunnels to carry electricity from source to substations. The substations are the national distribution company's interface with regional electricity companies, which step down the grid's transmission voltages to 33 Kv, 11 Kv, 400 v and 240 v for domestic and industrial consumption. Although flexibly manned, in operational terms, they are remotely manipulated from a central control centre to ensure that the capacity available in the grid is used in optimal and rational ways and that security of supply is maintained. Maintenance operations are also coordinated centrally from another centre, thereby separating operations from safety.

Scenario 1: Switching Operations Scenario (Barking)

Scenario 1 took place at a substation in East London handling voltages and circuits from 275 Kv down to 33 Kv and a Central Operations Control Room (COCR). It involved the switching out of three circuits relating to so-called 'Super Grid Transformers' (SGT), which convert incoming transmission voltages of 275 Kv down to 132 Kv or 33 Kv. Specifically, circuit SGT5 was being switched out for the installation of a brand-new transformer for a bulk electricity consumer, while SGT1A and 1B were being switched out for substation maintenance. Associated with such large pieces of high-voltage apparatus are several control circuits, large overhead line isolators, remotely operated air blast circuit breakers, other points of isolation, compressed air equipment and oil cooling apparatus. All of this disparate equipment had to be handled and made safe by qualified personnel in a highly prescribed manner. In addition, the work had to be centrally pre-planned by electrical engineers to ensure that other circuits were not affected and that the balance and capacity of the system was not compromised. Qualified personnel worked on-site to these plans and liaised with the COCR at key points during this process.

Scenario 2: Maintenance Scenario (Tottenham)

Scenario 2 took place at the COCR and a rural substation site. It involved the switching out of circuits and overhead lines in order to permit work to commence on pieces of the control equipment (Current and Voltage Transformers) used to provide readings and inputs into other automatic, on-site current, voltage and phase regulation devices. In addition, maintenance work was to be carried out on a Line Isolator (the large mechanical switching device that provides a point of isolation for a specific overhead line that departed from this substation [A] and terminated at another substation [B]), and major maintenance on a device called an *earth switch*. There were three main parties involved in the outage: a party working at Substation B on the outgoing substation A circuit, a party working at Substation A on the substation B circuit and an overhead line party working in between the two sites.

In both scenarios, the COCR operator, located a central control room (see Figure 4.3), took on the role of commander, planning the work to be undertaken and distributing work instructions to the Senior Authorised Persons (SAPs) and Authorised Persons (APs) located at the substations where the work was to be undertaken. In addition to overseeing the activities that were analysed for the purposes of this research, the COCR operator was also involved in other activities being undertaken elsewhere on the grid and so had other responsibilities and tasks to attend to during the study. The other agents involved in the scenarios included the Central Command (CC) Operator and Overhead Line Party (OLP) personnel working on the overhead

FIGURE 4.3 Central control room.

lines. The COCR operator communicated with the other agents via landline telephone and mobile phone. The COCR operator also had access to substation diagrams, work logs and databases and the internet. The structure of personnel for Scenario 1 is presented in Figure 4.4. The structure of personnel for Scenario 2 is presented in Figure 4.5.

METHODOLOGY

Design

The study was an observational study that involved direct observation of the activities undertaken during the scenarios analysed. Three researchers acted as observers, with one located in the field at the substation involved and two located at the central control room.

Participants

This study involved 11 participants who work for the energy distribution organisation in question. Scenario 1 involved the following four participants: a CC operator, a COCR operator and a SAP and AP. Scenario 2 involved the following seven participants: a CC operator, a COCR operator and a SAP, AP and Competent Person (CP) at one substation, and SAP at another substation and an overhead line party contact. Due to access restrictions and the nature of the study (observation during real work activities), it was not possible to collect demographic data for the participants involved.

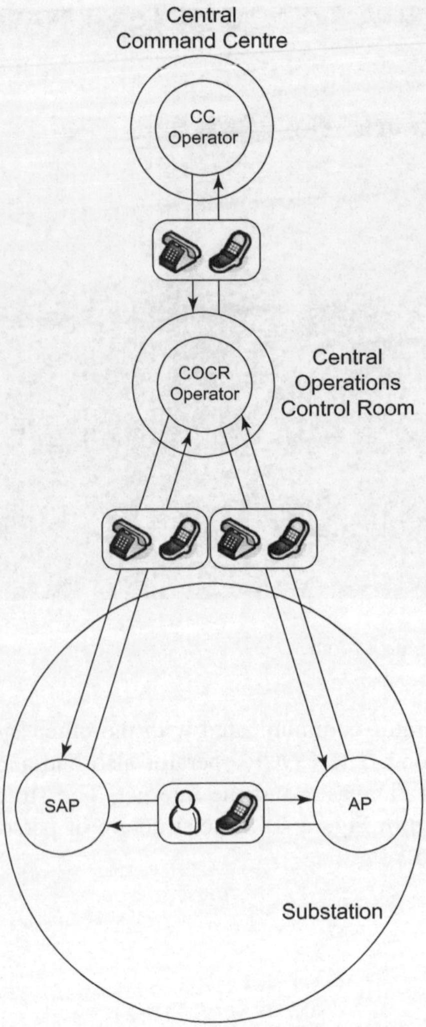

FIGURE 4.4 Scenario 1 network structure; CC Operator = Central Command Operator; COCR Operator = Central Operations Control Room Operator; SAP = Senior Authorised Person; AP = Authorised Person.

Materials

The observers used pen and paper, video and audio recording equipment to collect data during the observations. All observational and verbal transcripts were transcribed using Microsoft Word. A set of pre-defined interview probes were used for the CDM interviews, along with a CDM pro-forma to record participants' responses. A Dictaphone was also used to record the CDM interviews. Various other software programs were used to support the analysis presented, including the Agna network analysis and WESTT software packages and Microsoft Visio.

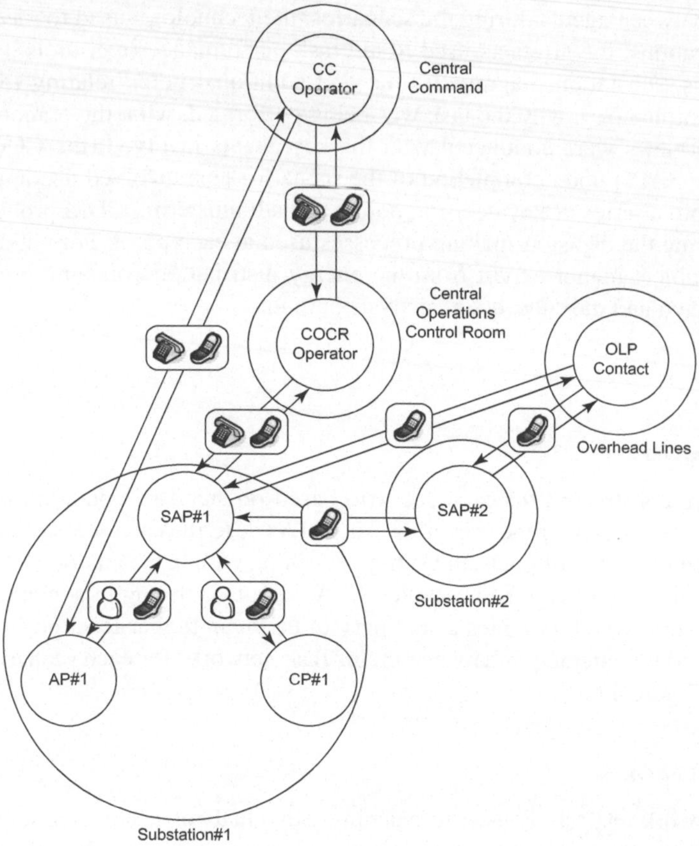

FIGURE 4.5 Scenario 2 network structure; CC Operator = Central Command Operator; COCR Operator = Central Operations Control Room Operator; SAP = Senior Authorised Person; AP = Authorised Person; CP = Competent Person; OLP Contact = Overhead Line Party Contact.

Procedure

The analyses were based on data collected during live observational study of the two scenarios. In each scenario, two observers were located at the COCR observing the COCR operator, and one observer was located in the field with the SAPs/APs at the substation involved. The analysts located at the COCR observed all of the COCR operator's activities and were able to discuss the activities being undertaken and to query different aspects of the scenarios as they unfolded. The analyst located at the substations observed the SAP/APs undertaking the work required and was also able to discuss aspects of the scenario with them. Observational transcripts were constructed, and audio recordings were used to record the communications between those involved. The data recorded via observational transcripts included a description of the activity (specific task steps, e.g. 'issue instructions to SAP at substation') performed by each of the agents involved, transcripts of the communications that occurred between agents during the scenarios, the technology used to mediate these communications, the artefacts used to aid task performance (e.g. tools, computers, instructions, substation diagrams), time and additional notes relating to the tasks being performed (e.g. why the task was being performed, what the outcomes were). CDM interviews were conducted with the key agents involved (the COCR operator and the SAPs) upon completion of the scenario. This involved decomposing the scenario into a series of key decision points and administering CDM probes in order to interrogate the decision-making processes used at each point. For validation purposes, a subject matter expert from the energy distribution company reviewed the data collected and the subsequent analysis outputs.

RESULTS

Task Networks

HTAs were constructed for each scenario based on standard operating procedures and direct observation of the scenarios. The HTAs describe each scenario in terms of a structured hierarchy of goals and sub-goals, along with feedback loops. One useful way of summarising large and complex HTA outputs is through the construction of a task network, which provides a summary of the main higher-level goals and tasks involved and the interaction between them. Task networks for each scenario are presented in Figure 4.6.

Social Networks

Social network analysis is used to examine individuals or teams who are linked to each other by communications and to subject that network to mathematical analysis using Graph Theory (Driskell and Mullen 2004). Social networks were constructed for each scenario based on the observed communications between agents during task performance. The structure of each network was then analysed using three network analysis metrics: density, sociometric status and centrality. Network *density* represents the level of interconnectivity of the network in terms of communications links between agents. Density is expressed as a value between 0 and 1, with 0 representing a network with no connections between agents and 1 representing a network in which every agent is connected to every other agent (Kakimoto et al. 2006; cited in Walker et al. 2011). *Sociometric status* provides a measure of how 'busy' each agent is relative to the total number of agents within the network under analysis (Houghton et al. 2006). *Centrality* is also a metric of the standing of each agent within a network (Houghton et al. 2006), but here this standing is in terms of its 'distance' from all other agents in the network. A central agent is one that is close to all other agents in the network, and a message conveyed from that agent to an arbitrarily selected other agent in the network would, on average, arrive via the least number of relaying hops (Houghton et al. 2006). The social network analysis for Scenarios 1 and 2 are presented in Figure 4.7.

The social network analysis outputs allow identification of the key communications 'hubs' within each network, as well as a comparison of the communications network apparent in each scenario. Key agents, specifically those that act as

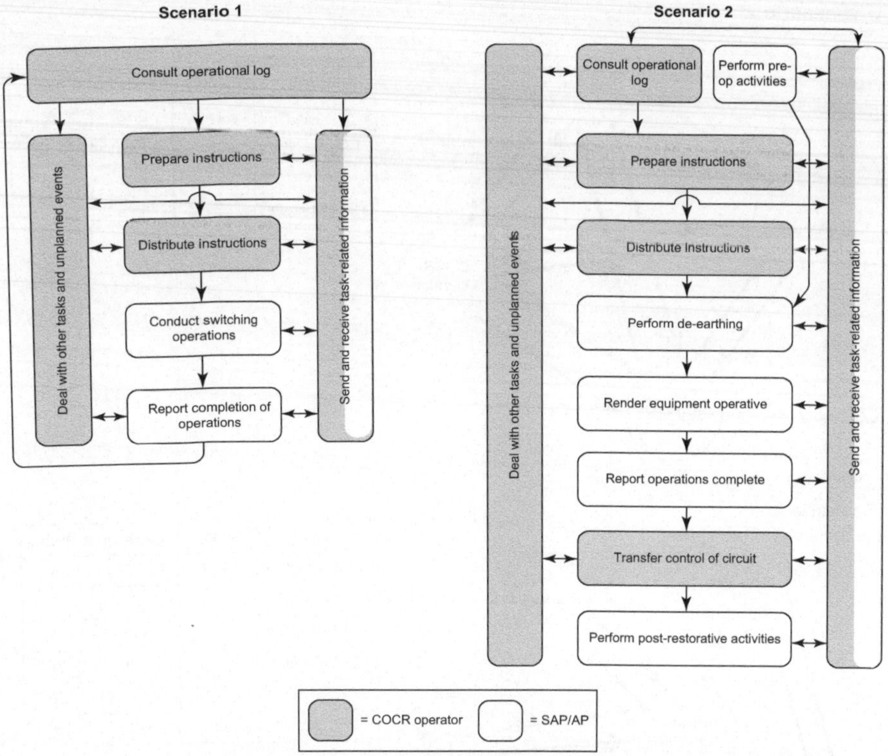

FIGURE 4.6 Task models for Scenarios 1 and 2; shading denotes which agents undertake activities related to each goal.

communications hubs within each network, are identified through examination of the centrality and sociometric status values. Agents with values above the centrality and sociometric status mean values for the overall network are defined as key agents. For Scenario 1, the key agents are the COCR operator and the SAP/AP at the substation. For Scenario 2, the COCR operator is the key agent. The network density values show that the communications network in Scenario 1 was denser than that of Scenario 2. That is, there are less agents involved in the scenario but a similar number of communications between them.

Information Networks

Information networks were constructed for each scenario phase (as defined by the task networks) using the observational transcripts, CDM interview responses and the HTA of the tasks performed. The information network for Scenario 1 is presented in Figure 4.8. Within Figure 4.8, each node represents a concept, and the directional arrows show the relationships between the concepts (e.g. substation 'has' location, display 'shows' circuit). The information networks were interrogated to identify the usage and ownership of concepts by each agent involved as well as the key concepts

Scenario 1

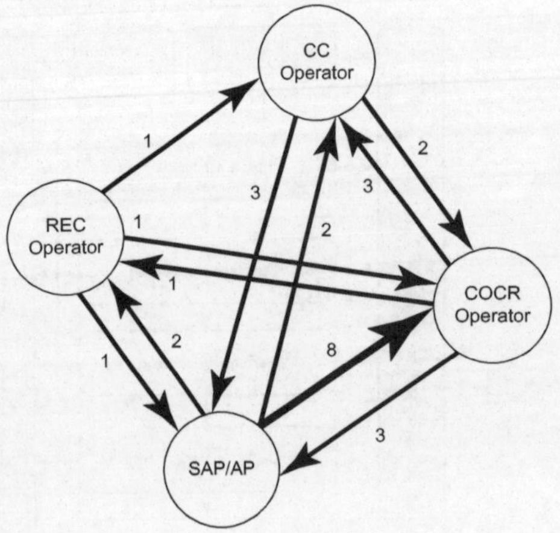

Agent	Density	Centrality	S. Status
COCR Operator	0.31	2.16	6
CC Operator		1.85	3.66
SAP/AP		2.16	6.3
REC Operator		1.85	2

Scenario 2

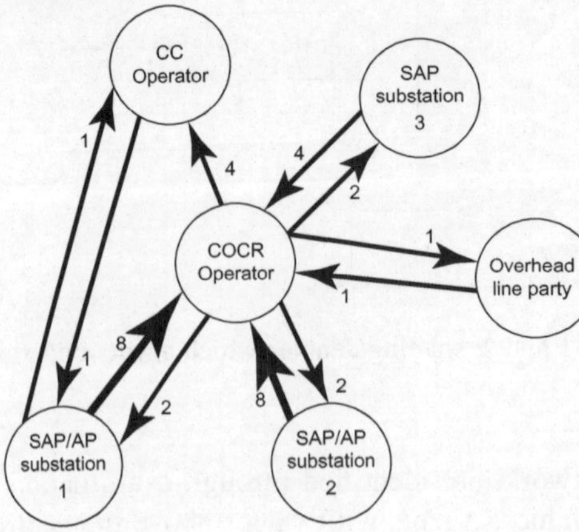

Agent	Density	Centrality	S. Status
COCR Operator	0.2	4.72	6.4
CC Operator		2.6	1.2
SAP/AP Sub 1		3.25	2.4
SAP/AP Sub 2		2.73	2
SAP/AP Sub 3		2.73	1.2
Overhead line party		2,73	0.4

FIGURE 4.7 Social network analysis outputs for Scenarios 1 and 2.

underpinning situation awareness. The usage of concepts by each agent is presented in Figures 4.9 (Scenario 1) and 4.10 (Scenario 2). Further, the usage of concepts by each agent per scenario phase is presented in Table 4.1.

The key concepts underpinning situation awareness were identified through the use of the centrality and sociometric status (see *social network analysis*). Key concepts are defined as those that have salience for each scenario phase, *salience* being defined as those concepts that act as hubs to other concepts. Those concepts with a sociometric status and centrality value above the mean values for the overall network are taken to be key concepts (Figure 4.11).

Additional EAST Analyses

A deeper level of analysis is provided through additional methods from the framework, including Operation Sequence Diagrams (OSDs), Coordination Demands

Analysis (CDA) (Burke 2004) and Communications Usage Diagram (CUD) (Watts and Monk 2000). In the present study, the task, social and knowledge network analyses were supplemented by OSD, CDA and CUD analyses. The purpose of these additional analyses was to examine levels of workload and teamwork during the scenarios and also to evaluate the technologies used to mediate communications.

CUD (Watts and Monk 2000) is used within the EAST framework to describe the communications between agents and evaluate the technologies used to mediate

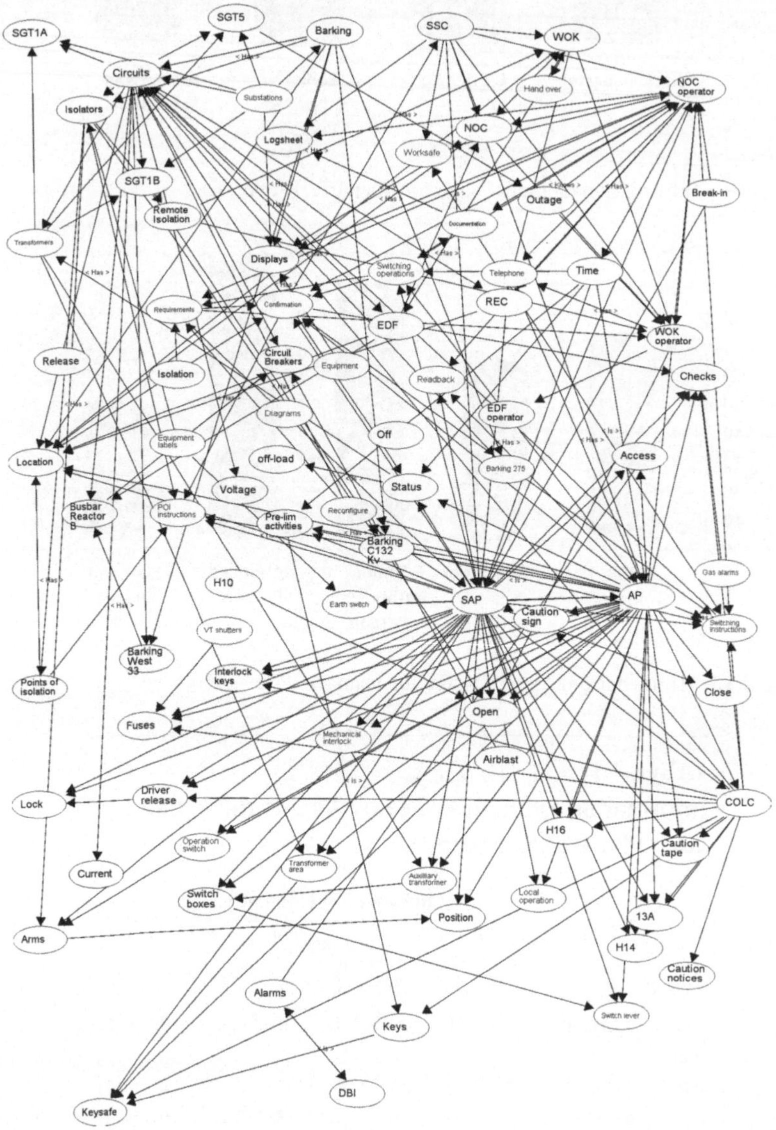

FIGURE 4.8 Propositional network for Scenario 1.

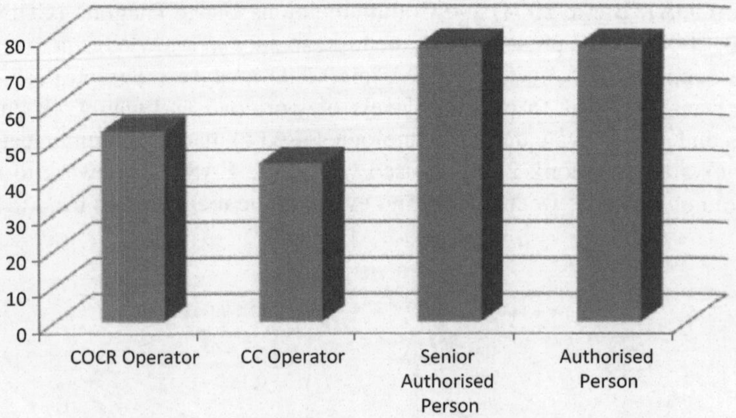

FIGURE 4.9 Information element usage for Scenario 1 (overall).

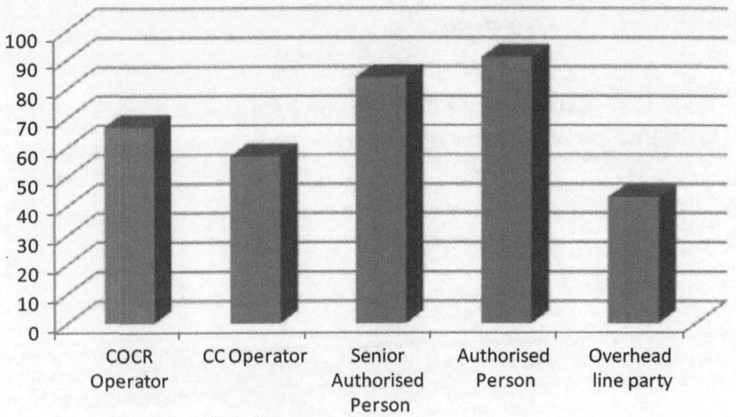

FIGURE 4.10 Information element usage for Scenario 2 (overall).

TABLE 4.1
Information Element Usage by Agents throughout Each Phase of Scenario 2

	Information Element Usage							
	Scenario Phase							
Agent	**1**	**2**	**3**	**4**	**5**	**6**	**7**	**Total**
COCR Operator	22	41	24	0	20	30	0	**67**
SAP	34	36	0	29	18	37	0	**84**
AP	33	29	0	24	18	3	14	**91**
CC Operator	26	0	0	0	0	0	0	**57**
OLP Contact	15	0	0	9	13	0	0	**43**

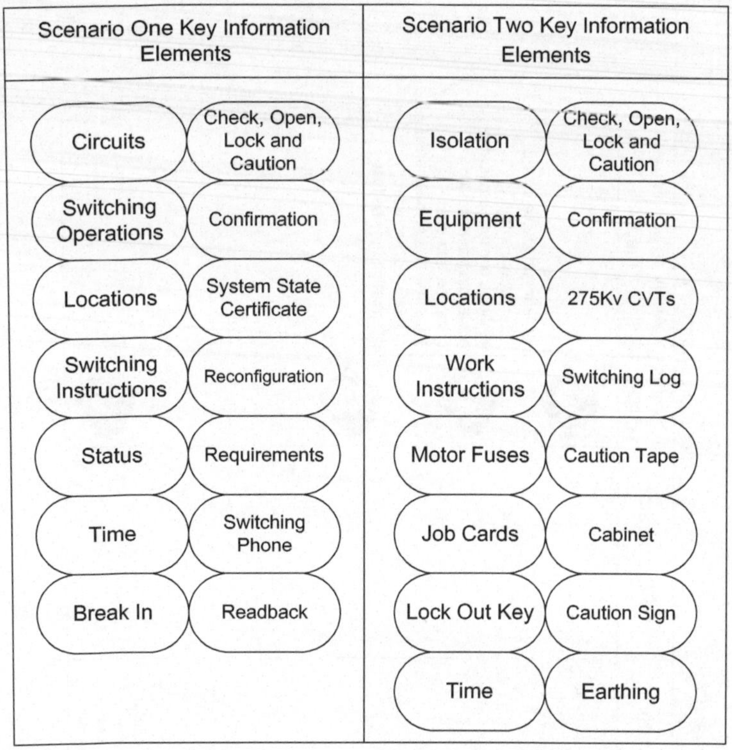

FIGURE 4.11 Key information elements for Scenarios 1 and 2.

these communications. For the present study, those bottom-level task steps from the HTAs involving communications between agents were examined. An extract from the CUD analysis from Scenario 1 is presented in Figure 4.12.

CDA (Burke 2004) is used to derive a rating of the level of coordination exhibited between team members during collaborative scenarios. Again, using the HTA as its primary input, teamwork tasks were first identified, following which coordination between team members on each teamwork task step was rated using a taxonomy of key teamwork behaviours. An extract of the CDA for Scenario 1 is presented in Table 4.2.

For Scenario 1, the CDA analysis revealed that 64% of the bottom-level HTA tasks were teamwork tasks. The mean level of coordination across all teamwork tasks was 1.57, which represents a medium level of coordination.

SUMMARY

The aim of this chapter was to demonstrate an integrated framework of Human Factors methods for analysing performance in complex sociotechnical systems. Although only a summary of the outputs derived were presented, the utility of taking a framework approach to the analysis of performance in complex sociotechnical systems is demonstrated. In this case, EAST enables description and examination of

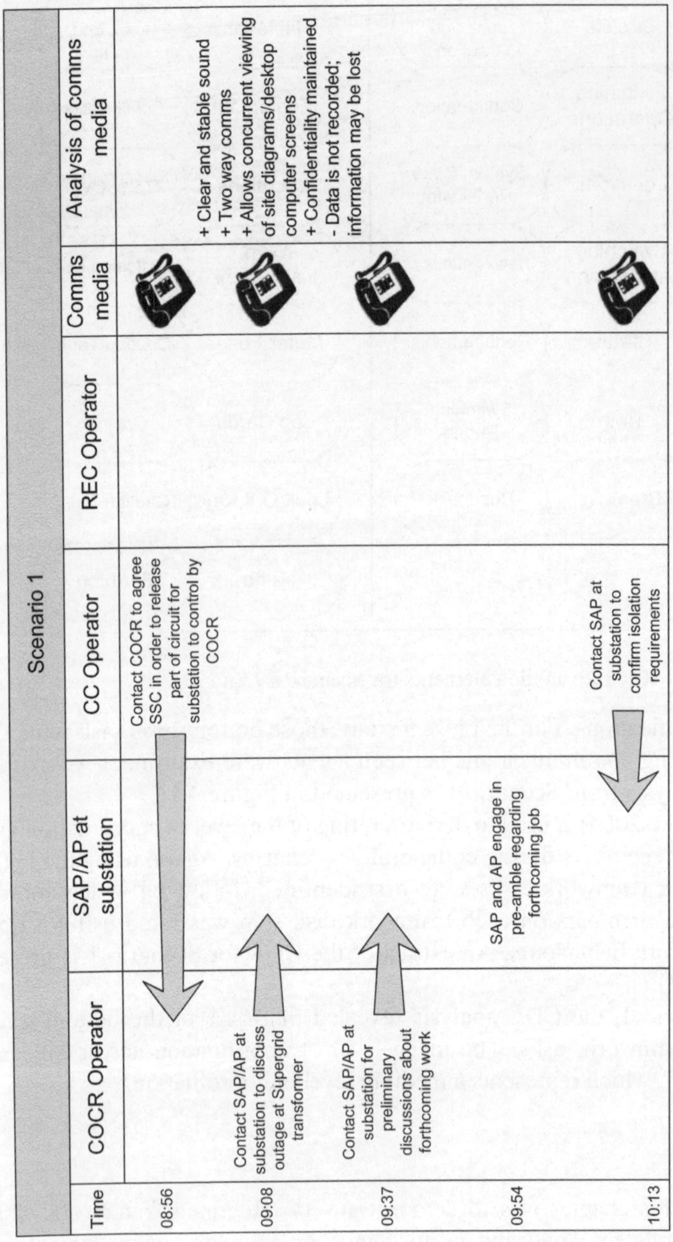

FIGURE 4.12 Communications Usage Diagram extract.

TABLE 4.2
CDA Extract

Teamwork = 64% of HTA Tasks										Task Work = 36% of HTA Tasks
			Mean Coordination Across all Teamwork Tasks = 1.57							
Task	**Agent**	**Task Type**	**Comms**	**SA**	**DM**	**MA**	**Lead**	**Adapt**	**Assert**	**Mean Co–ord**
1.1.1. Use phone to contact COCR	CC Operator	Task work	–	–	–	–	–	–	–	–
1.1.2. Exchange Ids	CC Operator & COCR Operator	Team work	3	3	1	1	1	1	1	1.57
1.1.3. Agree SSC documentation	CC Operator & COCR Operator	Team work	3	3	3	1	1	1	1	1.86
1.1.4.1. Agree SSC with CC Operator	COCR Operator	Team work	3	3	3	1	1	1	1	1.86
1.1.4.2. Agree time with CC Operator	COCR Operator	Team work	3	3	3	1	1	1	1	1.86
1.1.5.1. Record details onto log sheet	COCR Operator	Task work	–	–	–	–	–	–	–	–
1.1.5.2. Enter details into worksafe system	COCR Operator	Task work	–	–	–	–	–	–	–	–
1.2.1. Ask for isolators to be opened remotely	COCR Operator	Team work	3	3	1	2	2	1	1	1.86
1.2.2. Perform remote isolation	CC Operator	Task work	–	–	–	–	–	–	–	–
1.2.3. Check substation diagram on screen	COCR Operator	Task work	–	–	–	–	–	–	–	–
1.2.4. Terminate comms.	CC Operator & COCR Operator	Team work	3	1	1	1	1	1	1	1.29

TABLE 4.3
Elements of Command and Control Examined Through EAST

	HTA	Social Network Analysis	Propositional Networks	CDM	CUD	CDA	OSD
Tasks	■			■		■	■
Communications	■	■		■	■		■
Teamwork		■				■	■
Situation Awareness			■	■			
Decision Making	■			■			
Technologies	■	■	■	■	■		■
Workload	■			■			■

the task and social and knowledge networks underpinning complex sociotechnical system performance, along with a deeper analysis of various specific constructs. For example, in both scenarios, the methods with the EAST framework supported examination of the following elements of command and control.

Table 4.3 demonstrates how, when taken individually, the outputs offer insight into various components of command and control (e.g. situation awareness, teamwork, communications); for example, the propositional networks alone can be used to make judgements on the levels of situation awareness attained throughout the scenarios and also to determine what knowledge underpins situation awareness during energy distributed maintenance scenarios. Integration of the outputs, however, provides a much deeper insight. For example, viewing the task, social and knowledge networks together enables judgement to be made on how the procedures used and the organisation of the social network supported situation awareness during operations. In the case of the task networks, the standard operating procedures engendered high levels of closed loop communication between the agents involved throughout the scenario, including issuing of work instructions, regular updates on work progress and reporting of completed operations. In the case of the social networks, in both scenarios, each of the agents involved had one or more ways of communicating with all other agents (e.g. landline telephone, mobile phone and emails). Further, the hierarchical structure of the network meant that the COCR operator, effectively as network commander, would regularly contact (or be contacted by) agents in the field in order to gather or provide work progress updates, a process that served to update situation awareness throughout the activities. Combining the analysis outputs thus allows examination of not only *how* situation awareness was developed and maintained and *what* it comprised but also *why* efficient levels of situation awareness were achieved.

The EAST framework lends itself to in-depth evaluations of complex sociotechnical system performance, examination of specific constructs within complex sociotechnical systems (e.g. situation awareness, decision making, teamwork), and also system, training, procedure and technology design. Whilst not providing direct recommendations, the analyses produced are often highly useful in pinpointing specific issues limiting performance or generating system redesign recommendations.

REFERENCES

Burke, S. C. (2004). Team task analysis. In: N. A. Stanton, A. Hedge, K. Brookhuis, E. Salas and H. Hendrick (Eds.), *Handbook of Human Factors and Ergonomics Methods* (pp. 56.1–56.8). CRC Press: Boca Raton, FL.

Driskell, J. E. and Mullen, B. (2004). Social network analysis. In: N. A. Stanton, A. Hedge, K. Brookhuis, E. Salas and H. Hendrick (Eds.), *Handbook of Human Factors and Ergonomics Methods* (pp. 58.1–58.6). CRC Press: Boca Raton, FL.

Houghton, R. J., Baber, C., McMaster, R., Stanton, N. A., Salmon, P. M., Stewart, R. and Walker, G. H. (2006). Command and control in emergency services operations: A social network analysis. *Ergonomics*, 49, 1204–1225.

Stanton, N. A., Baber, C. and Harris, D. (2008). *Modelling Command and Control: Event Analysis of Systemic Teamwork*. Ashgate: Aldershot, UK.

Stanton, N. A., Salmon, P. M., Baber, C., Walker, G. H. and Jenkins, D. (2005). *Human Factors Methods: A Practical Guide for Engineering and Design*. Ashgate: Aldershot, UK.

Stanton, N. A., Salmon, P. M., Rafferty, L. A., Walker, G. H., Baber, C. and Jenkins, D. (2013). *Human Factors Methods: A Practical Guide for Engineering and Design* (second edition). Ashgate: Aldershot, UK.

Stanton, N. A., Stewart, R., Harris, D., Houghton, R. J., Baber, C., McMaster, R., Salmon, P. M. et al. (2006). Distributed situation awareness in dynamic systems: Theoretical development and application of an ergonomics methodology. *Ergonomics*, 49, 1288–1311.

Stewart, R., Stanton, N. A., Harris, D., Baber, C., Salmon, P. M., Mock, M., Tatlock, K., Wells, L. and Kay, A. (2008). Distributed situation awareness in an airborne warning and control system: Application of novel ergonomics methodology. *Cognition Technology and Work*, 10 (3), 221–229.

Walker, G. H., Gibson, H., Stanton, N. A., Baber, C., Salmon, P. M. and Green, D. (2006). Event analysis of systemic teamwork (EAST): A novel integration of ergonomics methods to analyse C4i activity. *Ergonomics*, 49, 1345–1369.

Walker, G. H., Stanton, N. A. and Salmon, P. M. (2011). Cognitive compatibility of motorcyclists and car drivers. *Accident Analysis & Prevention*, 43 (3), 878–888.

Walker, G. H., Stanton, N. A., Baber, C., Wells, L., Jenkins, D. P. and Salmon, P. M. (2010). From ethnography to the EAST method: A tractable approach for representing distributed cognition in air traffic control. *Ergonomics*, 53 (2), 184–197.

Watts, L. A. and Monk, A. F. (2000). Reasoning about tasks, activities and technology to support collaboration. In: J. Annett and N. Stanton (Eds.), *Task Analysis* (pp. 55–78). Taylor and Francis: London, UK.

5 EAST in a Submarine Control Room

ANALYSING DISTRIBUTED COGNITION

Distributed cognition is characterised by multiple individuals and teams working together in pursuit of a common goal (comprising multiple interacting sub-goals). High levels of communication and coordination are required, and there is often an onus placed on technologies to facilitate this. Hutchins' (1995) investigation into navigation on ships showed how distributed cognition worked in practice. In the analysis of a navigation task, the ship's crew were trying to determine two things: to identify the ship's current position and to estimate the ship's future position. Hutchins noted that this task was distributed among the crew and artefacts they use. No one individual was able to explain the entirety of navigation. Rather, to understand how ship navigation is performed, one needs to study the interactions within the wider sociotechnical system. Hutchins studied navigation in restricted waters as this is perhaps the most challenging task and requires up to 10 personnel: the navigator, assistant to the navigator, navigation plotter, navigation bearing recorder and timer, starboard pelorus operator (a starboard observer who takes bearing fixes from landmarks), port pelorus operator (a port observer who takes bearing fixes from landmarks), restricted manoeuvring helmsman, quartermaster of the watch, restricted manoeuvring helmsman in after steering and the fathometer operator. His explanation and analysis were presented in a mixture of transcripts of spoken voice, narrative, diagrams and photographs. The richness of the interpretation is derived by the reader through mental simulation of the scenario as they absorb the information presented. No single Human Factors method has convincingly covered all of this complexity in its entirety (Stanton et al. 2005) making the modelling of distributed cognition a significant challenge. Using an integrated suite of methods however, allows scenarios to be analysed from the perspectives described. More importantly, it allows the effects of one set of constructs on other sets of constructs to be considered, for example, how communications influence the way in which tasks are performed and, in turn, how the way in which the tasks are being performed influences the information being used (Stanton et al. 2008). There are also advantages associated with methods integration, because not only does the combination of existing methods bring reassurance in terms of a validation history, but it also enables the same data to be analysed from multiple perspectives. Assuming that the separate methods integrate on a theoretical level, then their application to the same data set offers a form of 'analysis triangulation'.

Various approaches have been put forward to analyse complex sociotechnical systems (Farrington-Darby et al. 2006; Furniss and Blandford 2006; Patrick et al. 2006; Stanton et al. 2006; Waterson 2009; Jenkins et al. 2011). The aim of the approaches is both to make the complexity of sociotechnical systems more explicit, so that the interactions between sub-system boundaries may be examined, and to reduce the complexity to a manageable level. This may seem at odds with the non-linear nature of complex systems and the unpredictable properties that may emerge from them (Walker et al. 2010a, b; Stanton et al. 2012). Nevertheless, the representations do enable inspection of the relationships between sub-systems and their boundaries, even if some of the fidelity is lost (Griffin et al. 2010). The co-evolution of systems design requires some occasional stock-taking before the next evolution (Stanton et al. 2012). The notion that problems become apparent at boundaries of layers, levels, structures and functions is not new, as it underpins general systems theory (von Bertalanffy 1950). Of particular interest here is the boundary between distributed cognitive sub-systems, as this is where problems may become apparent.

Walker et al. (2010a) proposed Event Analysis of Systemic Teamwork (EAST) as a method that could be used to show how distributed cognition could be modelled in air traffic control. The analysis showed how distributed cognition for complex systems could be represented by networks, with the distinct advantage being that the networks enabled both qualitative and quantitative investigations. Walker et al. argued that the multi-faceted nature of the different networks revealed the aggregated behaviours that emerge in complex sociotechnical systems. This representation was proposed as an alternative to the reductionistic approaches often used to understand systems, which present systems in their constituent parts but fail to capture the system as a whole. Walker et al. suggested that the insights gained by network modelling were superior to the traditional ethnographic narrative that has previously been used to describe distributed cognition because they present graphical models of systems. Griffin et al. (2010) go further to show how the EAST method offers insight into system failure. Again, the cited advantage of the approach was the non-reductionistic, non-taxonomic method for analysing non-normative behaviour of systems. Whilst EAST does not employ taxonomies in the analysis, the resultant network structures may be classified into archetypes. The systemic approach allows system interactions to be understood in their entirety (Plant and Stanton 2012).

EAST is underpinned by the notion that complex collaborative systems can be meaningfully understood through a network of networks approach (see Figure 5.1).

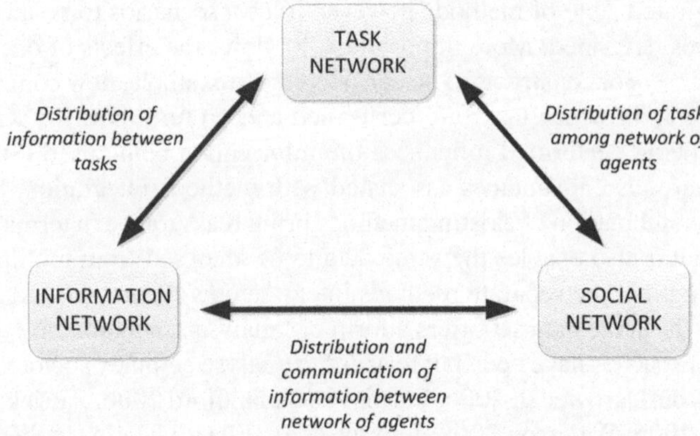

FIGURE 5.1 Network of networks approach.

Specifically, three networks are considered: task, social and information networks. Task networks describe the relationships between tasks and their sequence and interdependences. Social networks analyse the organisation of the system (i.e. communications structure) and the communications taking place between the actors working in the team. Finally, information networks describe the information that the different actors use and communicate during task performance (i.e. distributed situation awareness). Each of these approaches have been presented independently in other papers, such as Farrington-Darby et al.'s (2006) presentation of task diagrams in a study of railway controllers (an example of a task network); Furniss and Blandford's (2006) presentation of communication channels in emergency medical dispatch teams (an example of a social network); and Sanderson et al.'s (1989) analysis of verbal protocols for a process control task (an example of an information network). What EAST does is bring these three networks together into the same analysis framework.

The EAST framework lends itself to in-depth evaluations of complex system performance, examination of specific constructs within complex sociotechnical systems (e.g. situation awareness, decision making, teamwork) and also system, training, procedure and technology design. Whilst not providing direct recommendations, the analyses produced are often highly useful in identifying specific issues limiting performance or highlighting areas where system re-design could be beneficial.

This chapter presents a new shortened form of EAST that focuses on the analysis of collaborative performance (Stanton et al. 2008). This version of EAST does not rely upon the previously defined constituent methods of Hierarchical Task Analysis, Critical Decision Methods, Coordination Demand Analysis, Communications Usage Diagram and Operation Sequence Diagram. Rather, the task, social and information networks are developed directly from the raw data. It also extends the quantitative and qualitative network analysis methods to show how network statistics may be applied to all three networks and how the different networks' perspectives may be integrated. Since its conception, the framework has been applied in many domains, including naval warfare (Stanton et al. 2006), aviation (Stewart et al. 2008), air traffic control (Walker et al. 2010a), emergency services (Houghton et al. 2006), energy distribution (Salmon et al. 2008) and railway maintenance (Walker et al. 2006). The approach is gaining momentum as well as showing its domain independence.

SYSTEM PROPERTIES OF EAST

EAST is a Systems Ergonomics method and as such is able to be applied to all systems levels: micro (i.e. individual human–machine interaction), meso (i.e. organisations operating highly automated systems) and macro (i.e. multi-layered networked system). At each level of the systems analysis, it is possible to construct the triad of task, social and information networks as each is a distributed cognition system at different units of analysis. Further, the EAST representations could be nested such that the micro networks are contained with the meso networks which are also contained in the macro networks. So, whilst the networks might change in terms of the granularity of the analysis, the methodology would remain the same. Wilson (2012) identified six of the defining characteristics essential to Systems Ergonomics: systems

TABLE 5.1
Systems Characteristics of EAST

Characteristic	Property of EAST
Systems focus	Captures the whole sociotechnical system in the network analysis and does not favour one system over the other.
Context	Analyses system behaviour at work using observed and recorded from data from that context with input from subject matter experts. System boundaries are defined by subject matter of interest and may also emerge from the analysis conducted.
Interactions	The interacting parts of the system are revealed in the three networks and the relationships between the networks, as indicated in Figure 5.1. Thus, both interactions within and between networks can be analysed showing distributed cognition in terms of task-social, task-informational, social-informational and task-social-informational interactions.
Holism	The networks are analysed as a whole, both quantitatively (using SNA metrics) and qualitatively (using network archetypes). The networks are also superimposed upon each other to produced combined networks.
Emergence	The emergent properties of the system are revealed through the SNA metrics and the network archetypes.
Embedding	The method itself is embedded in the communications and systems engineering disciplines, so it offers familiarity to organisations wishing to scrutinise their sociotechnical systems. It has the benefit of representing the networks in graphical form as well as supporting metrics for detailed analysis.

Source: Wilson, J. R. *Work*, 41, 3861–3868, 2012.

focus, context, interactions, holism, emergence and embedding. Any method that lays claim to be a Systems Ergonomics, such as EAST, should embody these characteristics. Table 5.1 indicates how EAST encapsulates these properties.

As the contents of Table 5.1 attest, EAST passes the 'systems' test. By conceiving of systems as sets of interacting networks (i.e. task, social and information networks), it has the distinct advantage of enabling network analysis and classification of network archetypes. This means that the systems analysis methods can be framed in terms of quantitative and qualitative assessments. Networks do not differentiate between different types of node (e.g. artefacts and/or people) so that from a modelling perspective they are not constrained by existing structures of people and artefacts; rather, they are related to the tasks associated with a scenario. It is also possible to model the temporal aspects of networks by identifying critical moments in the sequence of activity. To do this, the scenario is divided into task phases allowing active and non-active elements to be specified and represented.

DATA COLLECTION AND ANALYSIS

EAST was undertaken as an 'information audit' of the relationship between the sound room and control room on board a submarine. The control and sound rooms have all the characteristics of distributed cognition problems (as described by

Hutchins 1995). The purpose was to understand how information flowed around the system in order to determine what requirements might be included in the next generation of tactical systems.

The data were collected over a series of observations in a control room and sound room layout in the Talisman Command Team Trainer. There is a strong communication link between the Officer of the Watch (OOW), Chief Petty Officer for Tactical Systems (CHOPS[TS]: sometimes called the OpsO or Operations Officer) and the Sound Room Controller (SRC, sometimes called the Sonar Controller or CHOPS[S]). The layout of the control room and sound room is fixed by the equipment layout but has been optimised over the decades to the current design. The photograph in Figure 5.2 shows a view into the control room with the periscope on the right-hand side.

All of the internal communications were recorded by plugging into the system and recording on a laptop computer using Audacity (version 1.3.14 software). The ambient communications were recorded using an external Yoga BM-26D boundary microphone.

The purpose of this chapter was to see if the networks could be analysed quantitatively using Social Network Analysis (SNA) metrics. SNA offers a means of analysing the network as a whole as well as the behaviour of individual nodes and their interactions. As such, SNA is potentially a very powerful tool for Systems Ergonomics. Whilst it has traditionally been applied to the analysis of social networks (as implied by the name of the method) Driskell and Mullen 2005; Houghton et al. 2006), there is no reason why it cannot be applied to other networks, such as task and information networks. This is a new application for the method but a potentially useful one. The method can also be applied to the design of anticipated networks, so that more effective task, social and information networks can be designed into new systems, which

FIGURE 5.2 View into the control room with periscope on right-hand side.

is another new avenue of research for Systems Ergonomics that would enable network resilience to be explored in a practical manner. The first step in a SNA involves defining the network that is to be analysed. Once the overall network type is specified, the tasks, agents or information should be specified. Once the type of network under analysis has been defined, the scenario(s) within which they will be analysed should be defined. For the purposes of this chapter, the scenario for return to periscope depth (RTPD) was considered. Once the network and scenario(s) under analysis are defined clearly, the data collection phase can begin. There are a number of metrics associated with the analysis of social networks, depending upon the type of evaluation that is being performed. The size of the network determines the number of possible relations, and the number of possible relations grows exponentially with the size of the network. This defines the network's complexity. An explanation of the metrics analysed using AGNA (version 2.1.1 – a software program for computing the SNA metrics) are provided below. The first set of metrics analyse the individual nodes:

- *Emission* and *reception* degree are the number of ties emanating from, and going to, each agent in the network.
- *Eccentricity* is defined by the largest number of hops an agent has to make to get from one side of the network to another.
- The *sociometric status* of each agent refers to the number of communications received and emitted, relative to the number of nodes in the network.
- Agent *centrality* is calculated in order to determine the central or key agent(s) within the network. There are a number of different centrality calculations that can be made. For example, agent centrality can be calculated using Bavelas-Leavitt's index.
- *Closeness* is the inverse of the sum of the shortest distances between each individual and every other person in the network. It reflects the ability to access information through the 'grapevine' of network members.
- *Farness* is the index of centrality for each node in the network, computed as the sum of each node to all other nodes in the network by the shortest path.
- *Betweenness* is defined by the presence of an agent between two other agents, which may be able to exert power through its role as an information broker.

The second set of metrics analyse the whole network:

- The *density* of a network is defined by the number of social relations that are actually observed and can be represented as some fraction of the total possible.
- *Cohesion* is defined as the number of reciprocal connections in the network divided by the maximum number of possible connections.
- The largest geodesic distance within a network defines its *diameter*, which can be thought of as another metric of the network's size, that is, the number of hops to get from one side of the network to the other.

The first step in the analysis was to calculate the statistics for each of the nodes in the network, of which a range can be produced to represent the metrics of distance (i.e. eccentricity), sociometrics (i.e. emission/reception and sociometric status) and centrality (i.e. centrality, closeness, farness and betweenness). The second step was to calculate statistics for the whole network (i.e. density, cohesion and diameter). The final step was to combine the networks for a qualitative assessment. Each of the networks is presented in turn followed by combinations of the networks.

TASK NETWORK ANALYSIS

The task network was constructed from the main phases of the RTPD activities involved in returning the submarine to periscope depth. Rather than represent the phases as a linear process, the task network portrays the relationships between the tasks that are non-sequential. The decision to RTPD is continually being assessed and reassessed. If an unexpected contact (i.e. another vessel) appears in the area (or going towards the area) where the submarine is heading, then the decision to RTPD will be cancelled and the submarine will head down to the safe depth (below that of the deepest ship's hull, i.e. greater than 30 metres). A depiction of the task network is shown in Figure 5.3. As Figure 5.3 shows, the focus of the task is on identifying all of the contacts surrounding the area where the submarine is intending to RTPD. This focus continues all the way up and when the 'look' (i.e. viewing the surface using the periscope) is established. At any point in the manoeuvre, the ship will return to a safe depth if a contact is detected. The decision to RTPD begins with the OOW calling an outstations briefing to ensure all people and systems on the submarine are in order and ready for the manoeuvre. This is followed by ballasting (to ensure the submarine is in trim) and clearing of stern arcs (so that the sound room can check for vessels behind the submarine). Then the sound room and control room are engaged in the activities to range all contacts within the local area to find a safe area of sea to RTPD. If no safe area is found, then the submarine will need to change area and continue to range contacts. When all contacts have been ranged and an area of sea has been identified, the OOW reports to the Captain for permission to RTPD. If permission is forthcoming, the OOW will request final reports from outstations to check that the people and systems on the submarine are in order and ready for the manoeuvre. A decision will be made by the OOW whether to conduct the standard routine (where range and bearing of all contacts are called out as the submarine returns to periscope depth) or a silent routine (where the submarine returns to periscope depth as stealthily as possible). At periscope depth (called PD in Figure 5.3) the OOW calls 'breaking' and conducts two sweeps with the periscope to check that the submarine is clear from potential collision with contacts and will then range the contacts manually to update the Submarine Maritime Command System (SMCS) operators. Then the mission intentions can be carried out. If the submarine is on a potential collision course with any of the contacts, the OOW will give the order to dive to the safe depth. As stated previously, the task network helps to identify interdependencies and relationships between tasks.

The network in Figure 5.3 was used to construct an association matrix to indicate the presence or absence of links between tasks. The number of links into and out of each task are shown in the 'reception' row and 'emission' rows respectively of Table 5.2. The network statistics (as described in Section 3 on data collection

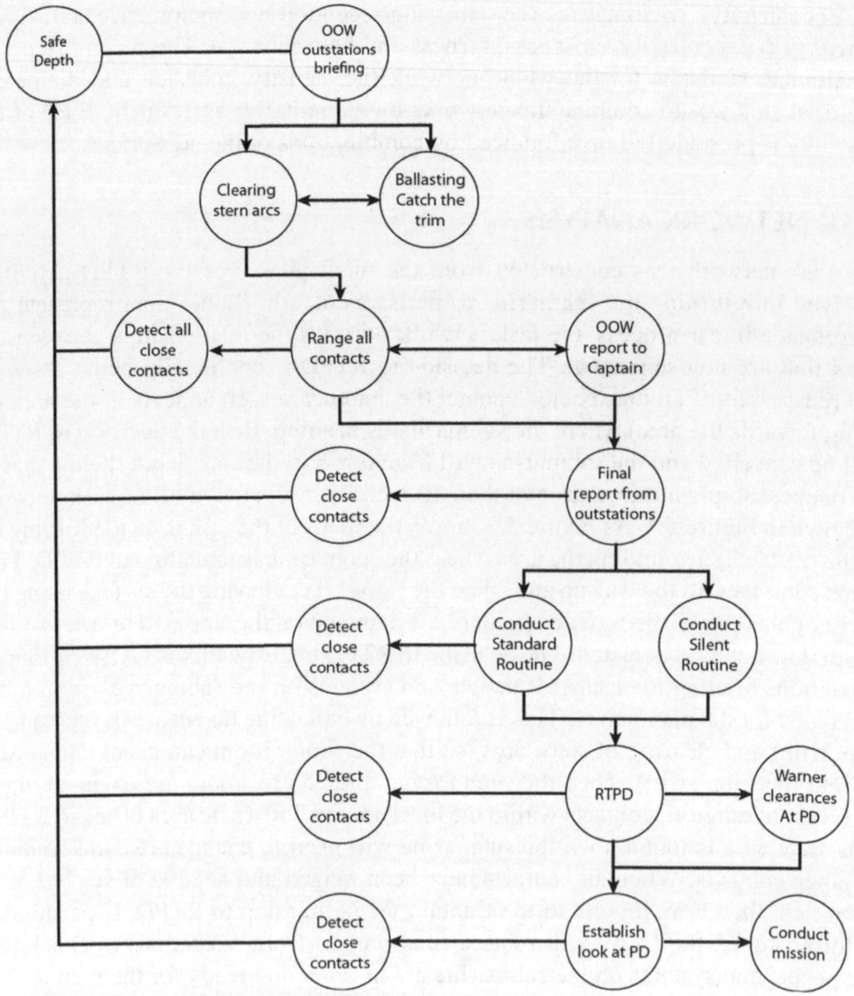

FIGURE 5.3 Task network for RTPD.

and analysis) were then computed using AGNA. Whilst using SNA to analyse task structures is somewhat new to ergonomics, it can give us some insights into the use of metrics. Obviously, the task 'detect close contacts' is critical in RTPD, which is reflected in most of the metrics. Some of the other tasks are less obvious. The two tasks of 'clear stern arcs' and 'range all contacts' score quite low on Sociometric Status but quite high on Centrality and Betweenness. The higher the value means the higher each task is scored on that metric.

The network density was 0.14 (i.e. a low distribution network) and cohesion was 0.02 (i.e. a very low level of reciprocal links – suggesting very strict dependences in the task network, as shown in the task network diagram) with a diameter of 9 (i.e. nine hops from one side of the network to the other). The network can also be described as weighted (i.e. non-uniform) and asymmetric (i.e. unbalanced).

TABLE 5.2
Analysis of Task Network

Task Network Analysis	Safe depth	OOW outstations briefing	Clear stern arcs	Ballasting	Range all contacts	Detect close contacts	OOW report to Captain	Final report from outstations	Conduct standard routine	Conduct silent routine	Return to periscope depth	Warner clearances at periscope depth	Establish look at periscope depth	Conduct mission
Reception	2	1	2	2	1	7	1	2	1	1	2	1	2	1
Emission	1	3	2	1	3	1	1	3	2	2	3	2	2	0
Eccentricity	8	7	6	7	5	9	6	6	7	7	7	8	9	0
Sociometric Status	0.2	0.3	0.3	0.2	0.3	0.6	0.2	0.4	0.2	0.2	0.4	0.2	0.3	0.1
Centrality (B-L)	7.7	7.8	7.9	6.9	7.7	7.5	6.5	7.3	6.4	6.4	6.3	5.7	5.8	10.4
Closeness	0.2	0.3	0.3	0.2	0.4	0.2	0.3	0.3	0.3	0.3	0.3	0.2	0.2	0
Farness	58	46	41	53	36	67	46	39	46	46	49	56	63	0
Betweenness	67	68	70	0	70	65	0	59	17	17	31	0	12	0

SOCIAL NETWORK ANALYSIS (SNA)

The social network analysed the relationships (e.g. communications and activity) between the agents during the scenario, that is, the Captain, OOW, Ships Control, WECDIS, Weapons Engineering Officer (WEO), OpsO, Sound Control and SMCS. The matrix represents the frequency of communications between each agent in the network that could be captured through the network communications system and the ambient communications (it is accepted that not all of the communications in the network could be captured, particularly those outside the network communications system). The association matrix for the RTPD scenario is presented in Table 5.3. This matrix shows whether an agent within the system can be associated with any other agent, specifically through frequency of communications.

Once the matrix of association is completed, the social network diagram can be created. The social network depicts each agent in the network and the communications between them. The communications are represented by directional arrows, and the frequency of communications is also presented, as shown in Figure 5.4. The thicker the line, the more communications have occurred, and vice versa.

Table 5.4 shows the metrics for the social network in the RTPD scenario. This highlights the OOW and Operations Officer (OpsO) as the highest on sociometric status and betweenness metrics, whereas the SMCS operator is highest on Centrality. The Captain, who has the most executive power in the ship, is modestly placed on the sociometric status metric for the RTPD scenario. The Captain's betweenness score hints at his executive power. The higher the value means the higher each task is scored on that metric.

Network density is equal to the total number of links between the agents in the network divided by the total number of possible links. Low network density figures are indicative of a broadly spread network with few links. High density figures indicate a tight network that has many links. In this case the network density was 0.4

TABLE 5.3
Association Matrix for the RTPD Scenario

From/To	Captain	OOW	OpsO	Sound Control	SMCS	WEO	Ships Control	Warner	WECDIS
Captain		2	1	1	1	2	1	1	1
OOW	9		9	4	4	4	21	4	4
OpsO	3	3		38	5	3	3	8	3
Sound Control	0	0	28		0	0	0	0	0
SMCS	0	0	0	0		0	0	0	0
WEO	2	0	0	0	0		0	0	0
Ships Control	0	31	0	0	0	0		0	0
Warner	0	0	4	0	0	0	0		0
WECDIS	0	1	0	0	0	0	0	0	

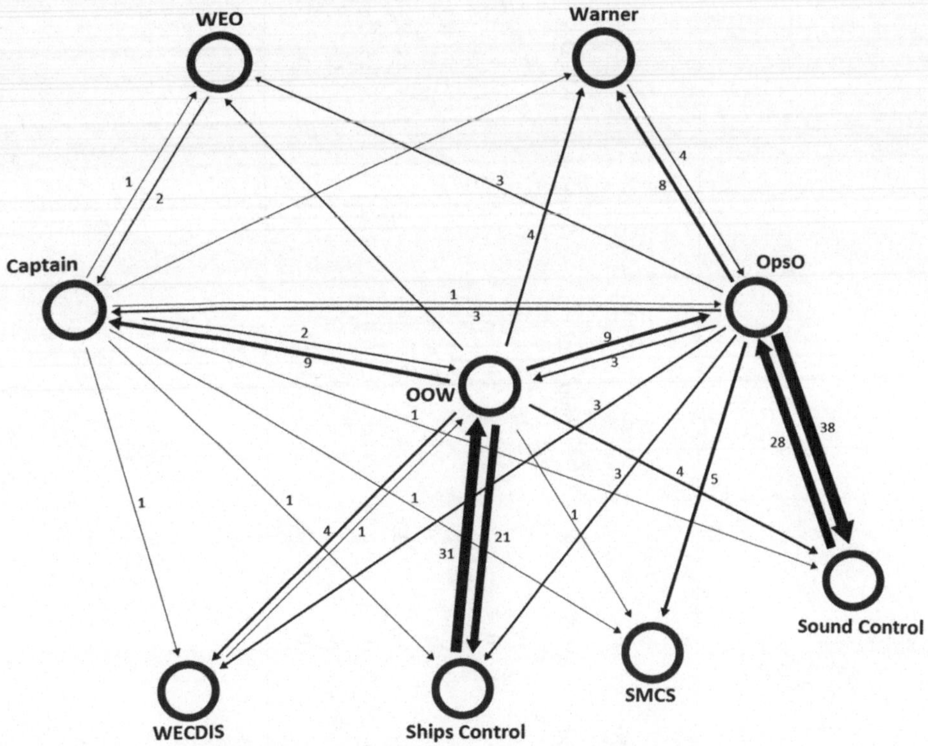

FIGURE 5.4 Social network diagram for the RTPD scenario.

(i.e. a medium distributed network – which is likely to have good resilience against failure of the network) and cohesion was 0.2 (i.e. a low level of reciprocal links) with a diameter of 2 (i.e. two hops from one side of the network to the other) and 29 edges (i.e. when an in-link to an agent had a corresponding out-link to another agent) for a network of nine nodes. The network can be described as weighted (i.e. non-uniform) and asymmetric (i.e. unbalanced). As a baseline study, this characterised the network for returning the submarine to periscope depth.

INFORMATION NETWORK ANALYSIS

The information networks were created by analysing the communications transcripts. Each 'concept' in the transcript was identified and paired with its nearest relation (i.e. concepts from the same sentence) to build up a network of information concepts in the control room related to the activities of returning the submarine to periscope depth. The networks were then presented to the subject matter experts (SMEs) for verification. The information may be related temporally or spatially to agent or tasks or both. In this way, it is possible to build up a picture of who knows what when and how this relates to task performance. This notion takes the activities of the sound room and control room into the realm of distributed cognition. In this

TABLE 5.4
Social Network Metrics for the Network for the RTPD Scenario

Social Network Analysis	Captain	OOW	OpsO	Sound	SMCS	WEO	Ship Cont	Warner	WECDIS
Reception	14	37	37	43	10	9	25	13	8
Emission	10	54	66	28	0	2	31	4	1
Eccentricity	1	1	1	2	0	2	2	2	2
Sociometric Status	3	11	12	9	1	1	7	2	1
Centrality(B-L)	5.2	5.5	5.5	3.8	7.6	3.8	3.8	3.8	3.8
Closeness	1	1	1	0.5	0	0.5	0.5	0.5	0.5
Farness	8	8	8	15	0	15	15	15	15
Betweenness	7	14	14	0	0	0	0	0	0

view, cognition is not an individual phenomenon but rather a systemic endeavour performed by the whole team working together. It is argued that cognition transcends the boundaries of individual actors and becomes a function that is achieved by coordination between the human and technological agents working within the collaborative system. Systems theory argues that the cognitive system needs to be analysed as a whole rather than on the basis of its constituent parts. To this end, the entire transcript is presented in the information network model, as the unit of analysis is not the individual person but the entire system under investigation. The information network for returning the submarine to periscope depth is presented in Figure 5.5 (note that the node 'cuts' at the bottom right of Figure 5.5 is duplicated merely as a convenience to disentangle the lines). By viewing the system as a whole, it does not matter if humans or technology own this information, just that the right information is activated and passed to the right agent at the right time. Individual human agents are not required to know everything, provided that the system has the information in some form or other.

As Figure 5.5 shows, the information network has 68 nodes. Central nodes in the network (informally defined on the basis of the centrality in the network) appear to be 'cuts', 'tracks', 'contacts', 'classification', 'reports', 'manoeuvre', 'depth' and 'trim'. The informational nodes coming off these more central key concept nodes provide the detail that would pertain to the situational specifics, such as the 'bearing' and 'range' for the 'fire control solution' or the type and nature of the vessel classification. The purpose of the information network was to identify the type of information required to return the submarine to periscope depth. A simplified version of the network was formed (abstracted from Figure 5.5) for the practical convenience of presenting the SNA metrics in this chapter, as is shown in Figure 5.6.

The metrics in Table 5.5 suggest three classes of information in the network. 'Contacts' has the highest sociometric status and is probably the most important piece of information in the control room. Next are two clusters of information. One refers to information about the activities of the submarine (i.e. depth, manoeuvre and course) and one refers to the activities of other vessels (i.e. tracks, picture and reports). The network statistics are useful in highlighting the important underlying features of the relationships between information elements from the control room. The higher the value, the higher each task is scored on that metric.

The network density was 0.18 (i.e. a low distribution network) and cohesion was 0.17 (i.e. a very low level of reciprocal links – suggesting very strict dependences in the information network, as shown in the task network diagram) with a diameter of 7 (i.e. seven hops from one side of the network to the other). The network can also be described as weighted (i.e. non-uniform) and asymmetric (i.e. unbalanced).

COMBINING NETWORK MODELS

As stated in the introduction to the chapter, the purpose of EAST is not only to consider the task, social and information networks in isolation of each other but also to consider the relationships between the three networks. These relationships are explored in the following sections. As proposed earlier, the method presents a distributed cognition perspective within a multi-person–machine system to show how

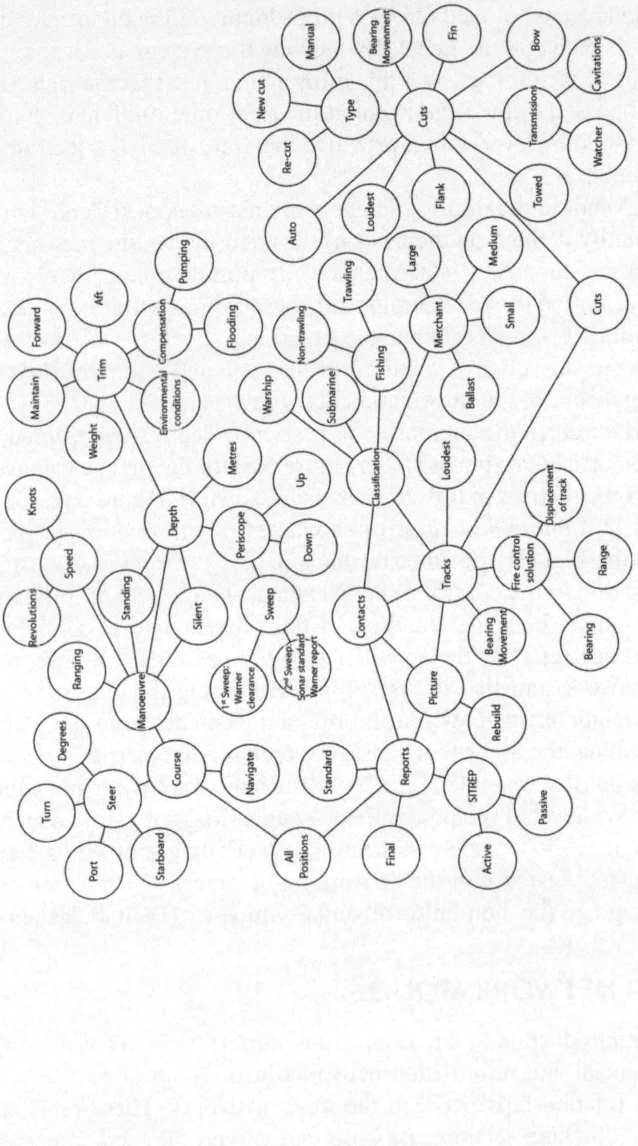

FIGURE 5.5 Information network for returning the submarine to periscope depth.

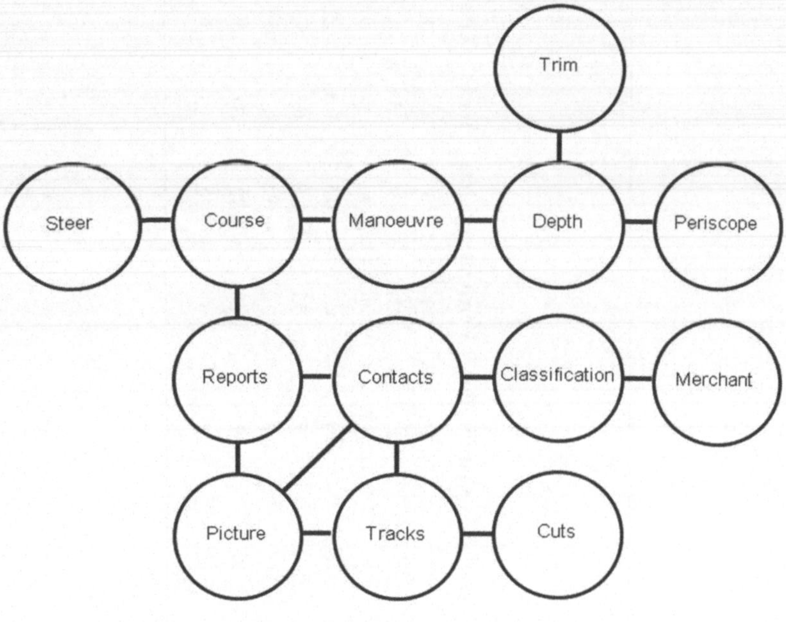

FIGURE 5.6 Simplified information network.

information is used and parsed between people performing tasks. First, the combined task and social networks are presented, wherein the people are mapped onto the task network. Next, the information and social networks are presented, wherein the people are mapped onto the information network. Finally, the three networks are integrated, wherein the information network is coded by the people and tasks.

Task and Social Networks

The relationships between the task and social networks show which agent is primarily involved in each aspect of the task network. These relationships are presented in Figure 5.7, in which the task network in Figure 5.3 has been colour coded by social network to indicate which role is primarily responsible for each task. As Figure 5.7 shows, the role specialists are involved in different tasks throughout the RTPD. The OOW calls the outstation briefing. The Ships Control Officer (SCO) and Planesman clear the stern arcs and ballast the submarine. The SRC, Sound Room Operators (SRO), Tactical Picture Supervisor (also called the Ops Officer, or OpsO) and SMCS are responsible for ranging contacts and reporting to the OOW. The OOW reports to the Captain that the submarine is ready to RTPD. The OOW receives final reports from the outstations. All are involved in the silent or standard routine. The OOW, SCO and Planesman are responsible for getting the submarine to periscope depth, whilst the SRC and SRO are listening for close contacts (which if detected would lead the OOW to order the submarine back to a safe depth). The Warner is responsible for giving clearances at periscope depth. Then the OOW and Periscope Watch Keeper (PWK) are responsible for establishing the look at periscope depth. The first

TABLE 5.5
Analysis of Simplified Information Network

Information Network Analysis	Periscope	Trim	Depth	Manoeuvre	Course	Steer	Report	Contacts	Classification	Merchant	Picture	Tracks	Cuts
Reception	1	1	3	2	3	1	3	3	3	1	3	3	1
Emission	1	1	3	2	3	1	3	4	2	1	3	3	1
Eccentricity	7	7	6	5	4	5	4	5	6	7	5	6	7
Sociometric Status	0.2	0.2	0.5	0.3	0.5	0.2	0.5	0.6	0.4	0.2	0.5	0.5	0.2
Centrality (B-L)	4.9	4.9	6.2	7.5	8.8	6.4	9.1	8.2	6.7	5.2	8.0	6.8	5.3
Closeness	0.2	0.2	0.3	0.3	0.4	0.3	0.4	0.4	0.3	0.2	0.4	0.3	0.2
Farness	52	52	41	34	29	40	28	30	39	50	32	37	48
Betweenness	0	0	42	54	78	0	72	42	23	0	22	23	0

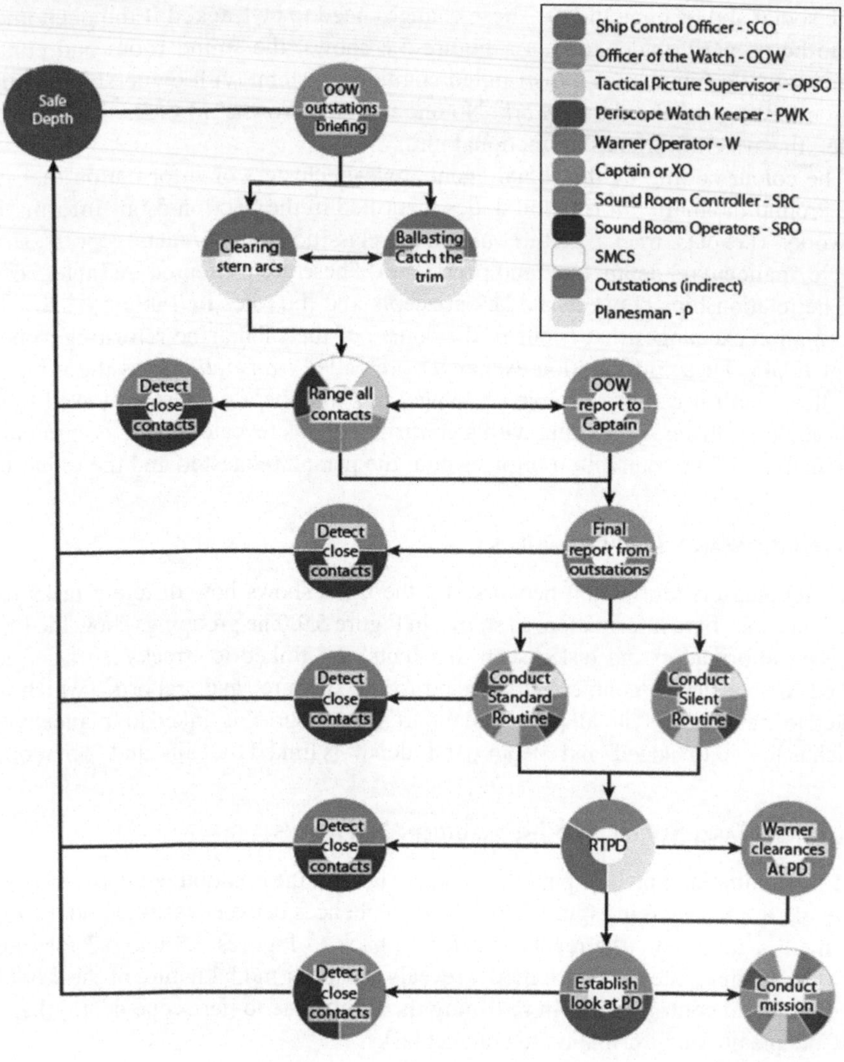

FIGURE 5.7 Task network coded by social agents.

sweep is to search for close in contacts and the second sweep for contacts at greater distance. Any contacts on a potential collision course would lead the OOW to order the submarine back to a safe depth. If the submarine is safe at periscope depth, then the mission intentions can be carried out, in which all of the specialist roles will be performing their respective tasks.

Information and Social Networks

The dynamic nature of information communication means it is subject to change moment by moment in light of changes in the task, environment and interactions (both social and technological). These changes need to be tracked if the phenomena are to be properly understood. As Figure 5.8 shows, the sound room and control room really do function as a distributed cognition system, with ownership of information flowing around the network. No one individual owns all of the information; rather, the system works as a functional unit.

The colour coding by the social agents reveals clusters of information that feature around the main information nodes identified in the Section 5 on 'information networks' (i.e. cuts, tracks, picture, merchant, classification, contacts, reports, steer, course, manoeuvre, depth, trim and periscope). These are presented in Table 5.6.

The relationships between the key concepts and the roles in Table 5.6 show the information exchanges that occur in the course of the submarine returning to periscope depth. These information exchanges are called *transactions*, as the information flows both ways. In a simple exchange, this may be a request, followed by an information transfer and ending with a confirmation (as revealed in the communication analyses). The transaction informs both the person requested and the requestor.

Information and Task Networks

The grouping of information networks by the tasks shows how different tasks use, distribute and share information, as shown in Figure 5.9. The groupings show the links between information and tasks, such that 'cuts' are linked to 'tracks', 'tracks' are linked to 'contacts', 'contacts' are linked to the 'picture' and 'reports', which are linked to the 'course' (via 'all positions' reports), the 'course' is linked to 'manoeuvre', which is linked to 'speed' and 'depth', and 'depth' is linked to 'trim' and 'periscope'.

Combined Task, Social and Information Networks

The final combined picture puts the task, social and information networks into one network, as presented in Figure 5.10. The dependences between tasks cannot be represented in this view (the reader is referred back to Figures 5.3 and 5.7 for those views). The integrated network model reveals the multi-modal nature of the work in the sound and control rooms in returning the submarine to periscope depth, that is, how people utilise information to conduct tasks.

As Figure 5.10 shows, no one individual possessed all of the information to perform the RTPD task. The tasks were distributed among the crew and artefacts they use. To understand how a submarine is returned to periscope depth, one needs to study the interactions within the sociotechnical system. In this study of the sound and control room, at least ten personnel had main roles in performing different tasks.

SUMMARY AND CONCLUSIONS

In summary, this analysis has reported on the application of the EAST to the analysis of the RTPD task in the control room of the Trafalgar class submarine. EAST acknowledges that systems are inherently complex, and multiple perspectives on the problem are required to more fully appreciate the relationships between the social

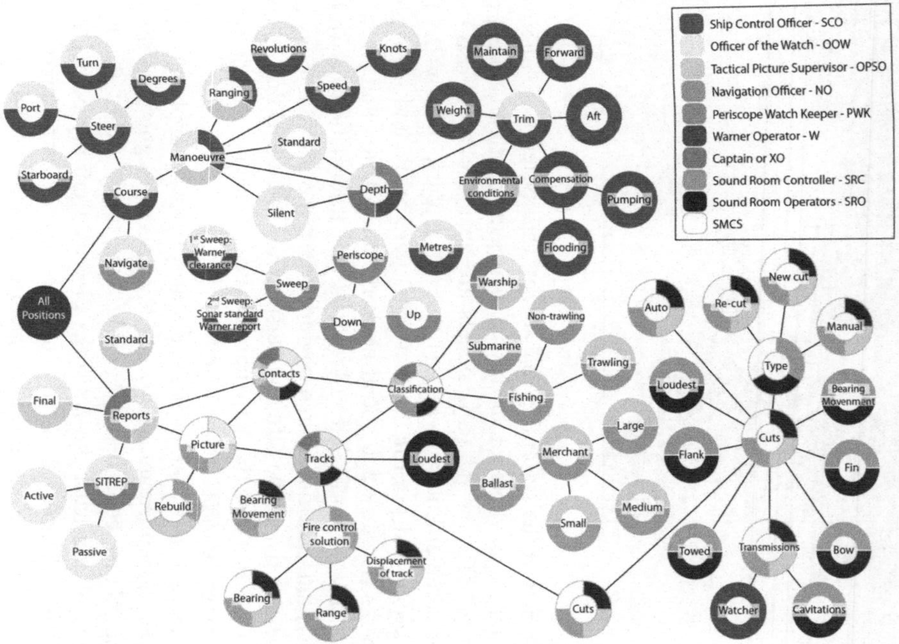

FIGURE 5.8 Information network coded by social agents.

TABLE 5.6
Analysis of Concepts by Social Agents

Key Concept	XO	OOW	SCO	OPSO	SMCS	SRC	SRO	PWK
Cuts				■	■	■	■	
Tracks	■	■		■	■	■	■	
Picture		■		■	■	■		
Merchant				■		■		
Classification	■	■		■	■	■	■	
Contacts	■	■		■	■	■	■	
Reports	■	■		■	■	■		
Steer		■	■					
Course		■	■					
Manoeuvre		■	■	■				
Depth	■	■	■					■
Trim		■	■					
Periscope		■						■

and technical aspects of the system. EAST accepts that systems are intertwined and analyses the system as a whole rather than on the basis of its constituent parts. The research has shown that the network models are able to characterise the domain in different, but complementary, ways. The seven outputs, namely the individual task networks, social and information networks (and associated metrics), and the combined networks (i.e. task and social network, information and social network, information and task network, and task, social and information network) offer a graphical representation of distributed cognition (e.g. the sound and control room) from different perspectives. The different perspectives offered by the representations are an attempt to characterise the activities between the sound and control rooms in returning the submarine to periscope depth.

This chapter extends the work of Walker et al. (2010a) by applying metrics to all three task, social and information networks to provide insights into the structural integrity and the relative contribution of each of the nodes. It has also been shown that it is possible to construct the networks directly from the observational data. Walker et al. (2010a) demonstrated the benefits of the network representations over the traditional ethnographic narratives and pictures (Hutchins 1995), which has been reinforced in the current chapter. It has been shown that the network models offer a useful way of considering distributed cognition in systems to reveal the interdependences between tasks, agents and information. There are some similarities between the approach taken by EAST and that of Actor Network Theory (ANT) (Engestrom 2000), but EAST represents the networks separately as well as together. Both EAST and ANT have, at their core, a conceptual triangulation between objects, actors and events. Both use networks of relationships to graphically display their analysis. Also, EAST goes further than ANT to apply statistical analysis to the networks as well as identify network archetypes. Nevertheless, there are important similarities in the

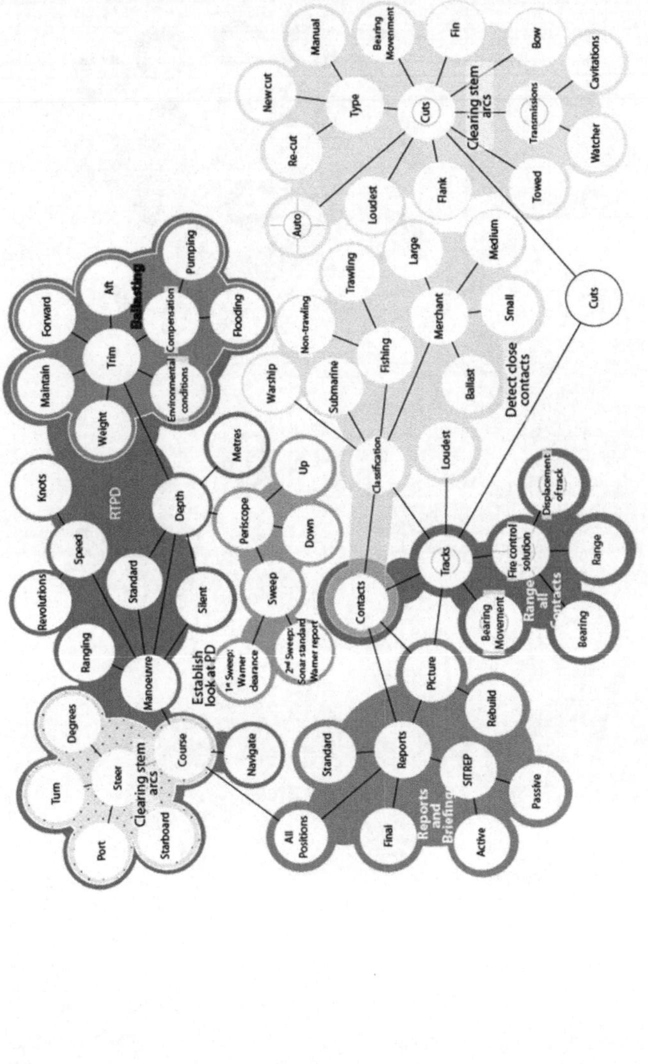

FIGURE 5.9 Combined information and task networks.

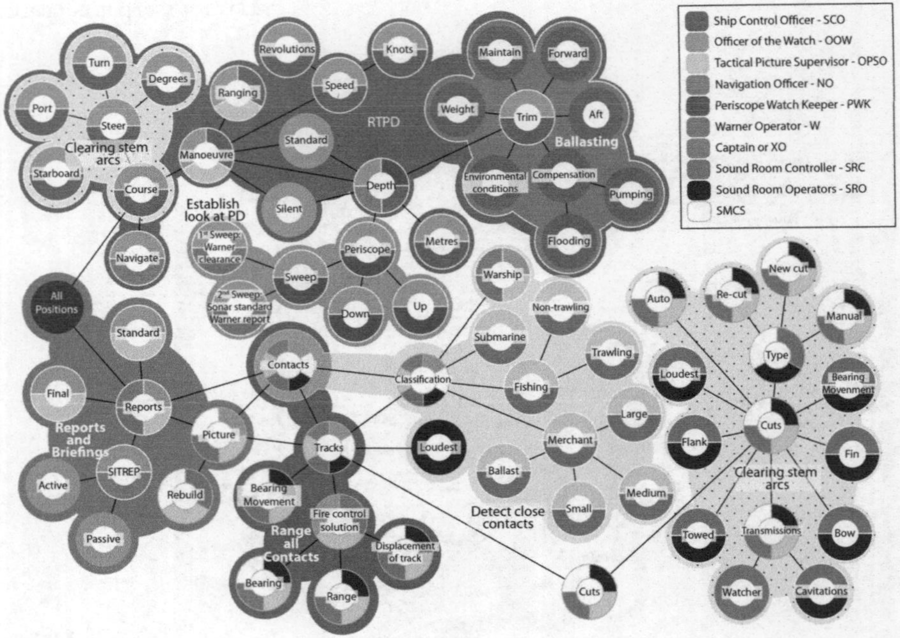

FIGURE 5.10 Integrated networks model.

approach and representation. Arguably, ANT relies more heavily upon the skill of the analyst to identify themes in the networks, whereas the use of verbal protocol data and SNA metrics by EAST has reduced this subjectivity. The use of social network statistics may be one way of examining the potential resilience of networks (Hollnagel et al. 2006) – which may be particularly useful as metrics of distributed cognition. This is a new concept and methodology, so further studies are required to test its efficacy.

The representations in this chapter revealed the clustering of tasks, social agents and information. These clusters show how the constituent parts of the networks have been bound together, either by chance or design. As the system has developed over a century of submarine operations, it reflects a high state of evolution, but that does not mean that the system cannot be improved upon. The representations afford both quantitative and qualitative structural analysis. The quantitative analyses have been presented at some length and offer insights into the potential integrity and resilience of the system. As a method for Resilience Engineering, EAST can be used to assess the potential weaknesses and points of failure in socio-technical structures. The qualitative analyses enable the network structures to be classified into archetypes (e.g. chain, circle, tree, star, mesh, small world and fully connected) (Stanton et al. 2008; Stanton et al. 2012). In these terms, the task network appears to be a hybrid of a chain and circle archetype, the social network appears to be a hybrid of a star and circle archetype and the information network appears to be a small world archetype. This analysis is somewhat speculative at this point, and further research is needed to understand the relationship between the metrics for network resilience and the archetypes of network structure. There are some early indications that the small world networks offer the greatest resilience and efficiency (Stanton et al. 2012).

EAST described the control room in terms of task, social and information networks as well as exploring the relationships between those networks. The individual networks were used to describe the respective relationships between the tasks (such as the task dependencies and sequences), between social agents (such as sociometric status of agents based on communications) and information (such as the interdependences between the concepts discussed). The combined task and social networks showed which roles were performing the tasks in series and parallel. The combined information and social networks showed which roles were communicating the information concepts. The three integrated networks described how information was used and communicated by people working together in the pursuit of tasks. Any new conceptualisation of the command system will need to consider the likely changes on these sociotechnical networked structures.

EAST was able to characterise the activities in the control room of the Trafalgar class submarine. Thus, it is capable of being applied to a complex sociotechnical 'system of systems' to present 'networks of networks'. EAST offered complementary descriptions of the requirements for the RTPD task using the current command system. The networks show multiple perspectives on the activities in the system, which is a necessary requirement for sociotechnical analysis. Further analysis should attempt to characterise future command systems so that the multiple perspectives can be compared and 'so-what' questions can be asked as ideas for design of the social and technical aspects of the system co-evolve (Clegg 2000; Walker et al. 2010b; Stanton et al. 2012). EAST has the potential to map the task, social and information networks and their interdependencies. Adopting this sociotechnical systems design approach would help to jointly optimise the whole system rather than the parts in isolation. To do this requires spending more time in initial modelling and prototyping, working with end-users and SMEs, with more focus on the social systems and ways of working (to redress the balance of focus on technical system development) than is currently the case. In the study reported in this chapter, the benefit of the EAST method is that it helped users understand what was going on between the sound and control room and how people share information related to the tasks they were performing. The ultimate goal of the work will be to model alternatives and provide metrics for choosing one alternative over another.

ACKNOWLEDGEMENTS

The author would like to thank Kevin Bessell of BAE Systems for his assistance with recording and transcribing the verbal communication data.

This work from the Human Factors Integration Defence Technology Centre (HFI DTC) was part-funded by the Human Capability Domain of the UK Ministry of Defence.

REFERENCES

Clegg, C. W. (2000). Sociotechnical principle for system design. *Applied Ergonomics*, 31, 463–477.

Driskell, J. E. and Mullen, B. (2005). Social network analysis. In: N. A. Stanton, A. Hedge, K. Brookhuis, E. Salas and H. Hendrick (Eds.), *Handbook of Human Factors and Ergonomics Methods*. Taylor & Francis: London, UK.

Engestrom, Y. (2000). Activity theory as a framework for analyzing and redesigning work. *Ergonomics*, 43 (7), 960–974.

Farrington-Darby, T., Wilson, J. R., Norris, B. J. and Clarke, T. (2006). A naturalistic study of railway controllers. *Ergonomics*, 49 (12–13), 1370–1394.

Furniss, D. and Blandford, A. (2006). Understanding emergency medical dispatch in terms of distributed cognition: A case study. *Ergonomics*, 49 (12–13), 1174–1203.

Griffin, T. G. C., Young, M. S. and Stanton, N. A. (2010). Investigating accident causation through information network modelling. *Ergonomics*, 53 (2), 198–210.

Hollnagel, E., Woods, D. D. and Leveson, N. (2006). *Resilience Engineering: Concepts and Precepts*. Ashgate: Aldershot, UK.

Houghton, R. J., Baber, C., McMaster, R., Stanton, N. A., Salmon, P. M., Stewart, R. and Walker, G. H. (2006). Command and control in emergency services operations: A social network analysis. *Ergonomics* 49 (12–13), 1204–1225.

Hutchins, E. (1995). *Cognition in the Wild*. MIT Press: Cambridge, MA.

Jenkins, D. P., Salmon, P. M., Stanton, N. A., Walker, G. H. and Rafferty, L. A. (2011). What could they have been thinking? How sociotechnical system design influences cognition: A case study of the Stockwell shooting. *Ergonomics*, 54 (2), 103–119.

McIlroy, R. C. and Stanton, N. A. (2011). Specifying the requirements for requirements specification: The case for work domain and worker competencies analyses. *Theoretical Issues in Ergonomics Science*, 13 (4), 450–471.

Patrick, J., James, N. and Ahmed, A. (2006). Human processes of control: Tracing the goals and strategies of control room teams. *Ergonomics*, 49 (12–13), 1395–1414.

Plant, K. L. and Stanton, N. A. (2012). Why did the pilots shut down the wrong engine? Explaining errors in context using Schema Theory and the Perceptual Cycle Model. *Safety Science*, 50 (2), 300–315.

Salmon, P. M., Stanton, N. A., Walker, G. H., Jenkins, D. P., Baber, C. and McMaster, R. (2008). Representing situation awareness in collaborative systems: A case study in the energy distribution domain. *Ergonomics*, 51 (3), 367–384.

Sanderson, P., Verhage, A. G. and Flud, R. B. (1989). State-space and verbal protocol methods for studying the human operator in process control. *Ergonomics*, 32 (11), 1343–1372.

Stanton, N. A., Baber, C. and Harris, D. (2008). *Modelling Command and Control: Event Analysis of Systemic Teamwork*. Ashgate: Aldershot, UK.

Stanton, N. A., Salmon, P. M., Walker, G. H., Baber, C. and Jenkins, D. P. (2005). *Human Factors Methods: A Practical Guide for Engineering and Design*. Ashgate: Aldershot, UK.

Stanton, N. A., Stewart, R., Harris, D., Houghton, R. J., Baber, C., McMaster, R., Salmon, P. M. et al. (2006). Distributed situation awareness in dynamic systems: Theoretical development and application of an ergonomics methodology. *Ergonomics* 49 (12–13), 1288–1311.

Stanton, N. A., Walker, G. H. and Sorensen, L. J. (2012). It's a small world after all: Contrasting hierarchical and edge networks in a simulated intelligence analysis task. *Ergonomics*, 55 (3), 265–281.

Stewart, R., Stanton, N. A., Harris, D., Baber, C., Salmon, P. M., Mock, M., Tatlock, K. and Wells, L. (2008). Distributed situational awareness in an airborne warning and control aircraft: Application of a novel ergonomics methodology. *Cognition, Technology and Work*, 10 (3), 221–229.

Vicente, K. J. (1999). *Cognitive Work Analysis: Toward Safe, Productive, and Healthy Computer-Based Work*. Lawrence Erlbaum Associates: Mahwah, NJ.

Von Bertalanffy, L. (1950). The theory of open systems in physics and biology. *Science*, 111, 23–29.

Walker, G. H., Gibson, H., Stanton, N. A., Baber, C., Salmon, P. M. and Green, D. (2006). Event analysis of systemic teamwork (EAST): A novel integration of ergonomics methods to analyse C4i activity. *Ergonomics*, 49 (12–13), 1345–1369.

Walker, G. H., Stanton, N. A., Baber, C., Wells, L., Gibson, H., Salmon, P. M. and Jenkins, D. P. (2010a). From ethnography to the EAST method: A tractable approach for representing distributed cognition in air traffic control. *Ergonomics*, 53 (2) 184–197.

Walker, G. H., Stanton, N. A., Salmon, P. M., Jenkins, D. P. and Rafferty, L. A. (2010b). Translating concepts of complexity to the field of ergonomics. *Ergonomics*, 53 (10), 1175–1186.

Waterson, P. (2009). A systems ergonomics analysis of the Maidstone and Tunbridge Wells infection outbreaks. *Ergonomics*, 52 (10), 1196–1205.

Wilson, J. R. (2012). Fundamentals of systems ergonomics. *Work*, 41, 3861–3868.

6 EAST in Railway Maintenance

With Huw Gibson and Chris Baber

In order to extract data from complex and diverse scenarios, a methodology called Event Analysis for Systemic Teamwork (EAST) was developed. With over 90 existing ergonomics methodologies already available, the approach taken was to integrate the following: a Hierarchical Task Analysis (Annett et al. 1971), a Coordination Demand Analysis (CDA_ (Burke 2005), a Communications Usage Diagram (CUD) (Watts and Monk 2000), a Social Network Analysis (Scott 1991; Driskall and Mullen 2005) and the Critical Decision Method (Klein et al. 1989). The outputs of these methods provide two summary representations in the form of an enhanced Operation Sequence Diagram and Propositional Network. These offer multiple and overlapping perspectives on key descriptive constructs, including who the agents are in a scenario, when tasks occur, where agents are located, how agents collaborate and communicate and what information is used and knowledge shared. The application of these methods to live data drawn from the UK rail industry demonstrates how alternate scenarios can be compared on key metrics, how multiple perspectives on the same data can be taken and what further detailed insights can be extracted. Applied across a number of scenarios in different civil and military domains, the ultimate aim of EAST is to provide data to develop generic models of C4i (command, control, communications, computers and intelligence) activity, and to improve the design of systems aimed at enhancing this management infrastructure.

INTRODUCTION

C4i is the management infrastructure for any large, complex and dynamic resource system (Harris and White 1987). An underlying mechanism for organisation and planning is needed where there are multiple individuals, or teams of individuals, who need to communicate and coordinate in order to achieve an efficient outcome to a common goal. Although the notion of command and control is one that might popularly be thought of as occurring in the military domain, examples of underlying organisational and planning infrastructures also abound in the civil domain.

C4i embodies people, teams, technology, knowledge and information. It is a multi-faceted endeavour that can be viewed from a variety of different and concurrent perspectives. This is can be challenging not just for those engaged in sustaining effective performance within a C4i scenario, but also for those engaged in analysing such a scenario. The multiple and concurrent perspectives include the micro-level descriptions of tasks, actors, and individual cognition, through the meso-level descriptions of team performance, through again to macro-level systems analysis. Add to this the notion that C4i also involves human and non-human elements and is comprised of activities that are highly related temporally, and it is clear that C4i research presents a considerable challenge. Smalley (2003) provides a contemporary perspective on these issues and an example of current research in this domain. Smalley's (2003) high-level analytical review of command and control systems also elucidates some of the different challenges inherent in C4i, which Harris and White (1987) summarise as being simply that

1. there is currently no theory of command and control;
2. there is informational chaos;
3. there is an ongoing technology revolution (Harris and White 1987).

The aim of this chapter is to present an analysis methodology that feeds into the development of generic models and theories of C4i. In order to do this, the method seeks to make sense of the informational chaos identified above by the structured and systematic extraction of data from C4i scenarios. The ultimate aim of this extraction and modelling process is for the Defence Technology Centre for Human Factors Integration (DTC HFI) to generate the level of understanding required to deploy novel forms of technology in ways that contribute optimally towards the effectiveness of achieving the goals of C4i scenarios. So, in summary, the context of the proposed descriptive methodology generates the following specific requirements to:

- generate the data necessary to understand key features of C4 scenarios
- use these data to develop theories and generic models
- use this understanding to design C4i systems that accurately meet user needs.

This chapter describes the development and application of the EAST methodology, which was designed to meet these requirements. EAST was developed specifically to analyse C4i in any domain (DTC HFI 2003).

DEVELOPMENT OF THE EAST METHODOLOGY

At the most basic level, the descriptive constructs in C4i can be distilled down to:

- *who*: (the agents participating in a C4i scenario are)
- *when*: (tasks are performed and which agents are associated with their performance)
- *where*: (agents are physically located)
- *how*: (agents collaborate and communicate to achieve scenario aims)
- *what*: (tasks are agents performing and what knowledge is used and/or shared).

There are over 90 different methods available to the ergonomist covering virtually all aspects of system design and description, including the constructs above

(Stanton et al. 2005). It would seem evident based on this that a high degree of circumspection is required before embarking on the development of yet more. This is especially so given the multi-faceted, and therefore multi-measurement, approach that would be required to analyse C4i scenarios. In other words, the task of developing and then subsequently validating the range of methods needed to access all the required perspectives on C4i is formidable. Instead, an approach based on method integration is proposed. This has a number of compelling advantages, because not only does the integration of existing methods bring reassurance in terms of a validation history, but it also enables the same data to be analysed from multiple perspectives. These multiple perspectives, as well as being inherent in the object that is being described and measured (C4i scenarios), also provides a form of internal validity. Assuming that the separate methods integrate on a theoretical level, then their application to the same data set offers a form of 'analysis triangulation'. Therefore, the internal structure of the EAST methodology can be broken down into three layers comprised of associated methods already established in the literature:

- Layer 1 – data collection methods
- Layer 2 – data analysis methods
- Layer 3 – representational methods

It is important to consider the theoretical issues surrounding the use of methods applicable to these three levels. This consideration is important not only from the point of view of validity in terms of the individual method but also in terms of compatibility 'between' methods.

Layer 1 – Data Collection Methods

Structured data collection methodologies are required to extract meaningful data from C4i scenarios. Annett (2003) argues that data collection should comprise of observation and interviews at the very least, and both Annett (2003) and Kieras (2003) argue for the least intrusive method of observation that circumstances permit. Therefore, a two-step process of observation and interview is proposed as a means to collect key information from within live C4i scenarios, where several people are likely to be working remotely from each other.

Observation itself takes the form of multi-site activity sampling (Stanton et al. 2005). Activity sampling is an aid to unobtrusively recording activity (or actual behaviour) in the field. This process involves the use of pre-defined categories or classes of action, which can then be used to create a checklist. The checklist could be completed at regular time intervals, for example, every 15 seconds (by ticking the appropriate action in the appropriate column for that observation), or it could be completed whenever an action occurs (by entering the time against the action). Activity sampling provides a means of relating the timing of actions to locations and to the decisions made by specific actors in context. It can be assumed that in a number of cases there will also be pre-existing material in the form of task analyses (or similar) that will enable task descriptions to be structured, sampled and even validated through observation.

In basic terms, whilst observational techniques provide information on the observable inputs and outputs of human information processing, they produce limited data on the process of decision making. Interviewing people enables the analyst to capture data on these unobservable processes, particularly if the interviewee is describing a recent event and how they dealt with it. In recent years, the study of decision making in real-world situations has received a great deal of attention, and there is a growing emphasis on the use of interviews to collect such information. The Critical Decision Method (CDM) (Klein and Armstrong 2005) is a contemporary example. According to Klein, 'The CDM is a retrospective interview strategy that applies a set of cognitive probes to actual non-routine incidents that required expert judgment or decision making' (Klein et al. 1989, p. 464). In this approach, the interview proceeds through a series of four stages: briefing and initial recall of incidents, identifying decision points in a specific incident, probing the decision points and checking. A slightly modified version of the CDM probes presented in O'Hare et al. (2000) are adopted within EAST and presented in Table 6.1. These permit elicitation of information on key decision points as well as non-routine 'incidents'.

The selection of data collection methods proceeds not just from the theoretical perspective but also the practical. Given that most C4i scenarios will involve the simultaneous observation of activities distributed geographically and among individuals, the proposed two-step process of activity sampling and CDM interview provides an expedient method of data capture in this context.

Layer 2 – Analysis Methods

Structured analysis methodologies take the data extracted earlier in the data collection phase and model them in terms of deeper, more fundamental concepts. These concepts relate to task and social structures, teamworking and mediating communications technology.

Hierarchical Task Analysis (HTA)

Task analysis is the activity of collecting, analysing and interpreting data on system performance (Annett and Stanton 2000; Diaper and Stanton 2004) and is one of the central underpinning analysis methods within EAST. According to Stanton (2004), task techniques can be broadly divided into five basic types: hierarchical lists (e.g. HTA and GOMS), narrative descriptions (e.g. the Crit and Cognitive Archaeology), flow diagrams (e.g. TAFEI [Task Analysis for Error Identification] and Trigger Analysis), hierarchical diagrams (e.g. HTA) and tables (e.g. Task-Centred Walkthrough, HTA, SGT [Sub Goal Template] and TAFEI). Some methods have multiple representations, such as HTA, which can be viewed as a hierarchical text list, a hierarchical diagram or in tabular format.

Hierarchical Task Analysis (HTA) is a means of describing a system in terms of goals and sub-goals, with feedback loops in a nested hierarchy (Annett 2005). Its enduring popularity and indeed its appropriateness within EAST can be put down to two key points. First, it is inherently flexible: the approach can be used to describe any system, even C4i. Second, it can be used for many ends, from person specification, to training requirements, to error prediction, to team performance assessment and to

TABLE 6.1
CDM Probes

Cognitive Cue	Sample Question
Goal specification	What were your specific goals at the various decision points?
Goal identification	What features were you looking at when you formulated your decision?
	How did you know that you needed to make the decision?
	How did you know when to make the decision?
Expectancy	Were you expecting to make this type of decision during the course of the event?
	Can you describe how this affected your decision-making process?
Conceptual model	Are there situations in which your decision would have turned out differently?
	Can you describe the nature of these situations and the characteristics that would have changed the outcome of your decision?
Influence of uncertainty	At any stage, were you uncertain about either the reliability or the relevance of information that you had available?
	At any stage, were you uncertain about the appropriateness of the decision?
Information integration	What was the most important piece of information that you used to formulate the decision?
Situation awareness	What information did you have available to you when formulating the decision?
Situation assessment	Did you use all the information available to you when formulating the decision?
	Was there any additional information that you might have used to assist in formulating the decision?
Options	Were there any other alternatives available to you other than the decision that you made?
	Why were these alternatives considered inappropriate?
Decision blocking	Was there any stage during the decision-making process in which you found it difficult to process and integrate the information available?
	Can you describe precisely the nature of the situation?
Basis of choice	Do you think that you could develop a rule, based on your experience, that could assist another person to make the same decision successfully?
	Do you think that anyone else would be able to use this rule successfully? Why?/Why not?
Generalisation	Were you at any time reminded of previous experiences in which a similar decision was made?
	Were you at any time reminded of previous experiences in which a different decision was made?

Source: O'Hare et al., *Task Analysis*, Taylor and Francis, London, UK, 2000, 170–190.

system design. The multiple perspectives available from HTA fit well with the multiple perspectives available from C4i scenarios. HTA has additional pragmatic benefits because it already underpins a number of subsequent analysis methodologies such as CDA (Burke 2005) and CUD (Watts and Monk 2000), making it an ideal candidate for method integration. Task analysts applying HTA are also required to understand both the ways in which people adapt to their environment and the ways that they adapt their environment to themselves. Thus, HTA has the further benefit of capturing and specifying the contextual conditions and precursors within C4i scenarios. These are represented within detailed task descriptions and as 'plans' within the hierarchy.

In its application within EAST, the definition of the HTA proceeds with reference both to what has been observed and to what may have been previously defined through any pre-existing task analyses. It is possible to meaningfully integrate these information sources on goal structure to produce a task analysis that accurately describes what has been observed, is consistent with what has already been pre-defined and also covers key decision points covered in the CDM interview.

Coordination Demand Analysis (CDA)

It might be assumed that C4i activity will be dominated by coordination activities, but this supposition needs to be checked. Individual tasks from the HTA can be assessed for the type of coordination that is required for successful performance using a method called CDA (Burke 2005). The method integrates with the HTA, where the tasks identified are assessed according to multi-dimensional aspects of teamworking, presented in Table 6.2.

Burke (2005) distinguishes between teamworking tasks (those comprised of the coordination dimensions in Table 6.2), and task work (individual tasks that are not

TABLE 6.2
Coordination Demand Dimensions

Coordination Dimension	Definition
Communication	Includes sending, receiving and acknowledging information among crew members
Situational Awareness (SA)	Refers to identifying the source and nature of problems, maintaining an accurate perception of the aircraft's location relative to the external environment and detecting situations that require action
Decision Making (DM)	Includes identifying possible solutions to problems, evaluating the consequences of each alternative, selecting the best alternative and gathering information needed prior to arriving at a decision.
Mission analysis (MA)	Includes monitoring, allocating and coordinating the resources of the crew and aircraft; prioritising tasks; setting goals and developing plans to accomplish the goals; creating contingency plans
Leadership	Refers to directing activities of others, monitoring and assessing the performance of crew members, motivating members and communicating mission requirements.
Adaptability	Refers to the ability to alter one's course of action as necessary, maintain constructive behaviour under pressure and adapt to internal or external changes
Assertiveness	Refers to the willingness to make decisions, demonstrating initiative and maintaining one's position until convinced otherwise by facts
Total coordination	Refers to the overall need for interaction and coordination among crew members

dependent on teamworking). For example, a teamwork task would be dealing with the issuing of an instruction to another individual, whereas a 'task work' task would be inputting data into the C4i system. It is argued by Burke that existing teamworking methodologies tackle only a few of the necessary skill aspects; therefore, CDA is proposed as a comprehensive and systematic descriptor. CDA also integrates readily within the EAST, where it provides a profile of teamworking skills according to the tasks identified earlier in the HTA.

Communications Usage Diagram (CUD)

The CUD (Watts and Monk 2000) is another task analysis technique that, in its current application, provides an opportunity to systematically critique the communications technology in use during a scenario. The critique is based upon the task flow. It identifies actors in specific locations, the communications in use within identified tasks and an approximate indication of the sequencing of events. The critique of communications technology currently in use is based upon this. It identifies the positive and negative points associated with a given communications media within the current context of use and enables the analyst to propose alternative solutions. The CUD is useful as a way of drawing out practical, design-related outputs as well as providing a more theoretical level of description.

Social Network Analysis (SNA)

SNA is a means to present and describe, in a compact and systematic fashion, the network structure underlying individuals or teams who are linked through communications with each other (Driskell and Mullen 2005). The relationships that are specified from this analysis can be used to determine what aspects of the network structure constrain or enhance the performance of agents in the network (Driskell and Mullen 2005). Unlike the other analysis methods within EAST, SNA is not directly based on the task analysis. However, foundation data on actors and communications are derived from the HTA and CUD.

SNA can be used to analyse the formal and informal relationships between people in a network, but there is no reason why they cannot also show technological mediation of communication and networks where some of the nodes are non-human. This is of particular relevance within C4i. In a practical sense, the approach might reveal sub-optimal networks and bottlenecks in communication. The approach also allows a complex network to be summarised on just a small number of key metrics and for these same metrics to permit easy comparison between different C4i scenarios. Another major advantage that Driskell and Mullen (2005) identify is that this 'network approach focuses on the relationships among actors embedded in their social context' (pp. 58). Again, another important consideration within C4i. Houghton et al. (2006) deal in more detail with this network perspective.

Layer 3 – Representational Methods

Representational methods take the extracted data modelled in the analysis section and provide a means of simplifying and presenting these deeper concepts in various integrated, graphical forms.

Operation Sequence Diagram (OSD)

Process charts offer a systematic approach to describing activities. They emphasise essential features using a graphical representation that is easy to follow and understand (Kirwan and Ainsworth 1992). Charting techniques such as this preserve the ability of preceding methods, such as SNA and HTA, to represent human and non-human elements of the C4i system (Drury 1990). Charting techniques are also capable of representing one of the key contextual factors in C4i scenarios, and that is the temporal structure and interrelations between and among processes.

Several enhancements to the OSD representation have been made to reflect the outputs of supporting analysis methods. The OSD presents a temporal overview of tasks, the outputs of the CDA, communications media from the CUD and links between agents from the SNA. Therefore, the approach summarises a large amount of supporting analysis in a fashion that is graphical and relatively easy to follow, whilst also being scalable to suit different 'sized' C4i scenarios. This enhanced version of OSD is therefore one of the key summary representations within EAST.

Two additional reasons underlie the selection of the OSD. First, OSD has long been a popular and useful tool in the Human Factors toolkit. Second, many approaches to systems analysis (such as Unified Modelling Language) make use of sequence diagrams in their representations, and so it ought to be possible to integrate a Human Factors OSD with the more technical sequence diagrams used by systems analysts. In this manner, we propose that the process-based analysis derived from EAST provides us with a convenient route into Human Factors Integration. This factor is consistent with the DTC HFI's overall aims.

Propositional Network (PN)

Propositional Networks (PNs) are like semantic networks in that they contain nodes (with words) and links between nodes. It is argued that the application of basic propositions and operators enables dictionary-like definitions of concepts to be derived (Ogden 1987). Stanton et al. (2006) take this basic notion and extend it to offer a novel way of modelling knowledge in any scenario. Knowledge relates strongly to SA. A systems view of SA (and indeed an individual view as well) can be understood as activated knowledge (Bell and Lyon 2000), and therefore propositional networks offer a novel and effective means of representing this 'systems level' view of SA. The theoretical background to this approach is described in detail in Stanton et al.'s DSA methodology (2006). However, an opportunity arises within EAST to apply the DSA methodology in a real sense to C4i scenarios.

The knowledge used in C4i activities is accessed via the CDM, where a systematic content analysis of the interview transcripts permits 'knowledge objects' to be extracted and subsequently linked. Knowledge objects are analogous to propositions and can be defined as an 'entity or phenomenon about which an individual requires information in order to act effectively' (Baber 2004). The resultant network of knowledge objects enables at least four powerful perspectives on SA.

1. Firstly, a major advantage of PAs is that they do not differentiate between different types of node (e.g. knowledge related to objects, people or ideas), and therefore, from a design perspective, they do not constrain assessments

to consideration of existing configurations of people and objects but rather to the required knowledge elements associated with a scenario.

2. Secondly, the network shows the totality of knowledge used in the scenario at a systems level, regardless of whether agents in the scenario are human or technical.
3. Thirdly, shared SA can be accessed from the CDM, where agents can be attributed to knowledge objects within the network.
4. Fourthly, Endsley (1995) states that SA occurs within a 'volume of time and space' (p. 36), and it is possible to illustrate this key temporal aspect of SA by animating the propositional network in terms of active and non-active knowledge objects (Figure 6.8) later in this chapter provides an illustration of the concept). To do this, the scenario is divided into tasks phases according to the higher-level goals of the HTA. The CDM relates to these phases and accesses information on decision making within them, allowing active and non-active knowledge objects to be specified and represented.

Summary of Methods

Whilst the methods above are tried and tested, the integration of them into EAST represents a new approach. Each method provides insight into the main descriptive constructs identified above. The HTA identifies the actors in the scenario, the temporal structure of tasks, where tasks (and associated actors) are taking place and what tasks are being performed. The HTA is one of the main foundation methods within EAST and drives the CDA analysis, which provides structured insight into the general question of how teamworking tasks are performed and what teamworking skills they require. The CUD is also founded on data from the HTA; it provides insight into who the actors are, the flow of tasks (i.e. when), where actors are located geographically, what items are being communicated and how (i.e. the communications technology used). The SNA extends the analysis of communications by considering the 'links' between actors rather than task flow. These two perspectives on 'the who' are complementary. Table 6.3 is a methods matrix that relates the component methods to the descriptive constructs identified above. The overlap between methods and the constructs they access is explained by the multiple perspectives provided on issues such as 'who' and 'what'. For example, the HTA deals with 'what' tasks, the

TABLE 6.3
Methods Matrix Mapping Descriptive C4i Constructs onto Component Methods of EAST

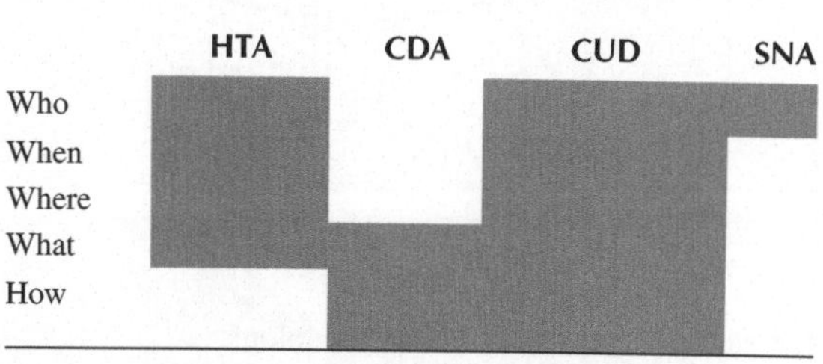

CDA deals with 'what' teamworking skills, and the CUD deals with 'what' communications technology is used. Each is a different but complementary perspective on the same descriptive construct and a different but complementary perspective on the same data derived from observation and interview, that is, analysis triangulation.

Structure of EAST Methodology

The internal structure of the methodology is illustrated in Figure 6.1. The live observation provides input into all of the Layer 2 analysis methods. The HTA serves to provide a definition of tasks and a goal structure for the remaining analysis methods. The analysis methods are then summarised and represented on an enhanced form of OSD (the SNA can also be used as a representational method, e.g. Houghton et al. 2006). The CDM interview data collected live are represented using a propositional network,

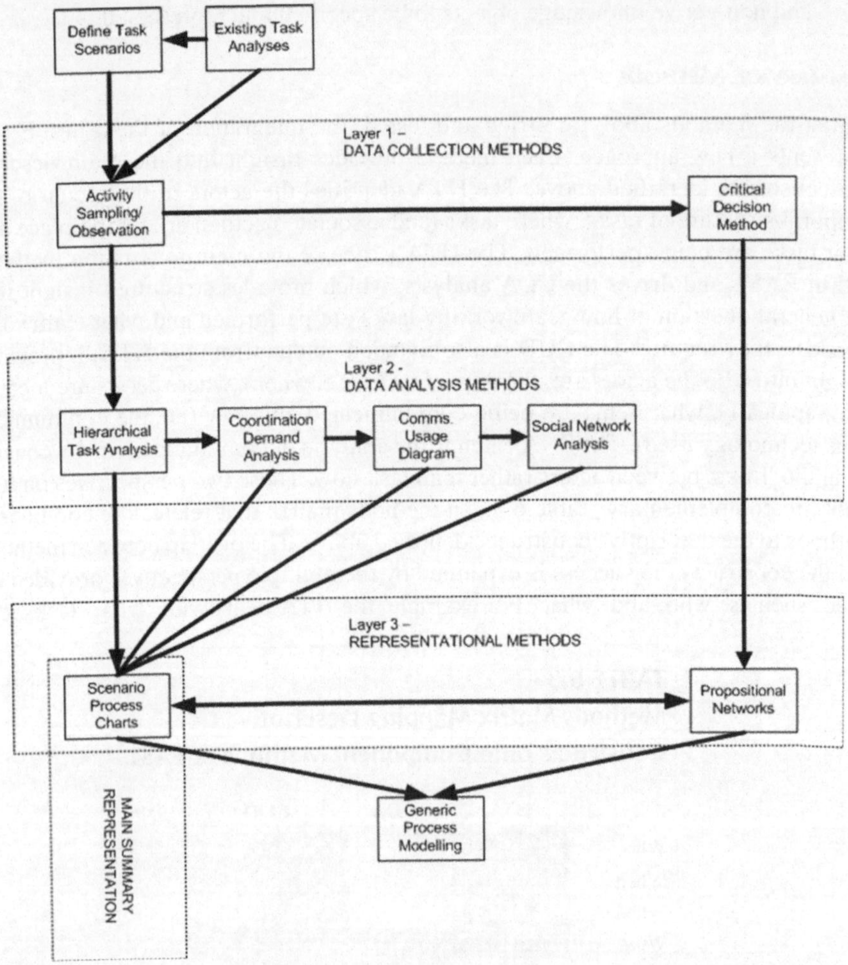

FIGURE 6.1 Internal structure of EAST methodology.

thus representing the structure of knowledge at key decision points in the scenario. The outputs of EAST provide an integrated picture of multiple perspectives; that is, the component methods are compatible with each other to the extent that they can be summarised within a common representation (e.g. the OSD), but the individual methods, and combinations of them, are also capable of generating insights at the core of C4i.

PROCEDURE

The high-level procedure for applying EAST can be summarised in Table 6.4 below as involving nine key steps. Obviously, within these overall headings, a degree of flexibility can be adopted to suit the particular domain or circumstances.

APPLICATION OF THE METHOD TO LIVE DATA

Three examples of complex and dynamic resource systems are taken from the UK rail industry. The scenarios serve as a source of live data to demonstrate the capability of the method. The data are sourced, in this case, from written transcripts based on communications between parties in the scenarios and interviews with subject matter experts.

Broadly speaking, the C4i activities under consideration are those involved in the setting up of safety systems required when carrying out maintenance of track. Safety systems are required so that workers on the track do not come into conflict with moving trains and that trains do not travel over railway infrastructure that is rendered unsafe by the maintenance work or the requirement for it. The strict procedures underpinning these systems are specified nationally in the UK railway industry Rule Book (RSSB 2003).

Railway operations are an example of civilian C4i, where a 'management infrastructure' is required and in place. Maintenance activities on the railway possess all the essential C4i ingredients, including

- that there is a common goal
- that there are individuals and teams coordinating to reach it
- that those parties are dispersed geographically
- that there are numerous systems, procedures and technology to support their endeavour

BACKGROUND

Under normal conditions, a signaller has the key responsibility for controlling train movements and maintaining safety for an area of railway line. This control occurs remote from the line at a control centre (a signal box or signalling centre). These can be located many miles from where activity could be taking place.

During maintenance, another person takes responsibility for an area of the line (sometimes referred to as a *possession*) and/or for preventing trains passing over the possession (measures referred to as *protection*). These individuals are normally termed the *person in charge of possession* (PICOP) or *controller of site safety*

TABLE 6.4
High-Level Procedure for EAST

STAGE	Action	Timing	Output
1	**Organise observation** It is proposed that many of the sessions can also be undertaken in simulators or training centres as an alternative to the field.	Prior to observation	Stakeholder involvement; details of scenario; number of personnel; key decision roles
2	**Conduct HTAs with SMEs** Existing task analysis material would be sourced during this stage and validated by SMEs. If required, new task analyses would be undertaken with SMEs.	Prior to observation	Definition of task scenario; details of task structure; activities and actors
3	**Define objectives** This stage will involve checking that the proposed technique is feasible (e.g. in terms of security, access, privacy etc.). If possible, define 'ideal' performance, perhaps in terms of a list of roles, responsibilities, decisions and timing of events.	Prior to observation	List of agreed objectives; agree observation posts; if possible, description of ideal performance
4	**Brief and train observers.**	Prior to observation 30–60 minutes.	Demonstrate sampling strategy; practice
5	**Arrive at scenario** On arrival at 'scenario', position observers at their posts. Synchronise watches.	On-site 5–10 minutes.	
6	**Observation** The incident is sampled using the pro-forma/experimental materials.	During observation Duration of incident or n × 20 minute observations.	Multiple activity samples
7	**Interview** Participants interviewed by observer(s) using CDM methodology.	After observation 1 hour per interviewee.	Interview transcript containing knowledge objects
8	**Define CDA and CUD outputs with SMEs** This involves engaging SMEs to assist in the rating of tasks along teamwork dimensions and the critique of communications technology in use during the scenario.	After observation (can be undertaken at any stage after Step 2 if required) 2 hours.	Validated CDA and CUD data
9	**Collation** The activity samples are collated to produce full EAST analysis.	After session	Analysis and Representations of C4i scenario

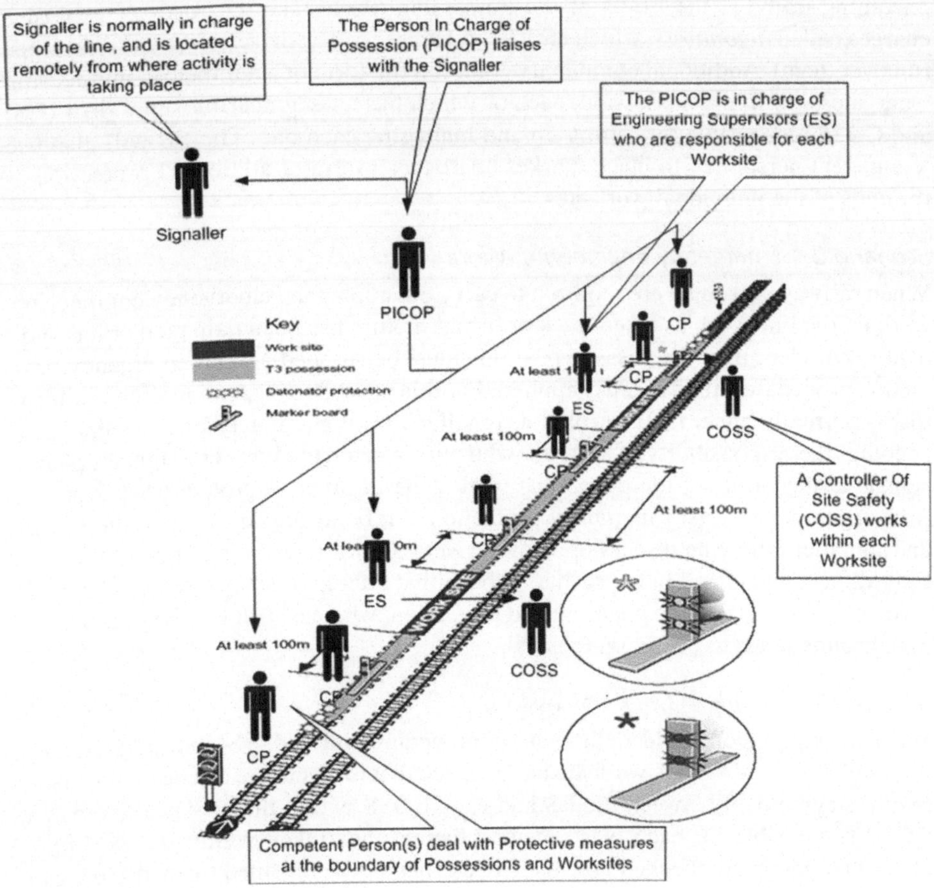

FIGURE 6.2 Overall diagram of the various track possession scenarios.

(COSS). Communication and coordination is required to transfer responsibility between the signaller and PICOP/COSS. The PICOP/COSS also has to communicate and coordinate with various other personnel, such as those carrying out maintenance within their areas of control, drivers of trains and on-track plant that may be in the possession and personnel implementing aspects of the possession (all of which may also be dispersed over a certain geographical area). The three specific maintenance scenarios are briefly described below, with Figure 6.2 providing additional clarity on the general layout and relative geographical positions of personnel.

Scenario 1 – Planned Maintenance Activities

This scenario describes the processes and activities for setting up a possession for a stretch of track so that planned maintenance can take place. This requires coordination between multiple parties, including communication between the signaller and PICOP (so that appropriate 'protecting' signals are set to danger) and the provision of instructions to a 'competent person' (CP) to place a form of protection against oncoming trains at the limits of the possession (these take the form of explosive charges called *detonators*, which emit a loud noise to alert drivers who may have just run over them). Additional complexity comes in the form of a number of engineering work sites within the possession, each of which has an engineering supervisor (ES) and COSS responsible for setting up and managing each one. The ES will also use 'competent personnel' to place marker boards as a form of additional protection at the ends of the individual worksites.

Scenario 2 – Emergency Engineering Work

When railway personnel are required to carry out unplanned emergency engineering work on the line, such as when track or infrastructure has been damaged or has suddenly degraded, then the passage of trains must be stopped and an emergency protection procedure called a *T2(X)* applied. For this procedure, a portion of the railway that is normally under the control of a signaller working remotely from a signal box becomes the responsibility of a COSS, who will work on the line. The signaler places signals at the limits of the dangerous work zone in order to protect track workers from train movements. Emergency protection can be arranged between the COSS and signaller following discussion with the Network Rail Area Operations Manager. Overall, it can be noted in the scenario above that *non*-emergency engineering work involves greater advanced planning and protection, whereas in emergency scenarios, organisation tends to occur 'on the day'.

Scenario 3 – Ending a Track Possession

When the possession is ended, the 'set-up' procedure outlined in Scenario 1 is largely reversed. First, the ES of a worksite has to check that the worksite can be closed. This requires agreement between the ES and each COSS within the worksite (there is a COSS responsible for each piece of work being undertaken within the worksite). Once this has been checked and the PICOP has been informed, then the ES can instruct a CP to remove the worksite marker boards. The PICOP is informed when this is completed and then, when all the worksites within a possession are closed, the possession itself can be closed. The PICOP can then instruct a CP to remove the possession protection. The PICOP will inform the signaller that lines are now safe and clear for trains to run. Control of the line is then passed from the PICOP to the signaller, and normal running of trains over the lines can resume.

RESULTS AND DISCUSSION

Analysis Methods

Task Networks

The first step in the EAST methodology, subsequent to collecting the data, is to model the goal structure of the scenario using HTA. The output of this step is represented in the form of task networks. These are graphical representations of the 'plan 0' within the HTA and depict the task structure in terms of how tasks relate to each other functionally and temporally. In the present case, the highly proceduralised and rigid nature of the activity is seen as a more or less linear task flow (Figure 6.3).

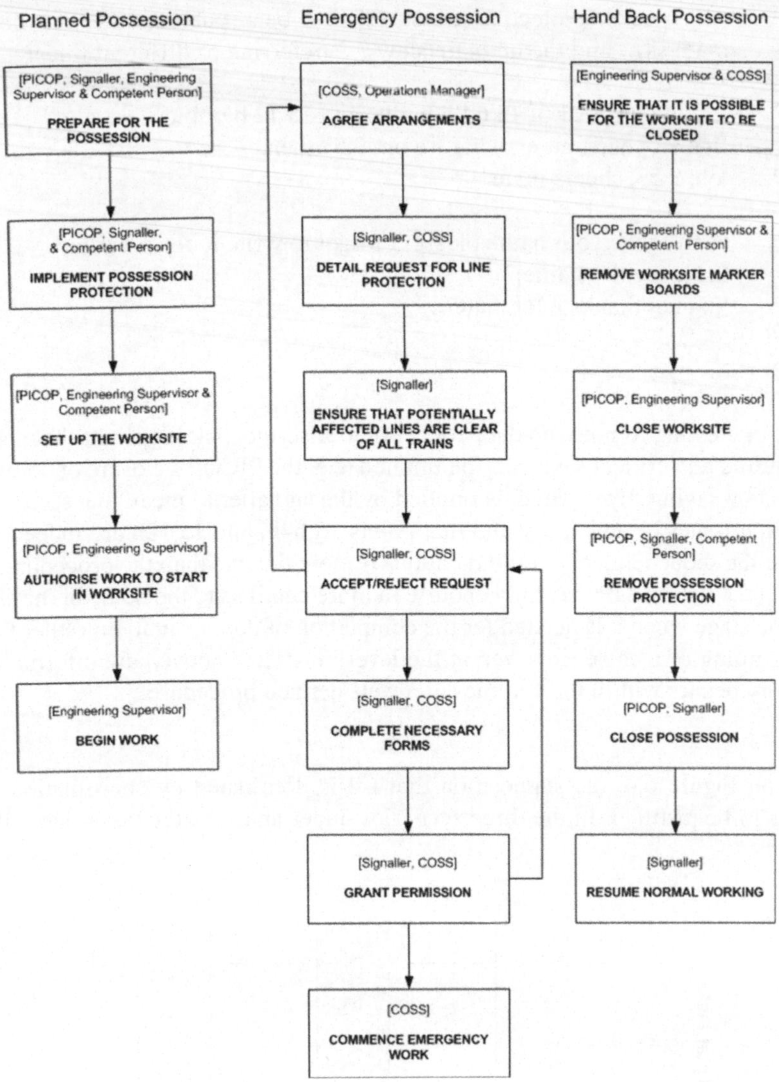

FIGURE 6.3 Task networks for each scenario.

The following is an extract from the verbal transcripts and shows a typical communication occurring between the signaller (S) and PICOP:

S 'Wimbledon …' [answers phone with name of signal box]

S '… panel 1' [signalling panel, and associated geographical area being worked by the signaller]

PICOP 'Hello Wimbledon, it's [name of PICOP] at Waterloo [station]'

S '… Yes …'

PICOP	'the blocks [protective measures] now been put out mate on the down main slow and the up main slow …' [referring to different lines]
S	'… right …'
PICOP	'… and its clear of 15 64 b points and 15 12 b points.'
S	'It's all yours, at, er, what we on, 9:53 then.'
PICOP	'oh, 9:53, cheers mate.'
S	'… ok …'
PICOP	'Can I take your name please, I forgot to write it down earlier'
S	'… [provides name].'
PICOP	'[name] thanks a lot mate …'
S	'ok …'
PICOP	'… bye.'

It is interesting to note the data collection transcripts that highlight that some of the required and critical steps may be implied (e.g. the PICOP's closure of communication is by saying 'bye', which is implied by the signaller to mean that there are no remaining issues or ambiguity and that points 15 64b, and 15 12b are indeed clear) and that the sequencing of communications is more flexible than the procedures may suggest (e.g. the PICOP may only enquire in more detail as to the name of the signaller at the stage when it is needed for the completion of documentation, rather than at the beginning of a call). However, at the level illustrated above, any informality or flexibility occurs within the confines of a well-defined procedure.

CDA

Based on Figure 6.4, the supposition that C4 is dominated by coordination tasks appears to be justified. In the three scenarios under analysis, the tasks that fall into

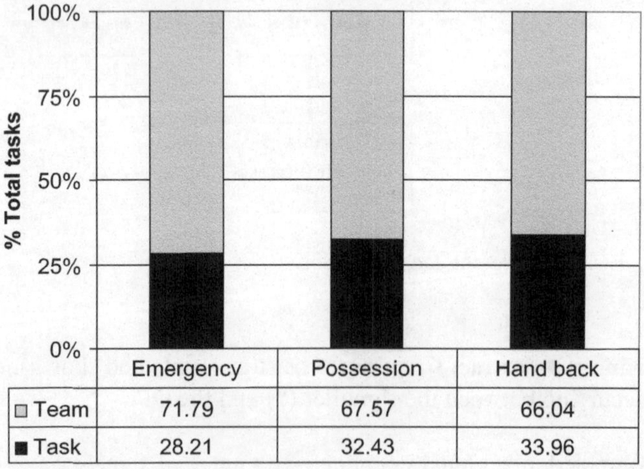

	Emergency	Possession	Hand back
Team	71.79	67.57	66.04
Task	28.21	32.43	33.96

FIGURE 6.4 Results of CDA analysis showing percentage of task/teamwork activities undertaken within each scenario.

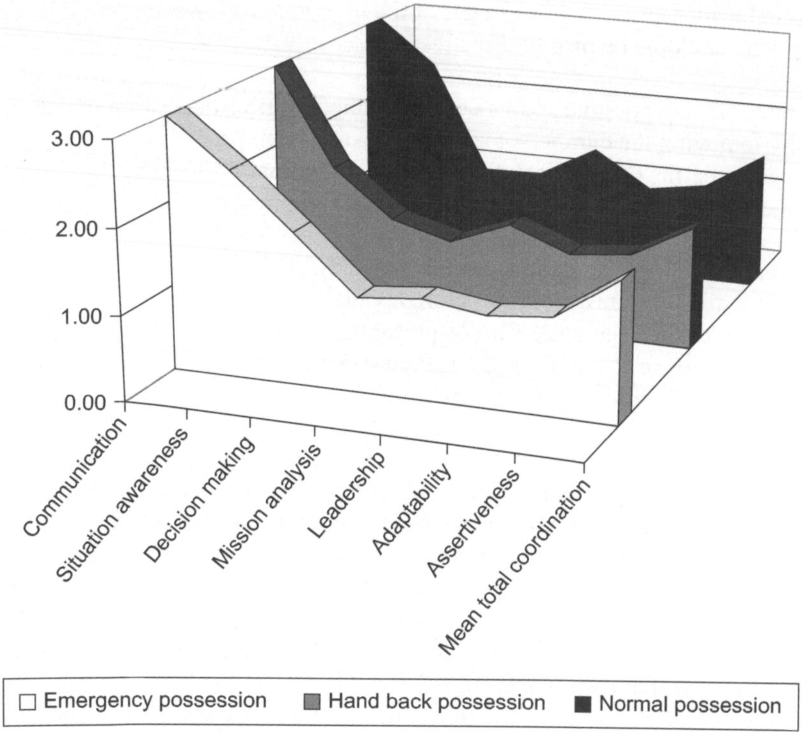

FIGURE 6.5 Results of CDA analysis showing profile of results on each of the coordination dimensions.

the 'teamwork' track, and that require coordination, form between 66% and 72% of total tasks undertaken.

Figure 6.5 extracts the teamwork tasks for further analysis. The analysis proceeds according to seven coordination dimensions and one summary total coordination score (based on the mean of the individual scores). This analysis reveals a broadly similar pattern of coordination activity within the scenarios; certainly the total coordination figures are comparable, falling within the mid-point of the rating scale. Of more interest is the pattern of results across the seven individual dimensions, where a distinctive footprint emerges. Communication, situational awareness (SA) and decision making are prominent dimensions, and there is also a smaller 'blip' for the leadership dimension. Leadership can be taken as a further indication perhaps of some decision-making activity. It can be further noted that the larger (and more complex) the scenario, the larger the leadership 'blip' is. Therefore, in summary, not only are the majority of total tasks dominated by coordination activities, but those activities are also dominated by communications and the creation and maintenance of SA.

CUD

General observations from the CUD analysis are that the communications are entirely verbal. Given the nature of the scenario, verbal and telephone communications appear to be appropriate at most stages of the interaction. However, possible technology options could be helpful in three respects.

1. by removing possible sources of error inherent in verbal communications
2. by removing the cumbersome nature of read-back procedures
3. by alleviating the physical disturbance to other tasks caused by the unscheduled and ad-hoc presentation of verbal communications.

Of course, any new approaches would require fuller risk justification and assessment within the wider task context before application. The key point is that the CUD method provides a systematic way of presenting the existing situation and considering alternatives to it based on data. Figure 6.6 summarises the communications technology in use within the scenarios.

SNA

Social Network Analysis is used in the EAST method to represent and summarise the communication/information links between agents in the scenario. Figure 6.6 presents a graphical representation of the networks derived from each of the scenarios and is also annotated with communications information drawn from the CUD analysis.

A range of mathematical metrics (derived from Graph Theory) can be applied. The results show that the PICOP, signaller and ESs have the highest levels of sociometric status and centrality. These metrics indicate that they are key agents in the scenarios. The notion of centrality is also borne out when considering 'betweenness'; that is, the PICOP and ES fall between pairs of other positions in the network the most frequently.

Having identified the key agents, it is also possible to view the network as a whole under the concept of *network density*. Density is the degree of interconnectivity between agents or the number of network links used compared to those that are

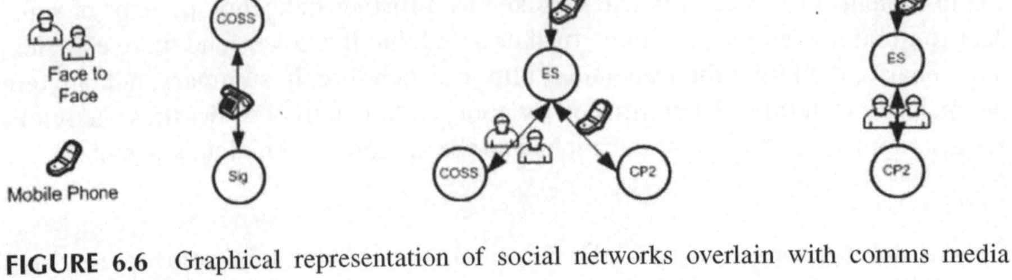

FIGURE 6.6 Graphical representation of social networks overlain with comms media drawn from CUD analysis.

TABLE 6.5
Comparison of Network Density between Scenarios

	Possession	Emergency Possession	Hand Back Possession
Density	0.4	0.67	0.33

theoretically available (the maximum being a case where all agents are linked to each other). In Table 6.5, it can be seen that the emergency possession scenario has the densest pattern of connectivity, with the remaining two scenarios being broadly comparable.

Although the metrics allow comparison between networks, intelligent interpretation is required. For example, a network with every agent connected to each other would permit easy dissemination of information, but it might also be inefficient. Similarly, having one central node may have advantages for coordination but offers the potential for an information bottleneck. Houghton et al. (2006) discuss the implications of several prototypical networks and their comparison to live scenario networks. The main point is that the interpretation and then subsequent comparison of networks has to take into account a range of contextual factors. For the time being, the networks derived from the scenarios above appear to be relatively well matched to the procedures being undertaken, with a mix of central agents and interconnectivity.

Representational Methods

Scenario Process Charts (OSD)

Figure 6.7 presents a sample of an enhanced OSD from scenario 1 (the planned maintenance activities), and highlights how the preceding methods are integrated with it.

The operations loading is presented in Table 6.6, showing the PICOP, signaller and CP as the most heavily loaded individuals in the network in terms of tasks. The operations loading table provides a further level of summarisation in being able to capture, in a relatively compact manner, the process-based aspects of what is often a large OSD.

Propositional Networks (PN)

The PN provides an overview, for each scenario, of all the knowledge elements and their relationships. It also allows the knowledge elements related to a specific phase in the scenario to be described, as well as the history of previously used knowledge points. Figure 6.8 displays the network elements related to taking a possession for emergency maintenance. Shaded cells denote knowledge objects that are currently active within the task phase. Faded cells indicate previously active knowledge. The main image in Figure 6.8 is intended to provide information on knowledge objects, whereas the smaller networks alongside are merely illustrative of changing activation.

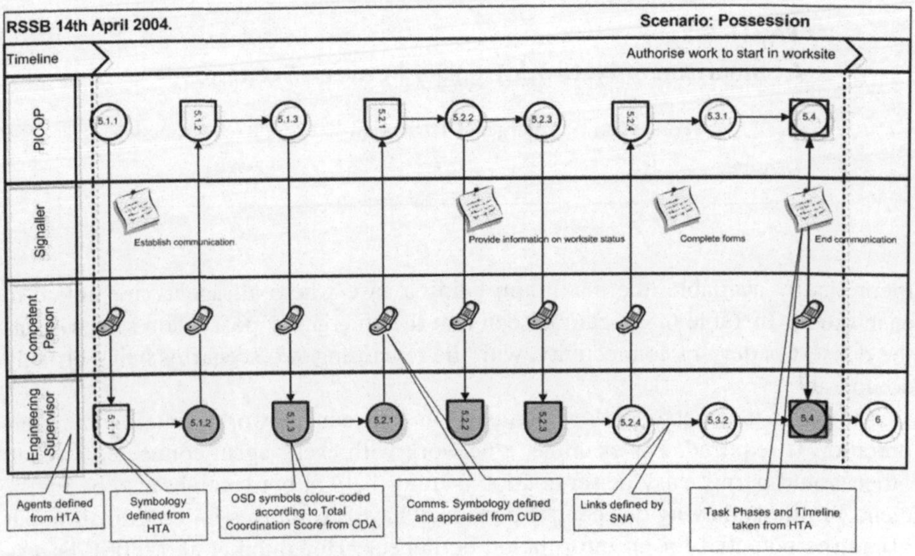

FIGURE 6.7 Enhanced OSD summary representation.

Despite the emergency procedure scenario being a relatively simple proceduralised task, the PN appears complex. It should be recognised that this complexity does not necessarily reflect complexity in the task itself. For example, some of the knowledge elements are internalised skills (e.g. participant knowledge of the railway rule book or a signaller's knowledge of the current status of the railway) or simple objects (e.g. telephones) that would not provide load or add to the complexity perceived by a skilled participant. It does, however, demonstrate that a large number of related objects are used to accomplish relatively simple tasks. In the context of task redesign, the propositional network provides a prompt that can allow system designers to question the necessity for knowledge objects, the form those objects can take and how they are communicated. This is a novel perspective. The network could also be considered as a tool for designing training, which describes the knowledge elements and relationships that someone undertaking the task must have available.

TABLE 6.6
Task Loading Table

	Operations				
Agent	**Operation**	**Receive**	**Decide**	**Transport**	**Total**
PICOP	54	30	1		85
Signaller	17	16	1		34
Eng. Supervisor	13	17		1	31
Competent Person 1	17	20		1	38

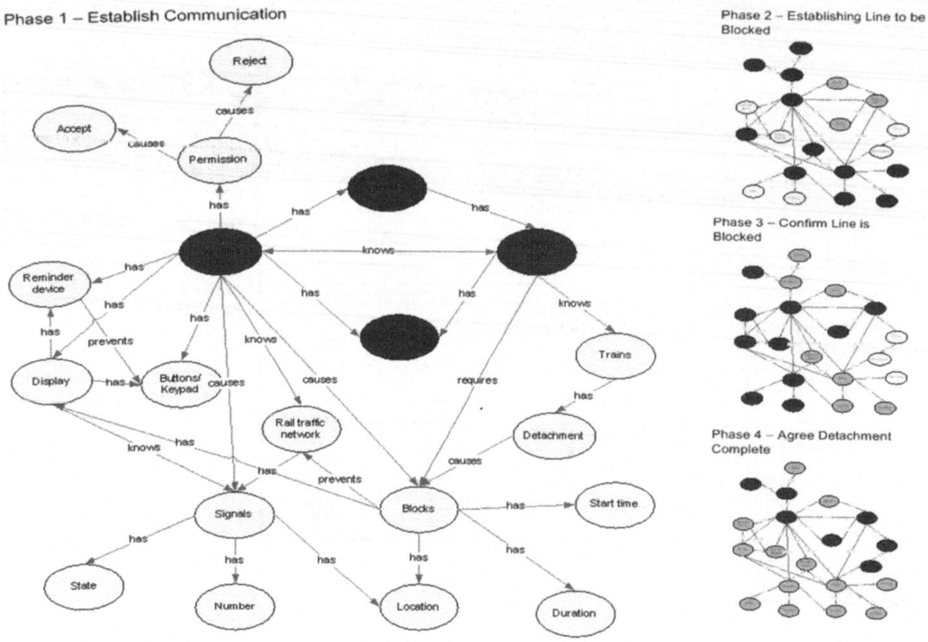

FIGURE 6.8 Illustration of propositional networks for phases within Scenario 3.

Summary

The summary and individual outputs derived from EAST have been presented. It should be evident that the data generated from an EAST analysis are extensive, so in Figure 6.9, an attempt is made to convey an impression of what the total EAST representation looks like. It is from this representation that a C4i scenario can be surveyed relatively quickly according to the key constructs of who, when, where, how and what. In turn, areas and themes that require further detailed insights can be extracted and examined.

CONCLUSIONS

Further Insights

The individual outputs available from EAST present a number of distinct but overlapping perspectives on the railway maintenance scenarios. This structured collection of data is required for the subsequent generic modelling and C4i theory development that is the main purpose of, but is separate from, the EAST method. However, along with contributing to theoretical data, a number of more practical themes can be drawn from each EAST analysis. The main practical themes related to this EAST analysis of railway data have been identified as follows:

1. The HTA and OSD highlight the sequential nature of the tasks involved. There is relatively limited decision making, and therefore the complexity comes from a requirement on the personnel involved to time and precisely sequence their actions.

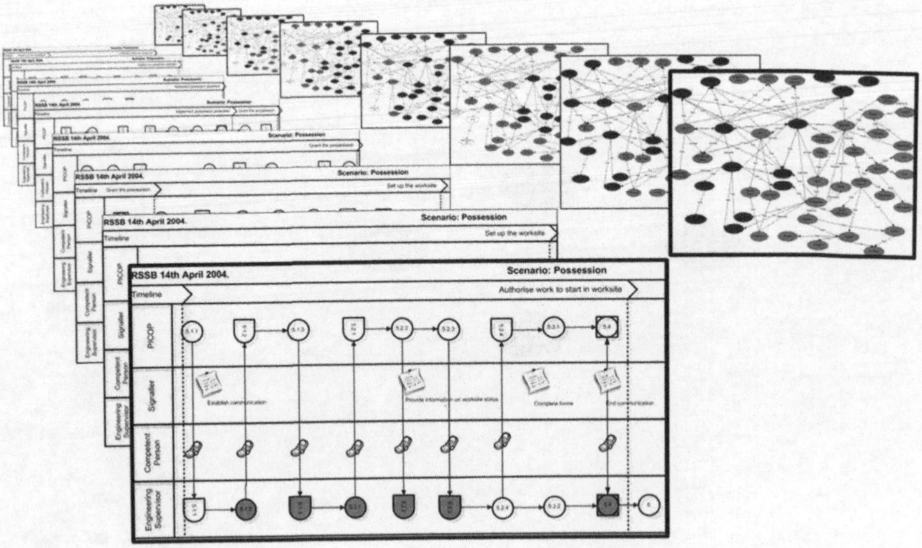

FIGURE 6.9 Summary of application of EAST to live railway data.

2. Pursuant to this, the analysis highlights a routine requirement in each communication to identify the people involved and the purpose of the call. This is due to the verbal nature of the telephone technology in use.
3. The analysis highlights the key role of communication in this process and that communication plays a part in, and is dependent on, SA.
4. Finally, the signaller (and other key agents) are required to develop situational awareness about the requirements of the possession communicated by various parties physically present at the trackside. They then have to consider the feasibility of carrying out the requirements in the context of planned information (where available) and the current status of train movements in the location.

The creation and maintenance of SA based on communications between actors is a key issue and can be examined further. One strategy to enable this is to extract the knowledge objects that the signaller requires, and that different actors possess. The relevant propositional networks and companion CDMs can be used for this purpose. The sharing of information can then be related to the sequence and timing of actions, and actors. Where knowledge objects are shared, consideration can also be given to the means by which this sharing occurs, for example, the communications technology that facilitates this sharing. This particular approach enables the analyst to proceed from a theoretical level to one that is much closer to practical design outputs. Figure 6.10 below illustrates this example of an approach to a more detailed analysis.

In describing this example, the intention is to convey how the general representations provided by EAST can be drilled down into to provide very specific insights from multiple perspectives. These multiple perspectives can be theoretical and practical.

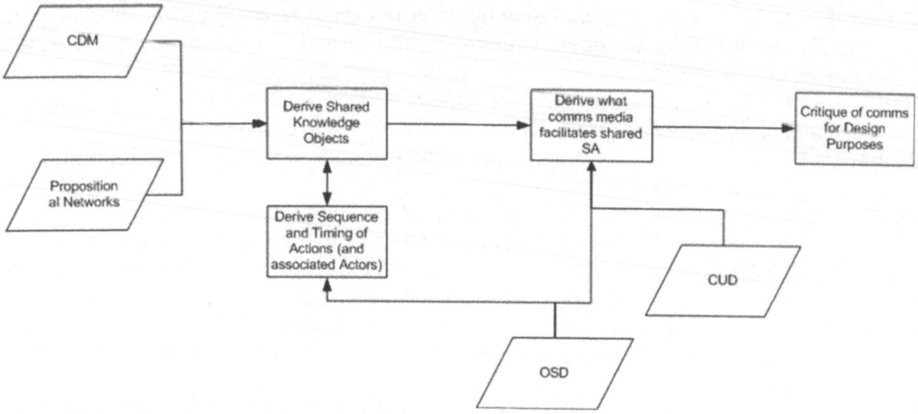

FIGURE 6.10 Plan for detailed analysis of communications and SA within railway scenarios

Wider Context

In conclusion, the application of the EAST methodology to live railway data illustrates the design and descriptive capability of the method. The method summarises the task structure, network structure and operational context into a form that enables ready comparisons on key metrics to be drawn between scenarios and for individual practical themes and insights to be analysed further. The method is also scalable, and its application by the DTC HFI to a host of military and civil domains is providing the data necessary to provide input into the development of generic models and theories of C4i.

ACKNOWLEDGEMENTS

The assistance of the Rail Safety and Standards Board is gratefully acknowledged, and the data presented here are used with their kind permission.

This work from the Human Factors Integration Defence Technology Centre (HFI DTC) was part-funded by the Human Sciences Domain of the UK Ministry of Defence Scientific Research Programme.

REFERENCES

Anderson, J. R. (1980). *Cognitive Psychology and its Implications.* Freeman: San Francisco, CA.

Annett, J. (2003). Hierarchical task analysis. In D. Diaper and N. A. Stanton (Eds.), *Handbook of Task Analysis in Human–Computer Interaction* (pp. 67–82). Lawrence Erlbaum Associates: Mahwah, NJ.

Annett, J. (2005). Hierarchical task analysis. In N. A. Stanton et al. (Eds.), *Handbook of Human Factors and Ergonomics Methods* (pp. 33.1–33.7). CRC: London, UK.

Annett, J., Duncan, K. D., Stammers, R. B. and Gray, M. (1971). *Task Analysis.* Her Majesty's Stationery Office: London, UK.

Annett, J. and Stanton, N. A. (2000). *Task Analysis.* Taylor and Francis: London.

Baber, C. (2004). Personal communication.

Bell, H. H. and Lyon, D. R. (2000). Using observer ratings to assess situation awareness. In M. R. Endsley (Ed.), *Situation Awareness Analysis and Measurement* (pp. 129–146). Lawrence Erlbaum Associates : Mahwah, NJ.

Burke, S. C. (2005). Team task analysis. In N. A. Stanton et al. (Eds.), *Handbook of Human Factors and Ergonomics Methods* (pp. 56.1–56.8). CRC: London, UK.

Collins, A. M. and Loftus, E. F. (1975). A spreading-activation theory of semantic processing. *Psychological Review*, 82, 407–428.

Defence Technology Centre for Human Factors Integration (2003). *Methodology for DTC HFI WP1 Field Trials* (Report: Draft 2.1.). Baber, C. & Stanton, N. A: London, UK.

Diaper, D. and Stanton, N. A. (2004). *Handbook of Task Analysis in Human–Computer Interaction*. Lawrence Erlbaum Associates: Mahwah, NJ.

Driskell, J. E. and Mullen, B. (2005). Social network analysis. In N. A. Stanton et al. (Eds.), *Handbook of Human Factors and Ergonomics Methods* (pp. 58.1–58.6). CRC: London, UK.

Drury, C. G. (1990). Methods for direct observation of performance. In J. R. Wilson and E. N. Corlett (Eds.), *Evaluation of Human Work – A Practical Ergonomics Methodology* (pp. 45–68). Taylor and Francis: London, UK.

Endsley, M. R. (1995). Toward a theory of situation awareness in dynamic systems. *Human Factors*, 37(1), 32–64.

Harris, C. J. and White, I. (1987). *Advances in Command, Control & Communication Systems*. Peregrinus: London, UK.

Houghton, R. J., Baber, C., McMaster, R., Stanton, N. A., Salmon, P., Stewart, R. and Walker, G. (2006). Command and control in emergency services operations: A social network analysis. *Ergonomics*, 49 (12–13), 1204–1225.

Kieras, D. E. (2003). GOMS models and task analysis. In D. Diaper and N. A. Stanton (Eds.), *Handbook of Task Analysis in Human–Computer Interaction* (pp. 83–116). Lawrence Erlbaum Associates: Mahwah, NJ.

Kirwan, B. and Ainsworth, L. K. (1992). *A Guide to Task Analysis*. Taylor and Francis: London, UK.

Klein, G. and Armstrong, A. A. (2005). Critical decision method. In N. A. Stanton et al. (Eds.), *Handbook of Human Factors and Ergonomics Methods* (pp. 35.1–35.8). CRC: London, UK.

Klein, G. A., Calderwood, R. and MacGregor, D. (1989). Critical decision method for eliciting knowledge. *IEEE Transactions on Systems, Man, and Cybernetics*, 19(3), 462–472.

Ogden, G. C. (1987). Concept, knowledge and thought. *Annual Review of Psychology*, 38, 203–227.

O'Hare, D., Wiggins, M., Williams, A. and Wong, W. (2000). Cognitive task analysis for decision centred design and training. In J. Annett and N. A. Stanton (Eds.), *Task Analysis* (pp. 170–190). Taylor and Francis: London, UK.

Quillian, R. (1969). The teachable language comprehender: A simulation program and theory of language. *Communications of the ACM*, 12459–12476.

Rail Safety and Standards Board. (2003). *Master Rule Book*. Newark: Willsons.

Scott, J. (1991). *Social Network Analysis: A Handbook*. Sage: London, UK.

Smalley, J. (2003). Cognitive factors in the analysis, design, and assessment of command and control systems. In E. Hollnagel (Ed.), *Handbook of Cognitive Task Design* (pp. 223–253). Lawrence Erlbaum Associates: Mahwah, NJ.

Stanton, N. A. (2004). The psychology of task analysis today. In D. Diaper and N. A. Stanton (Eds.), *Handbook of Task Analysis in Human–Computer Interaction* (pp. 567–584). Lawrence Erlbaum Associates: Mahwah, NJ.

Stanton, N. A., Baber, C. and Young, M. S. (2005). Observation. In N. A. Stanton et al. (Eds.), *Handbook of Human Factors and Ergonomics Methods* (pp. 28.1–28.7). CRC: London, UK.

Stanton, N., Hedge, A., Brookhuis, K., Salas, E., and Hendrick, H. (2005). *Handbook of Human Factors and Ergonomics Methods*. CRC: London, UK.

Stanton, N. A., Stewart, R., Harris, D., Houghton, R. J., Baber, C., McMaster, R., Salmon, P., Hoyle, G., Walker, G., Young, M. S., Linsell, M., Dymott, R. and Green, D. (2006). Distributed situation awareness in dynamic systems: Theoretical development and application of an ergonomics methodology. *Ergonomics*, 49(12–13), 1288–1311.

Watts, L. A., and Monk, A. F. (2000). Reasoning about tasks, activities and technology to support collaboration. In J. Annett and N. Stanton (Eds.), *Task Analysis* (pp. 55–78). Taylor and Francis: London, UK.

7 EAST at Road Intersections

With Michael G. Lenné and J. Ashleigh Filtness

STATEMENT OF RELEVANCE

Intersection safety currently represents a key road safety issue worldwide. In line with recent calls for a systems approach to road safety, this chapter presents a novel application of a framework of ergonomics methods for studying intersection 'system' performance. The findings support development of interventions that consider all road users as opposed to one group in isolation.

INTRODUCTION

Road transport–related trauma continues to be one of the leading causes of death and disability throughout the world (World Health Organisation 2009). Although significant reductions in injuries and fatalities have been made over the last four decades in most motorised countries (Elvik 2010), a number of complex intractable issues remain. One of these is collisions between different types of road users at intersections. In Australia, for example, the majority of urban crashes and a substantial proportion of rural crashes occur at intersections (McLean et al. 2010). Safety at intersections, therefore, represents a key road safety issue.

The prevalent approach to understanding and preventing collisions at intersections has tended to be road user centric and component driven, focusing on specific road user types (e.g. drivers) and/or fixing a component of the problem (e.g. making motorcycles more conspicuous; Gershen and Shinar 2013). The fundamental limitation of this reductionist approach is that there is a focus on only a component of the problem (e.g. rider conspicuity). In recent times, researchers have made a strong case for the 'systems' approach when considering road user behaviour and safety (e.g. Larsson et al. 2010; Salmon et al. 2012). The systems approach has important implications for how safety outcomes are achieved. Are they achieved through focusing on one component of the system in isolation (a reductionist approach) or can they, instead, be achieved by looking at the wider interacting system of human and technical elements (a systems approach)? In other words, does safety arise from 'adding' together interventions performed on individual elements of the system, such as the road environment and intersection infrastructure, training, road rules and regulations, environmental conditions and individual road users, or does it multiply or 'emerge' from these interactions? If it does, then a focus on individual elements will fail to capture important road safety interventions that could have disproportionately large effects.

Unsurprisingly, the systems approach has long been discussed in the context of road traffic crashes (e.g. Wagenaar and Reason 1990), but despite this, very few studies go beyond a focus on just one component of the system. Previous studies of road user behaviour at intersections have largely been road user centric, focusing on individual road user factors such as driving errors (e.g. Gstalter and Fastenmeier 2010; Sanding 2009) and pedestrian behaviours (e.g. King et al. 2009). Moreover, the majority of road safety research focusing on vulnerable road users, such as cyclists and motorcyclists, has been undertaken from the perspective of drivers (e.g. Hole and Tyrrell 1995; Schepers and Brinker 2011) or engineering (e.g. Highways Agency 2012). The consequence is that other factors influencing behaviour are left untouched and so diminish the effectiveness of the reductionist intervention (Salmon et al. 2012). For this reason, the reductionist, component driven approach to road safety has been criticised for failing to take into account the inherent complexity in transportation systems (e.g. Larsson et al. 2010; Salmon et al. 2012). This article is a response to this criticism.

This article argues that the long-standing problem of collisions between distinct road users at intersections cannot be solved through a reductionist approach that does not consider the overall sociotechnical intersection system. That is, consideration of individual road user behaviours in isolation, whilst yielding important information, will not necessarily lead to appropriate interventions that support the behaviour of all road users. This creates a challenging methodological problem. Existing approaches for studying road user behaviour, such as simulation (e.g. de Winter et al. 2009), microsimulation (e.g. Bell et al. 2012) and questionnaires (e.g. Reason et al. 1990) typically focus on one component only (e.g. 'driving' simulators, 'driver' behaviour questionnaire). To go beyond mere lip service to the 'systems approach' requires new approaches that embed individual road users within a wider joint cognitive system.

The analysis presented in this chapter moves towards systems analyses in road transport by considering the intersection 'system', comprising different road users (e.g. drivers, riders, cyclists and pedestrians), vehicles and the road environment. Using data derived from a recent on-road study, the Event Analysis of Systemic Teamwork framework (EAST) (Stanton et al. 2013; Walker et al. 2006, 2010) was used to evaluate behaviour at three major signalised intersections. The aims were to examine the differences in behaviour of different road users within the intersection system, to identify the factors underpinning conflicts between distinct road users at intersections and to inform the development of new more holistic intersection interventions that are appropriate for all road users.

EAST Intersection Case Study

The key benefit of a systems approach is that it offers a universal language for describing any type of system. Entities and artefacts are represented as nodes and the relationships between them as links. As such, there is no conceptual reason why a systems approach cannot be applied to intersections. Indeed, previous studies of road user behaviour at intersections suggest that two 'systemic' issues may be particularly problematic: intersection design and infrastructure that does not meet users' expectations (e.g. Cornelissen et al. 2013) and incompatibilities between different road users, such as drivers and motorcyclists (Salmon et al. 2013; Walker et al. 2011). The aim of the study was, through considering the entire intersection system as the unit of analysis, to investigate these issues further by applying the EAST framework to real-world, on-road data. The EAST framework lends itself to an in-depth evaluation of intersection performance of this kind. It facilitates examination of specific constructs relevant to the two key issues mentioned above, such as situation awareness, decision making, teamwork, training and road design. Whilst not providing direct recommendations, the analyses will

1. help to reduce complexity
2. identify specific issues limiting performance
3. highlight areas where system redesign could be beneficial.

The data were derived from an on-road study that covered a pre-defined 15km urban route in the south-east suburbs of Melbourne, Australia. The analysis focused on three major signalised intersections located along the route (see Figure 7.1). Each intersection required a right-hand turn in order to pass through the intersection and remain on the study route.

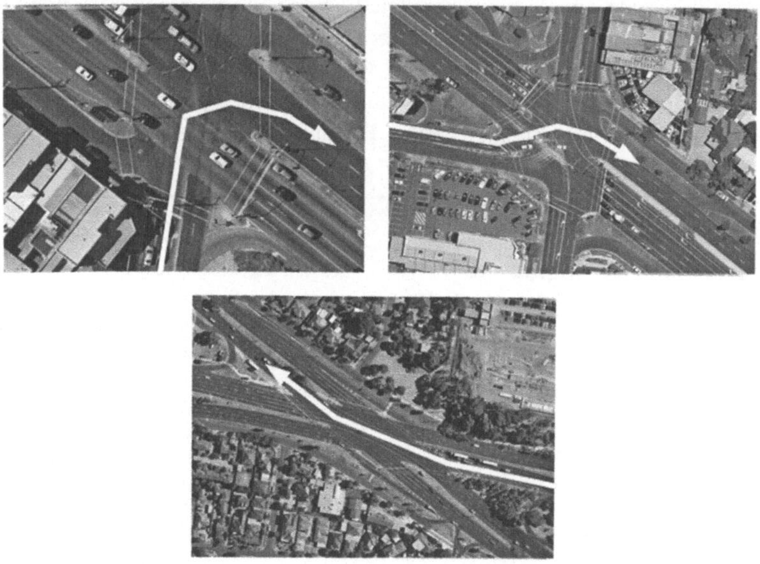

FIGURE 7.1 The three intersections studied (white arrow denotes route though intersection).

METHODOLOGY

Design

The work reported in this chapter is based around an on-road study investigating the influence of road design on the behaviour of different types of road user. The study used a semi-naturalistic paradigm whereby participants drove an instrumented vehicle around a pre-defined urban route. Drivers drove an instrumented test vehicle, whilst motorcyclists and cyclists completed the route using their own motorcycle or bicycle, which was equipped with video and audio recording equipment. Pedestrians negotiated the three intersections on foot whilst wearing video recording glasses and a microphone connected to a Dictaphone. All participants provided concurrent verbal (think aloud) protocols as they negotiated the route.

Participants

Seventy-eight participants (52 male, 26 female) aged 21–64 years (mean = 35.81, SD = 13.03) took part in the study. The sample comprised 20 car drivers, 18 motorcyclists, 20 cyclists and 20 pedestrians. An overview of the participants in each group is presented in Table 7.1.

All participants were recruited through a weekly on-line university newsletter and were compensated for their time. Prior to commencing the study, ethics approval was formally granted by the Monash Human Ethics Committee.

Materials

A demographic questionnaire was completed using pen and paper. A desktop driving simulator was used to provide verbal protocol training. A 15 km urban route, located in the south-eastern suburbs of Melbourne, was used for the on-road study component. Drivers drove the route in a 2004 Holden Calais sedan equipped to collect various vehicle, driving scene and driver-related data. A Dictaphone was used to record participant's

TABLE 7.1
Participant Demographics

Road User Group	Mean Age (SD)	Gender	Hours Using Respective Mode of Transport per Week
Drivers	34.9yrs (12.53)	10 males 10 females	11.5 hours
Cyclists	32.4yrs (10.42)	15 males 5 females	6.85 hours
Motorcyclists	45.5yrs (12.87)	17 males 1 female	7 hours
Pedestrians	30.5yrs (11.86)	10 males 10 females	8.92 hours

verbal protocols. Each motorcyclist participant's motorcycle was fitted with an Oregon Scientific ATC9K portable camera, which, depending on the motorcycle model, was fixed either to the handlebars or front headlight assembly. The ATC9K camera records the visual scene, speed and distance travelled. A microphone was fitted inside each motorcyclist participant's helmet and attached to the Dictaphone to record their verbal protocols. For cyclist participants, the ATC9K portable camera was fitted to their helmet and they wore Imging HD video cycling glasses. Pedestrians negotiated the intersections on foot whilst wearing Imging HD video sunglasses and a microphone linked to a Dictaphone. Motorcyclist, cyclist and pedestrian participants also carried a Garmin GPS unit to record aspects such as speed, distance travelled, route taken and so on. All verbal protocols were transcribed using Microsoft Word. During data analysis, the Leximancer™ content analysis software was used to construct knowledge networks.

Procedure

Individual participants negotiated the intersections on separate occasions. In order to control for traffic conditions, all trials took place at the same pre-defined times on weekdays (10 am or 2 pm Monday to Friday). Participants first completed an informed consent form and demographic questionnaire and were then briefed on the research and its aims (expressed generally as a study of driver behaviour). Following this, they were given training in providing verbal protocols, which included instruction from an experimenter followed by a desktop driving simulator task where they were asked to complete the drive whilst providing a verbal protocol. The experimenter monitored the drive and provided feedback to the participant regarding the quality of their verbal protocol. Participants were then shown the study route and were given time to memorise it. When comfortable with the verbal protocol technique and route, participants were taken to their vehicle and asked to prepare themselves for the test. They were then given a demonstration of the video and audio recording equipment, which was also set to record at this point. Following this, the experimenter instructed the participant to begin the study route. For the drivers, an experimenter was located in the vehicle and provided route directions if necessary. For the motorcyclists and cyclists, an experimenter followed behind (in a car for the motorcyclists, on a bicycle for the cyclists), ready to intervene if the participants strayed off route. Pedestrians were taken by car to the first intersection and instructed to negotiate the intersection and walk to a set point following the intersection. Once the participant reached this point, they were picked up by the experimenter and driven to the next intersection. This process was repeated until all three intersections had been negotiated.

Participants' verbal protocols were transcribed verbatim using Microsoft Word. For data reduction purposes, extracts of each verbal transcript for each intersection were taken from the overall transcripts. The extracts were taken based on the video data and pre-defined points in the road environment (e.g. beginning and end of intersection). The three network representations were constructed as follows:

1. *Task networks*: The task networks were built based on experimenter observations, a review of the video data and the verbal transcripts. This led to the construction of generic task networks for each road user group,

incorporating the range of participant behaviours observed within each group at the three intersections.

2. *Social networks*: In the present analysis, the social networks represent each participant's interaction with the intersection 'system'. Construction of the social networks involved reviewing the verbal transcripts and recording all instances in which the participant in question verbalised an interaction with part of the intersection system (e.g. 'I'm just checking the traffic lights ahead', 'I've just noticed a car pulling up behind me', 'I'm just going to press the pedestrian crossing button'). The frequency of interactions with different parts of the intersection system were calculated for each road user group and represented using standard social network diagrams.
3. *Situation awareness networks*: Situation awareness networks were constructed for each participant group at each intersection through analysis of participants' verbal transcripts using the Leximancer content analysis tool. Leximancer uses text representations of natural language to interrogate verbal transcripts and identify themes, concepts and the relationships between them. The software does this by using algorithms linked to an in-built thesaurus and by focusing on features within the verbal transcripts such as word proximity, quantity and salience. Leximancer thus automates the content analysis procedure by processing verbal transcript data through five stages: conversion of raw text data, concept identification, thesaurus learning, concept location and mapping (i.e. creation of network). The output is a network representing concepts derived from the verbal transcript and the relationships between them reflected within the verbalisations. The software has previously been used for situation awareness network construction and analysis (e.g. Salmon et al. 2013; Walker et al. 2011) and is important to analyses of this kind since it provides a reliable, repeatable process for constructing the networks.

Each form of network was analysed using selected network metrics that have previously been used in EAST analyses. Although predominately used for the analyses of social networks, Stanton (2014) demonstrated the utility of applying selected network analysis metrics to the task, social and situation awareness networks generated through EAST analyses. In the present study, the following network analysis metrics were applied to each network:

1. *Network density* – Network density represents the level of interconnectivity of the network in terms of relations between nodes. Density is expressed as a value between 0 and 1, with 0 representing a network with no connections between nodes and 1 representing a network in which every node is connected to every other concept (Kakimoto et al. 2006; cited in Walker et al. 2011).
2. *Sociometric status (individual nodes).* Sociometric status provides a measure of how 'busy' a node is relative to the total number of nodes within the network under analysis (Houghton et al. 2006). In the present analysis,

nodes with sociometric status values greater than the mean sociometric status value plus one standard deviation are taken to be 'key' (i.e. most connected) nodes within each network.

RESULTS

TASK NETWORKS

The 'negotiate intersection' task networks provide a summary of the main goals and tasks involved in negotiating signalised intersections for each road user group. The task networks for each road user group are presented in Figures 7.2 through 7.5.

The task networks highlight the differences, in terms of goals and tasks performed, between the four road user groups when negotiating intersections. As shown in Figures 7.2 through 7.5, when negotiating intersections, the goals and subsequent tasks are markedly different across the four road user groups studied. The only task common across the four road user groups is that of 'maintaining situation awareness', and this task in itself entails maintaining awareness of different aspects of the intersection depending on road user type (e.g. cyclists and motorcyclists require awareness of debris on the road, whereas drivers do not).

One key issue highlighted by the task networks is the differences in the level of flexibility afforded to each road user by the intersections studied. Cyclists have a significant level of flexibility in that they can proceed through the intersection in a variety of different ways. Depending on traffic conditions and perceived level of risk, they can either turn right on the road within the flow of traffic, turn right via the pedestrian crossings and along the footpath or turn right using a 'hook' turn, whereby they proceed straight on through the intersection, join the traffic queue to the left-hand side

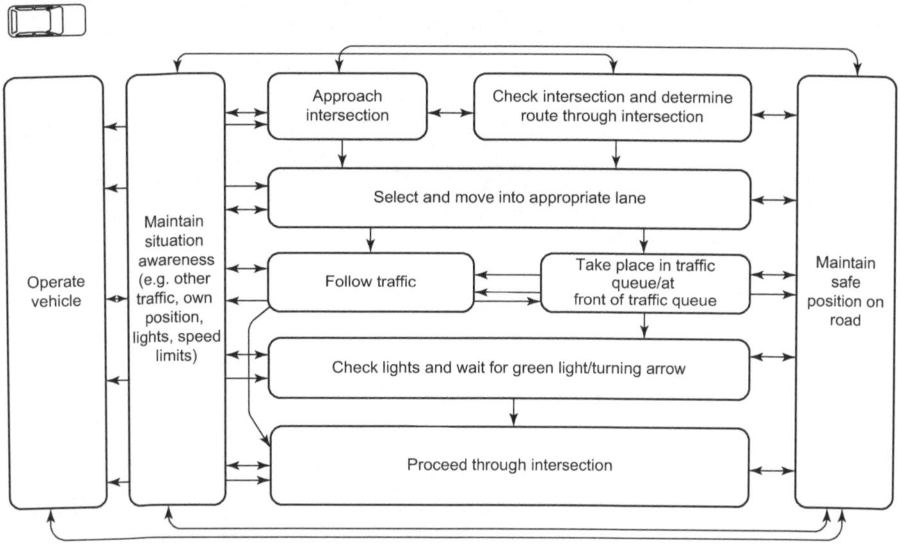

FIGURE 7.2 Driver intersection task network.

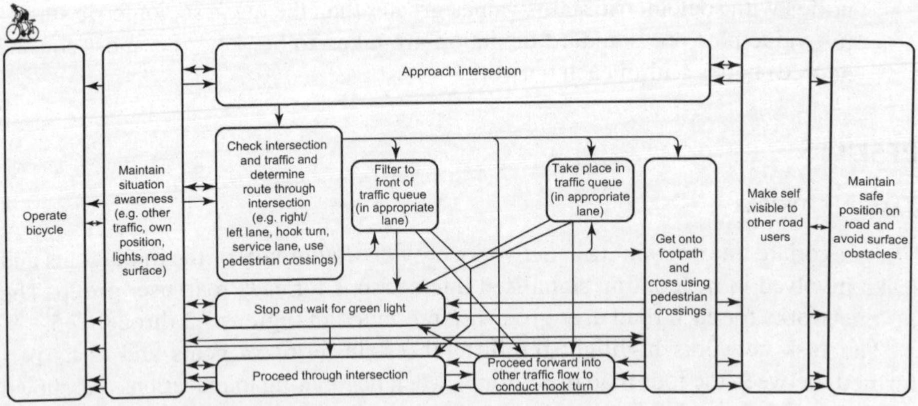

FIGURE 7.3 Cyclist intersection task network.

and then wait for a green light and proceed straight through the intersection (achieving the originally desired right-hand turn). This flexibility is also not restricted to a certain point within the intersection; cyclists can change their route through at any point prior to the intersection and even when passing through the intersection. For pedestrians, a degree of flexibility is extant through the ability to either cross via the pedestrian crossing or the road itself. Motorcyclists' behaviour is more restricted; however, they still have some flexibility in that they can choose to use the stay in the normal traffic queue or to filter up between the traffic to the front of the queue. Of the four road user groups, drivers are the most restricted in terms of behaviour since they can only select one of two turning lanes and progress through the intersection.

Other important differences across the task networks are present. These include the task of filtering up through the traffic queue undertaken by both cyclists and motorcyclists in pursuit of the goal to get to the front of the traffic queue, the

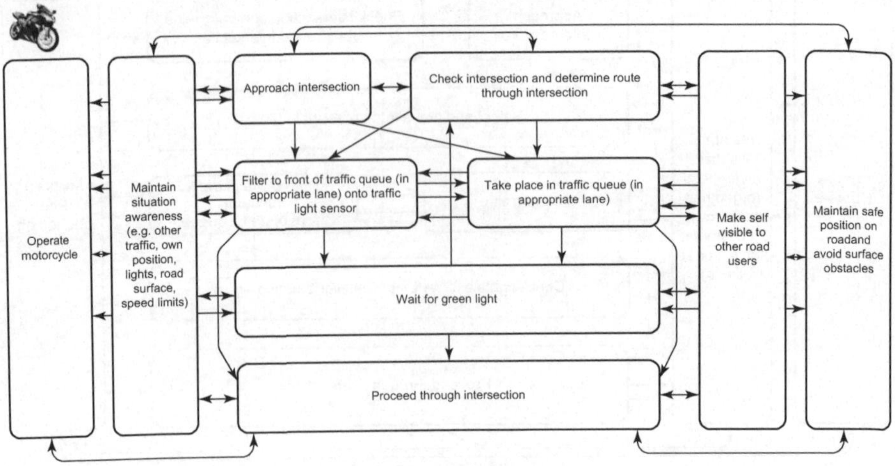

FIGURE 7.4 Motorcyclist intersection task network.

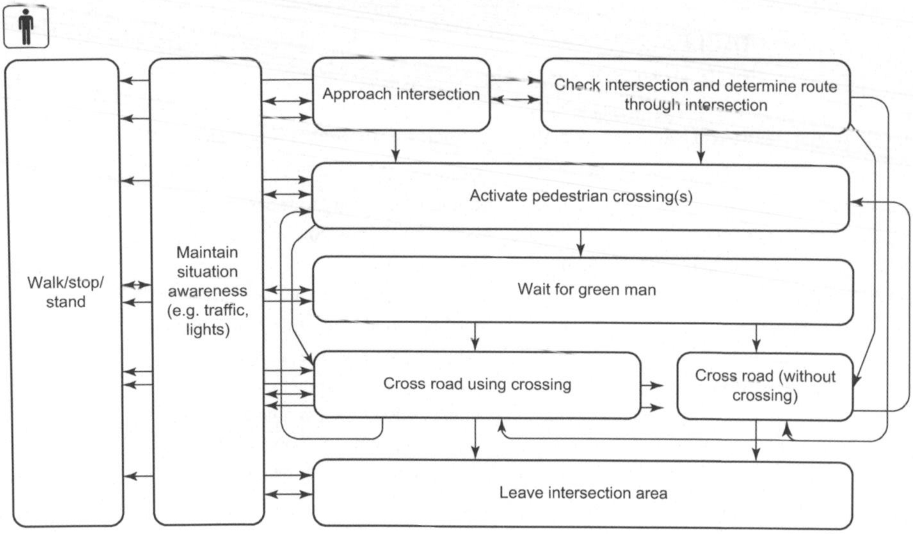

FIGURE 7.5 Pedestrian intersection task network.

requirement for cyclists and motorcyclists to undertake activities in order to make themselves visible to other road users (e.g. wiggling, flashing lights, eye-balling drivers, revving engine) and the requirement for cyclists and motorcyclists to check the road surface for debris/hazards/obstacles and to avoid debris on the road as they turn through the intersection.

The task models confirm that there are key differences in the goals being pursued and tasks being performed across the different road user groups. Moreover, they reveal differences that raise concerns regarding potential conflicts between distinct road users. For example, the level of flexibility afforded to cyclists, motorcyclists and pedestrians ostensibly gives them a degree of unpredictability in and around intersections. A cyclist in the right-hand turn lane may decide at a late stage to cross the intersection via the pedestrian crossing and thus manoeuvre across three lanes of traffic. Depending on how alert drivers are, this unpredictability could lead to conflicts. The different means of negotiating the intersection employed by different road users also raise potential conflicts, for example, cyclist and motorcyclist filtering behaviour coupled with drivers changing lanes and following the traffic ahead, or cyclists negotiating the intersection on the footpath and via pedestrian crossings.

The density and sociometric status calculations for the task networks are presented in Table 7.2.

Table 7.2 shows that the driver, cyclist and pedestrian task networks had similar levels of density; however, the motorcyclist task network is the most dense (0.81). This indicates that, of the four road users, motorcyclists have a more connected set of tasks when negotiating intersections. The key tasks, as identified through the sociometric status metric, demonstrate the importance of 'operate vehicle' and 'maintain situation awareness' tasks for all road users but also shows the importance of the 'make self visible' and 'maintain safe position on road' for cyclists and motorcyclists.

TABLE 7.2
Task Network Analysis

Road User	Density	Sociometric Status Key Tasks
Drivers	0.62	Operate vehicle Maintain situation awareness Maintain safe position on road
Cyclists	0.69	Operate bicycle Maintain situation awareness Make self visible to other road users Maintain safe position on road
Motorcyclists	0.81	Operate motorcycle Maintain situation awareness Make self visible to other road users Maintain safe position on road
Pedestrians	0.63	Walk/stop/stand Maintain situation awareness

Social Networks

The social networks depict each road user group's interaction with the intersections studied. For example, the social networks for each road user group at Intersection 1 are presented in Figure 7.6. This shows the total number of interactions with different parts of the intersection for each road user group. The thickness of the arrows represents the frequency of interactions, with the thicker arrows representing a higher frequency across each road user group.

The social network analyses highlight the fact that intersections are characteristic of joint cognitive, or distributed cognition, systems. The networks show a high interaction between human and non-human agents and also the key role of non-human agents within the right turn at intersection task. For each form of road user, the turn right task involves multiple non-human agents, such as the traffic lights, road and road markings, other vehicles and road signage. The importance of this finding is that compatibility between human and non-human agents is known to be a key element of efficient distributed cognition system performance (Salmon et al. 2008; Stanton et al. 2006; Walker et al. 2010).

Density was calculated for each network, and the sociometric status metric was used to identify the most connected nodes within each network (outside of the road users themselves). These network analysis results are presented in Table 7.3.

Table 7.3 demonstrates that the pedestrians were the most connected in terms of their interactions with the intersection environments. They interacted with the greatest number of intersection components (e.g. lights, button, traffic), had the highest number of interactions and achieved the highest sociometric status values across all three intersections. The drivers, on the other hand, were the least connected of the four groups, with smaller networks, the lowest number of interactions and the smallest sociometric status values across the three intersections.

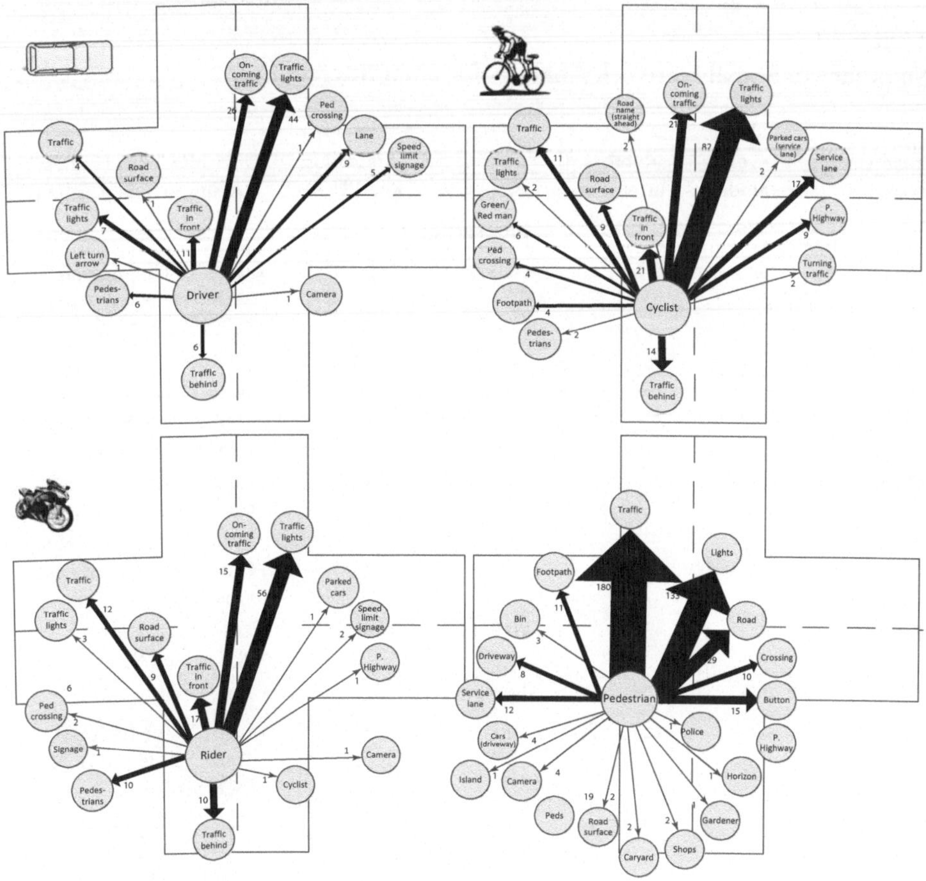

FIGURE 7.6 Social networks for Intersection 1.

The social networks also reveal important differences in how the different road user groups interact with intersection systems. Again, differences across the four road user groups are apparent. The networks show that the drivers also interacted less with certain parts of the intersection than did the other road users. For example, across the three intersections, the most frequent interaction for drivers was observing the oncoming traffic, traffic lights ahead and the traffic in front. Although cyclists and motorcyclists had frequent interactions with these parts of the intersections, they tended to have more interactions with these components, and other interactions were also prominent, such as checking for traffic behind and observing the traffic generally (i.e. travelling across the intersection). The low number of driver interactions focusing on the traffic generally, the traffic behind and adjacent traffic is a concern, particularly when coupled with the task network finding that the other road user groups may be unpredictable in and around the intersection.

The other road user groups also had interactions not found in the driver group. Unsurprisingly, pedestrians had various interactions related to the activity of interacting with the pedestrian crossing infrastructure, such as pressing the 'button'.

TABLE 7.3
Summary of Social Network Analysis for Each Intersection

Road User Group	Number of Nodes	Number of Interactions	Density	Sociometric Status of Road User	Nodes with Highest Sociometric Status Values
			Intersection 1		
Drivers	12	116	0.083	11.09	Lights, Oncoming traffic, Traffic (In front), Lane, Pedestrians, Traffic (Behind)
Cyclists	13	204	0.08	17.17	Lights, Oncoming traffic, Traffic (In front), Service lane, Traffic (Behind)
Motorcyclists	14	208	0.07	11.31	Lights, Traffic (In front), Oncoming traffic, Traffic (General), Traffic (Behind), Pedestrians, Parked cars, Road, Camera, Cyclist
Pedestrians	20	423	0.05	22.78	Traffic, Lights, Road, Pedestrians, Service lane
			Intersection 2		
Drivers	10	127	0.1	14.11	Lights, Traffic (In front), Traffic (General), Traffic (Behind), Sign
Cyclists	20	208	0.05	10.95	Lights, Traffic (General), Service lane, Traffic (In front), Road
Motorcyclists	13	116	0.08	9.66	Lights, Traffic (In front), Road, Traffic (General), Traffic (Behind)
Pedestrians	28	388	0.036	14.37	Traffic (General), Lights, Road, Intersection, Button
			Intersection 3		
Drivers	14	152	0.071	11.69	Lights, Traffic (General), Traffic (In front), Intersection, Lane
Cyclists	20	243	0.05	12.78	Lights, Traffic (General), Traffic (Behind), Lane, Traffic (In front)
Motorcyclists	18	113	0.053	6.28	Lights, Traffic (General), Lane, Road, Intersection
Pedestrians	30	767	0.033	26.44	Traffic (General), Lights, Road, Pedestrians, Intersection

Interestingly, cyclists also had a number of interactions with the footpath and pedestrian crossing infrastructure, which reflects the decision by selected cycling participants to cross the intersection using the footpath and pedestrian crossing. Other notable interactions specific to road user groups were motorcyclists and cyclists checking the 'road surface' for debris and oil, cyclists and motorcyclists monitoring traffic adjacent to them, motorcyclists checking the traffic lights facing other traffic (to work out when other traffic would be stopping and going) and cyclists monitoring pedestrian behaviours in and around the intersection.

Situation Awareness Networks

Leximancer was used to construct overall driver, cyclist, motorcyclist and pedestrian situation awareness networks for each intersection. For example, Figure 7.7 depicts each road user group's situation awareness network, overlaid on one another, for Intersection 1. Within Figure 7.7, the nodes and links are shaded to depict each road user group's situation awareness. Figure 7.7 shows how situation awareness differed across the distinct road user groups whilst negotiating Intersection 1, both in terms of the concepts underpinning situation awareness (i.e. nodes in the network) and in the way in which the concepts were linked together (i.e. links between the nodes in the network). Moreover, the network demonstrates that, even when the different road users were using the same concept, they were doing so in conjunction with other different concepts. This pattern is repeated over the other two intersections studied. The multi-road user situation awareness networks therefore confirm that driver, cyclist, motorcyclist and pedestrian situation awareness was different when negotiating the three intersections studied.

A summary of the situation awareness network analysis is presented in Table 7.4.

Table 7.4 highlights some important differences in situation awareness across the four road user groups. Pedestrian situation awareness networks comprised more situation awareness concepts at each intersection, whereas the driver networks were the smallest. The driver situation awareness networks were the most dense at Intersections 1 and 2, whereas the motorcyclists were most dense at Intersection 3. The same pattern was found for the mean sociometric status values, which indicates that the driver networks were most connected in terms of links between concepts at Intersections 1 and 2, whereas the motorcyclists were most connected at Intersection 3.

The key concepts underpinning each road user groups' situation awareness also reveal important differences. At Intersection 1, they key concepts are broadly similar across all four road user groups, with situation awareness predominantly underpinned by the lights, the lights' status (red or green) and the traffic. Notable differences at Intersection 1 include a focus of cyclists on the 'intersection' itself and a focus of motorcyclists on the 'lane' within which to progress through the intersection. At Intersection 2, more differences in key concepts are apparent. Again, the drivers' focus is on the lights; however, the cyclists are also focused on the 'service lane', which 'lane' to use, 'turning' traffic and the 'road' itself. The motorcyclists' key concepts include 'looking', 'behind', and again the 'turning' traffic and the 'lane' to use. At Intersection 3, the key concepts are again different. The drivers' key

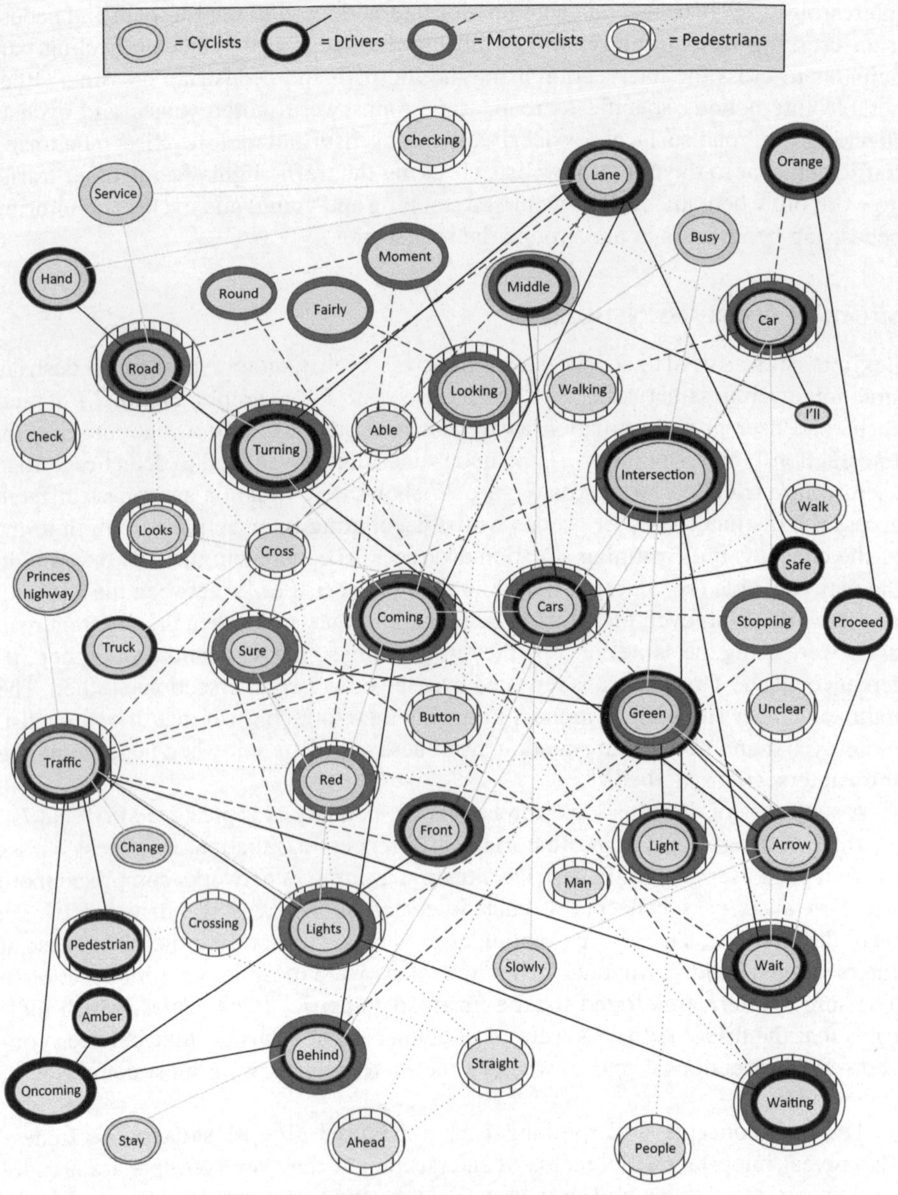

FIGURE 7.7 Intersection 1 situation awareness network showing driver, cyclist, motorcyclist and pedestrian situation awareness networks mapped onto each other.

concepts now include distinct areas around the car, such as 'front' and 'behind', and also the 'lanes'. The cyclists' key concepts also include 'ahead' and 'behind', 'wait' and 'crossing'. The motorcyclists' key concepts relate to the lights (i.e. green) and the 'lane' within which to progress through the intersection. Finally, the pedestrians have a larger set of key concepts at Intersection 3, with key concepts covering the

TABLE 7.4
Summary of Situation Awareness Network Analysis for Each Intersection

Road User Group	Number of Concepts	Density	Mean Sociometric Status	Key Concepts
			Intersection 1	
Drivers	26	0.075	0.151	Green, Oncoming, Traffic
Cyclists	28	0.061	0.122	Coming, Cars, Intersection, Red, Traffic
Motorcyclists	28	0.066	0.132	Lane, Traffic, Cars, Arrow
Pedestrians	29	0.069	0.138	Green, Red
			Intersection 2	
Drivers	27	0.074	0.148	Lights, Light, Lane
Cyclists	29	0.069	0.138	Cars, Green, Road, Take, Service, Lane, Turning
Motorcyclists	28	0.071	0.143	Looking, Behind, Green, Cars, Turning, Lane
Pedestrians	31	0.064	0.129	Lights, Light, Green, Turning, Coming, Road, Crossing
			Intersection 3	
Drivers	27	0.074	0.148	Front, Red, Dandenong road, Making, Behind, Lanes
Cyclists	28	0.071	0.143	Turning, Stay, Behind, Ahead, Wait, Lane, Lights, Crossing
Motorcyclists	26	0.075	0.151	Green, Lane
Pedestrians	34	0.055	0.110	Road, Cars, Coming, Cross, Wait, Traffic, Lights, Green, Red, Light,

'road', the traffic and its behaviour (e.g. 'cars', 'traffic', 'coming') and the lights and their status.

The differences in the situation awareness networks were explored further by examining the 'common' concepts (i.e. those found in all road user groups' situation awareness networks) along with the concepts unique to each road user group at each intersection (See Figure 7.8). Figure 7.8 shows that the concepts common across all four road user groups at the three intersections were mainly related to the cars, traffic, the road, the lights and their status (e.g. green), the intersection and the act of turning. This reflects a high focus of all road users on cars, the traffic lights and the intersection environment itself.

Important differences in situation awareness concepts are shown in Figure 7.8. At Intersections 1 and 2, the cyclist networks include the concepts 'service', 'stay' and 'route', all of which reflect the key decision that they face regarding whether or not to use the service lane on approach to the intersection and then cross via the footpath and pedestrian crossing, or to stay on the road and go through the intersection with the normal traffic flow. Moreover, prior to the intersection,

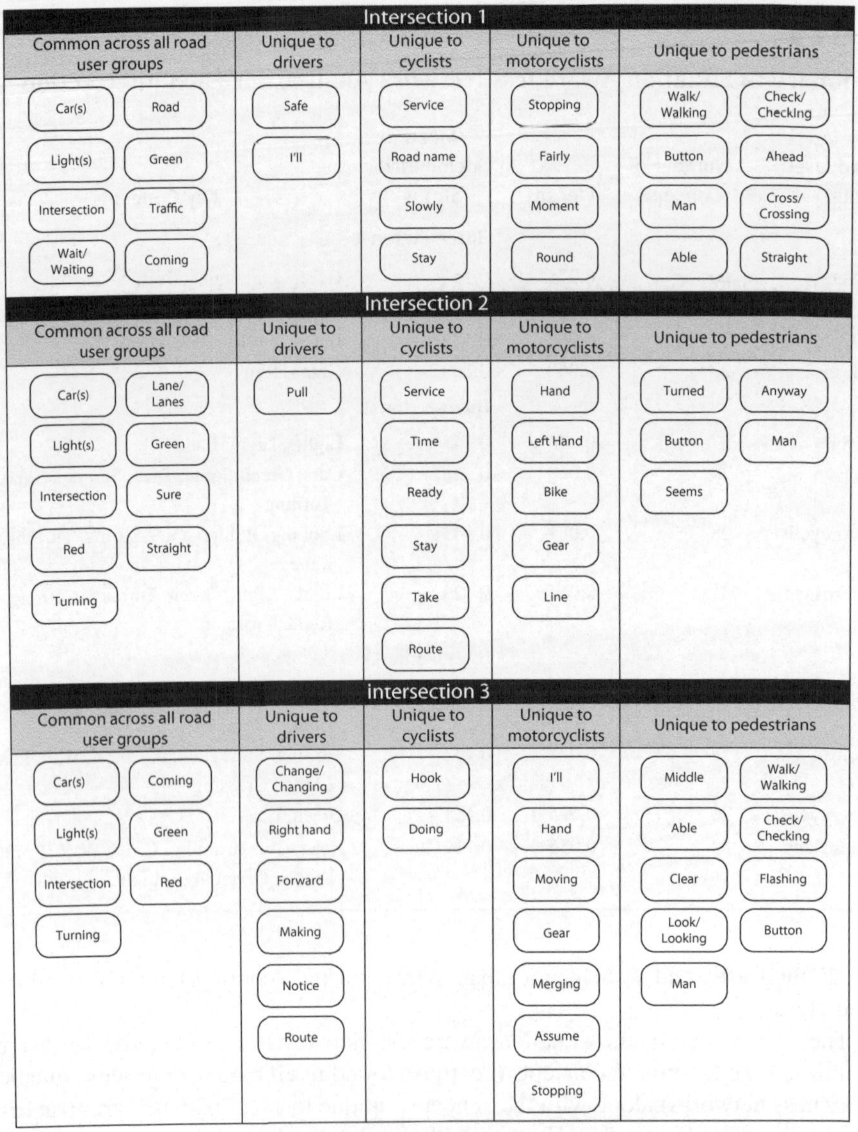

FIGURE 7.8 Common and unique concepts across road user groups at each intersection.

the cyclists also decide whether they will leave the road and get back into the service lane once they have passed through the intersection. At Intersection 3, the cyclists' network included the 'hook' concept, which refers to their decision regarding whether or not to use a hook turn in order to turn right at the intersection. Again, this reflects a key decision whereby cyclists try to work out whether it is safe enough to pass through the intersection on the road within the flow of traffic or whether they need to perform a hook turn to avoid conflict with other traffic also turning right.

For the motorcyclists, the unique concepts relate primarily to the selection of the left or right-hand lane to negotiate the intersection (e.g. 'hand', 'left hand', 'merging'), the motorcycle itself (e.g. 'bike', 'gear') and the 'line' that they should take through the intersection. The 'stopping' concept refers not only to motorcyclists' own braking behaviour but also to them checking that other traffic approaching from behind is stopping when the traffic lights are on red. For the pedestrians, the unique concepts were primarily related to the physical acts of walking (e.g. 'walk/walking') and crossing the road (e.g. 'cross/crossing') and also the crossing infrastructure (e.g. 'button', 'green man'). Interestingly, only the pedestrian networks included the concepts 'check/checking' and 'look/looking', which indicate that the other road users placed less emphasis on checking other traffic and the road environment when negotiating the intersections.

DISCUSSION

The aim of this article was to demonstrate how systems analysis methods such EAST can be used to better understand the complex behaviour of intersection systems. In doing so, we wished to examine the differences in behaviour across distinct road users and to identify factors that potentially create conflicts between them. The study aimed to demonstrate that the EAST method provides a means to reduce complexity, identify specific issues limiting performance and highlight areas where system redesign could be beneficial.

Reducing Complexity

The analysis demonstrates that, even when faced with the same intersection and right-hand turn task, the goals and tasks, interactions and knowledge used by distinct road users is different. This might not be particularly ground-breaking; however, identifying a structured methodology and associated metrics that are capable of revealing these differences using on-road, real-world data is a substantive contribution to the current state of the art. Further, the framework moves towards the much-heralded use of systems analysis techniques rather than reductionist, individual road user techniques for road transport analyses. Notably, although a systems approach has been called for by many (e.g. Larsson et al. 2010), to date there has been a dearth of appropriate methods proposed for supporting systems analyses in road transport (Salmon et al. 2012).

The EAST analysis also confirms the notion that road transport environments represent joint cognitive systems in which cognition is distributed across both human and non-human agents and is achieved through coordination between agents. This not only confirms the requirement for systems analysis techniques in road transport applications but also highlights the pressing need for studies of distributed cognition in the road transport context. In addition, this suggests that a distributed situation awareness approach (e.g. Salmon et al. 2008) in which the distribution of awareness across the intersection system is focused on, rather than the awareness of one road user group alone (e.g. drivers in motorcycle conspicuity studies), is required. Given the rapid advances in the capability of in-vehicle and roadside technologies and their likely increased role in road systems (e.g. intelligent infrastructure, car-to-car communications), distributed cognition and distributed situation awareness studies in support of future road system design efforts are recommended.

Performance Limitations

If tasks, interactions and knowledge are different across road users, and intersections are joint cognitive systems, then the key to intersection safety is ensuring that the differences are compatible and that road user activities and cognition connect together. This raises two lines of inquiry for the present study findings. First, do the differences revealed by EAST create conflicts between road users, or are they compatible? And second, does intersection system design (e.g. road environment design, training, road rules and regulations) support these differences in a way that ensures compatibility between distinct road users? It is in synthesising the tripartite EAST perspective to answer these questions that the framework becomes most powerful. Unfortunately, examination of the networks produced indicates a level of disharmony that suggests compatibility is under threat in some instances; in some circumstances, intersections may behave more like disjoint cognitive systems. For example, the networks demonstrate that the limited latitude for behaviour for drivers at intersections is in fact in conflict with the highly flexible and thus unpredictable behaviour of cyclists, motorcyclists and pedestrians. Here, flexibility, an often-sought-for characteristic in complex sociotechnical systems, coupled with poor intersection design, is creating collisions between distinct road users at intersections.

In the case of cyclists, for example, the combined networks clearly demonstrate their high level of flexibility since they can proceed through the intersection in multiple ways (i.e. hook turn, in normal flow of traffic, filtering, on the footpath and via pedestrian crossings). The networks also show that the result of this is that cyclists' behaviour and situation awareness are focused on working out which way through is appropriate, and that their behaviour is generally unpredictable since their chosen route through can change on a moment-by-moment basis. Examination of the driver social and situation awareness networks, however, shows that they may not be expecting cyclists, and indeed motorcyclists, to make major manoeuvres in close proximity to the intersection. Consequently, they do not behave or think in a manner that is compatible with other cyclists and motorcyclists. The driver situation awareness networks embodied a high focus on cars, the traffic lights and the act of turning right and an absence of 'check/checking', 'look/looking', and left- or right-'hand' side concepts. The driver social networks showed a limited interaction in which the oncoming traffic, traffic lights and traffic directly in front of the vehicle represent the majority of their interaction with the intersection system. Drivers also had only limited interactions with the areas adjacent to and behind their own vehicle.

This 'flexibility problem', as we will call it, is likely brought about by the fact that it represents flexibility not designed into the system per se but flexibility derived from emergence. For example, cyclists' range of possible behaviours at intersections is brought about by the lack of support for, and high risk associated with, turning right in the normal flow of traffic. Working within the constraints of intersection systems, cyclists and motorcyclists have come up with new ways of taking a safer path through the intersection. The design of the three intersections is therefore both creating this flexibility and then limiting compatibility between the cyclist and the driver. There is currently no signage warning drivers of the presence of cyclists or motorcyclists or the likelihood that they will be manoeuvring in and around the intersection. Also, where present, the cycling lanes preceding the intersections end a significant distance before the intersection itself, and there are no dedicated cycle or motorcycle lanes proceeding through the intersections. In addition, the road rules prohibit cyclists from cycling on footpaths and motorcyclists from filtering up the traffic queue. The effect of this is that drivers and pedestrians have limited experience of the prohibited behaviours and often may not be expecting them.

System Redesign

One of the aims of the analysis presented was to identify opportunities to create interventions designed to support safer interactions between distinct road users at intersections. The findings suggest that there are various avenues that can be pursued; however, the key is that all road users and non-human agents be considered, as opposed to one group in isolation. Intersection design should strive to create joint cognitive intersection systems.

To solve the flexibility problem, it would appear that the key is not only to ensure that different road users have some degree of flexibility, but also to ensure that the way in which the intersection is designed supports this flexibility and ensures that the resulting behaviours are compatible. For example, it is appropriate to maintain the level of flexibility for cyclists; however, other measures should be taken to support it. For example, avenues for making other road users (pedestrians, drivers) more expectant of cyclists and their range of behaviours will be fruitful. The use of cross-mode training has previously been raised as a way of developing anticipatory schema that support perception of other distinct road users (e.g. Magazzu et al. 2006; Walker et al. 2011). Cross-mode training incorporating both cyclist and motorcyclist training for drivers is likely to increase their expectancy levels regarding cyclist and motorcyclist behaviours at intersections. There is also a clear role for road design, with dedicated cycling lanes (on the road and on the footpath) and signage warning drivers of the presence of cyclists and the likelihood that they will make major manoeuvres in close proximity to the intersection.

As an exploratory study, this research did have some limitations. First, road users negotiated the same intersections under similar traffic conditions but not at the same time as one another (i.e. they were not negotiating the intersections together at the same time). Analysis of intersection system performance when road users are interacting together at the same intersection at the same time would provide more valid data on the potential conflicts between them. Second, the social networks were based on verbal protocols describing interactions rather than eye-tracking data showing actual interactions with objects in the road environment. Future studies in which eye-tracking data are used to construct the social networks are planned by these authors.

CONCLUSION

It is concluded that systems analysis methods, such as EAST, provide a useful avenue for understanding and eradicating road safety issues. From the analysis presented, it is concluded that there was a significant level of disharmony in the intersections studied. Although intersections are characteristic of joint cognitive systems, the intersections studied often behaved like disjointed cognitive systems. Differences were found in the way in which components (or road users) act, think and interact with the environment; however, it appears that intersections are designed from the viewpoint that all road users need to act and think in the same manner. Moreover, the analysis suggests that the way in which the intersections are currently designed may be restricting compatibility between different road user groups. It is argued that a failure to consider the overall intersection system during intersection design, including distinct road users' physical and cognitive behaviours and their differences, is facilitating the problem of intersection collisions. Moreover, a failure to consider these differences in future will render intersection collisions an intractable problem. Only a systems-based approach will be able to make substantive and sustainable inroads into accident reduction.

REFERENCES

Bell, M. C., Galatioto, F., Giuffrè, T. and Tesoriere, G. (2012). Novel application of red-light runner proneness theory within traffic microsimulation to an actual signal junction. *Accident Analysis and Prevention*, 46, 26–36.

Cornelissen, M., Salmon, P. M., McClure, R. and Stanton, N. A. (2013). Using Cognitive Work Analysis and the Strategies Analysis Diagram to understand variability in road user behaviour at intersections. *Ergonomics*, 56 (5), 764–780.

de Winter, J. C. F., de Groot, S., Mulder, M., Wieringa, P. A., Dankelman, J. and Mulder, J. A. (2009). Relationships between driving simulator performance and driving test results. *Ergonomics*, 52 (2), 137–153.

Elvik, R. (2010). Why some road safety problems are more difficult to solve than others. *Accident Analysis and Prevention*, 42, 1089–1096.

Farrington-Darby, T., Wilson, J. R., Norris, B. J. and Clarke, T. (2006). A naturalistic study of railway controllers. *Ergonomics*, 49, 1370–1394.

Furniss, D. and Blandford, A. (2006). Understanding emergency medical dispatch in terms of distributed cognition: A case study. *Ergonomics* 49 (12–13), 1174–1203.

Gershon, P. and Shinar, D. (2013). Increasing motorcycles' attention and search conspicuity by using Alternating-Blinking Lights System (ABLS). *Accident Analysis and Prevention*, 50, 801–810.

Gstalter, H. and Fastenmeier, W. (2010). Reliability of drivers in urban intersections. *Accident Analysis and Prevention*, 42 (1), 225–234.

Highways Agency (2012). *Design Manual for Roads and Bridges*. Highways Agency: London, UK.

Hole, G. J. and Tyrrell, L. (1995). The influence of perceptual 'set' on the detection of motorcyclists using daytime headlights. *Ergonomics*, 38 (7), 1326–1341.

Houghton, R. J., Baber, C., McMaster, R., Stanton, N. A., Salmon, P. M., Stewart, R. and Walker, G. H. (2006). Command and control in emergency services operations: A social network analysis. *Ergonomics*, 49, 1204–1225.

King, M. J., Soole, D. and Ghafourian, A. (2009). Illegal pedestrian crossing at signalised intersections: Incidence and relative risk. *Accident Analysis & Prevention*, 41 (3), 485–490.

Larsson, P., Dekker, S. W. A. and Tingvall, C. (2010). The need for a systems theory approach to road safety. *Safety Science*, 48 (9), 1167–1174.

Magazzu, D., Comelli, M., and Marinoni, A. (2006). "Are Car Drivers Holding A Motorcycle License Less Responsible for Motorcycle–Car Crash Occurrence? A Non-Parametric Approach." *Accident Analysis & Prevention* 38 (2): 365–370.

McLean, J., Croft, P., Elazar, N. and Roper, P. (2010). *Safe Intersection Approach Treatments and Safer Speeds Through Intersections: Final Report, Phase 1, AP–R363/10*. Austroads: Sydney, Australia.

Reason, J., Manstead, A., Stradling, S., Baxter, J. and Campbell, K. (1990). Errors and violations on the roads: a real distinction?. *Ergonomics*, 33, 1315–1332.

Salmon, P. M., McClure, R. and Stanton, N. A. (2012). Road transport in drift? Applying contemporary systems thinking to road safety. *Safety Science*, 50 (9), 1829–1838.

Salmon, P. M. Young, K. L. and Cornelissen, M. (2013). Compatible cognition amongst road users: The compatibility of driver, motorcyclist, and cyclist situation awareness. *Safety Science*, 56, 6–17.

Salmon, P. M., Stanton, N. A., Walker, G. H., Jenkins, D. P., Baber, C. and McMaster, R. (2008). Representing situation awareness in collaborative systems: A case study in the energy distribution domain. *Ergonomics*, 51 (3), 367–384.

Sandin, J. (2009). An analysis of common patterns in aggregated causation charts from intersection crashes. *Accident Analysis & Prevention*, 41 (3), 624–632.

Sanderson, P., Verhage, A. G. and Flud, R. B. (1989). State-space and verbal protocol methods for studying the human operator in process control. *Ergonomics*, 32 (11), 1343–1372.

Scheper, P. A. and den Brinker, B. B. (2011). What do cyclists need to see to avoid single-bicycle crashes? *Ergonomics*, 54 (4), 315–327.

Stanton, N. A. (2014). Representing distributed cognition in complex systems: How a submarine returns to periscope depth. *Ergonomics*, 57:3, 403–418.

Stanton, N. A., Salmon, P. M., Rafferty, L., Walker, G., Baber, C., and Jenkins, D. P. (2013). *Human Factors Methods: A Practical Guide for Engineering and Design*. Second Edition. Ashgate: Aldershot, UK.

Stanton, N. A., Stewart, R., Harris, D., Houghton, R. J., Baber, C., McMaster, R., Salmon, et al. (2006). Distributed situation awareness in dynamic systems: theoretical development and application of an ergonomics methodology. *Ergonomics*, Vol. 49, pp. 221–229.

Stewart, R. J., Stanton, N. A., Harris, D., Baber, C., Salmon, P. M., Mock, M., Tatlock, K., Wells, L., and Kay, A. (2008). Distributed situational awareness in an airborne warning and control aircraft: Application of a novel ergonomics methodology. *Cognition, Technology and Work*, 10 (3), 221–229.

Wagenaar, W. A. and Reason, J. T. (1990). Types and tokens in road accident causation. *Ergonomics*, 33 (10–11), 1365–1375.

Walker, G. H., Stanton, N. A. and Salmon, P. M. (2011). Cognitive compatibility of motorcyclists and car drivers. *Accident Analysis & Prevention*, 43 (3), 878–888.

Walker, G. H., Gibson, H., Stanton, N. A., Baber, C., Salmon, P. M., and Green, D. (2006). Event analysis of systemic teamwork (EAST): A novel integration of ergonomics methods to analyse C4i activity. *Ergonomics*, 49, 1345–1369.

Walker, G. H., Stanton, N. A., Baber, C., Wells, L., Jenkins, D. P. and Salmon, P. M. (2010). From Ethnography to the EAST method: A tractable approach for representing distributed cognition in air traffic control. *Ergonomics*, 53 (2), 184–197.

World Health Organisation (2009). Global status report on road safety: Time for action. World Health Organization: Geneva, Switzerland. Available at: www.who.int/violence_injury_prevention/road_safety_status/2009.

8 EAST in Elite Women's Cycling Teams

With Clare Dallat and Amanda Clacy

INTRODUCTION

Situation awareness remains one the most ubiquitous of all ergonomics concepts, with applications in an ever-expanding list of domains (Stanton et al. 2015). One area in particular that has been receiving increasing attention is elite sport, with recent studies considering situation awareness among Australian Rules Football umpires (Neville et al. 2016), handball teams (De Keukelaere et al. 2013) and elite hammer-throwing and rowing athletes and coaches (Macquet and Stanton 2013). Whilst applications have included testing of theory and methods, the primary function has been to inform the development of coaching strategies and interventions designed to optimize performance (e.g. Macquet and Stanton 2014).

Despite exhibiting many features that are characteristic of the domains in which ergonomics practitioners work, elite cycling has not yet been studied from this perspective. One interesting feature is the critical role that situation awareness apparently plays at the individual rider level, at the cycling team level and at the overall cycling system level. To date, however, the focus of research in elite cycling has largely been on the development of quantitative models to describe the behaviour of cycling pelotons (e.g. Trenchard 2015; Trenchard et al. 2014). What rider, team and system situation awareness comprises, and how it can be optimised, remained unexplored. Such knowledge could potentially provide teams with a performance edge.

Much has been written about the contention surrounding situation awareness and the appropriate theoretical and methodological approach to adopt in ergonomics studies (see Salmon et al. 2008 and Stanton et al. 2017). In a recent review of situation awareness models and their utility for studying and optimising elite sport systems, Neville and Salmon (2016) identified Stanton et al.'s Distributed Situation Awareness (DSA) model as highly suitable for sporting applications. Prior to this, Salmon et al. (2010) had noted that the Event Analysis of Systemic Teamwork (EAST) framework offers a powerful framework for understanding elite sports systems. Salmon et al. (2010) concluded from test applications that the EAST framework was highly suited to describing and analysing sports system performance and to supporting the development of interventions designed to improve sports performance.

This chapter brings together both of these arguments, presenting an EAST analysis of DSA and teamwork in an elite women's cycling team during two road race stages. The study was exploratory in nature and aimed to describe and understand DSA and teamwork to assist the team in optimising performance in future events.

A secondary aim of the analysis was to further examine the utility of applying both DSA and EAST in elite sports. In presenting this study here, the intention is to encourage further applications of EAST in the sporting context.

Elite Women's Cycling

The analysis presented in this chapter focuses on two elite women's road races undertaken as part of the 2016 Australian Subaru National Road Series (NRS). NRS road races typically range in length from 50 kilometres (km) to 130 km and involve teams of four to five riders. Each team has a so-called 'protected' or 'general classification' (selected based on overall NRS points standing), with the remaining riders acting in a 'domestique'* or support capacity.

The team's goal is to ensure that their protected rider finishes each stage of the race in the best position and time possible. Accordingly, the domestiques employ various tactics to assist their protected rider, such as allowing the protected rider to ride behind them ('on their wheel') to reduce their physical workload or 'leading them out' to initiate a sprint finish.

The riders from each team form a *peloton* of cyclists that negotiate the race route together. The peloton has been formally defined as 'a group of cyclists that are coupled together through the mutual energy benefits of drafting, whereby cyclists follow others in zones of reduced air resistance' (Trenchard et al. 2014, p. 92). Throughout the race, riders may attempt to 'attack' and break away from the peloton, and other teams' domestiques may attempt to chase them down and bring them back to the peloton or let them go until a later point in the race. Alternatively, the peloton may stay together, resulting in a 'bunch sprint' finish. Throughout the race, slower riders drop out of the peloton and often work together to attempt to re-join the peloton. Likewise, domestiques who have fulfilled their role of putting the protected rider in a position to finish the race often drop out of the peloton due to fatigue. An additional feature of the NRS women's road races are shorter sprint and Queen of the Mountain (QOM) sub-sections, whereby bonus points are offered for the top three cyclists within each sub-race.

A convoy of vehicles comprising the Commissaires; race officials, medical support, neutral spares, the media and team support vehicles follows behind the peloton. The Commissaires controls the race and following convoy, communicating with the cyclists via loudspeaker and the vehicles in the convoy via UHF radio or in person by driving alongside them. A medical support vehicle follows behind the Commissaires to provide assistance in the event of a crash or injury. Each team then has their own specific support vehicle, typically containing a Director Sportif (DS), a mechanic (in some cases this may be the DS), spare parts (e.g. wheels), bicycle tools and food and drink for the riders. Each support vehicle is permitted to provide mechanical or nutritional support to team riders and can offer tactical guidance and encouragement through verbal exchanges. The positioning of the support vehicles in the peloton convoy is based on the order of riders in the race (e.g. the team with a rider in first place is given the first support vehicle spot within the convoy).

Together the peloton and convoy form a complex, highly dynamic system comprising multiple human (riders, DS, mechanic) and non-human agents (bicycles, vehicles,

* In French, *domestique* translates as 'servant'

bicycle computers), which exhibits emergent properties, non-linear interactions, multiple control and feedback loops, loose and tight coupling and rapid decision making. As such, it is amenable to ergonomics inquiry whilst offering relevant examples of sociotechnical system behaviour that could inform theoretical, methodological and practical contributions in other safety critical domains. It is also characteristic of precisely the kind of system that EAST was developed to describe and analyse.

Event Analysis of Systemic Teamwork

The EAST framework provides an integrated suite of ergonomics methods for analysing the behaviour of complex sociotechnical systems. Heavily underpinned by network theory and analysis, EAST takes the premise that system performance can be meaningfully described via a 'network of networks' approach, and it uses three interlinked network-based representations to describe and analyse behaviour. Task networks are used to describe the goals and subsequent tasks being performed within a system (i.e. which agents, both human and non-human, do what). Social networks are used to analyse the structure of the system and the communications taking place between agents (i.e. who/what interacts and communicates with whom/what). Situation awareness networks show how information and knowledge is distributed across different agents within the system (i.e. who/what knows what at different points in time). By illustrating the relationships between task, social and situation awareness networks and then by interrogating these networks, an in-depth understanding of system behaviour is achieved.

Salmon et al. (2010) first discussed the potential utility of using EAST to analyse sports performance, presenting an EAST analysis of a fell running race that examined runner goals, situation awareness, decision making and workload. Whilst others have since used components of EAST to examine elite sports performance (e.g. Neville et al. 2016), the three forms of EAST network are yet to be used in conjunction with one another to analyse elite sports performance.

EAST Analysis of Elite Women's Cycling

The study was a naturalistic study whereby the research team collected data from an elite women's road racing team during two Australian NRS events from the 2016 NRS season. Both events were elite cycling road race events incorporating a series of stages, including a time trial, road races and a criterium race. The analysis presented in this chapter focuses on one of the races from the first NRS event: stage 1 of the Battle on the Border, which took place in the Tweed Coast region, New South Wales. The full analysis will be included in the conference presentation.

METHODS

Participants

The participants were members of an elite women's Australian NRS cycling team. For the present analysis, the sample included five riders, one DS and one mechanic.

Due to the naturalistic nature of the study, it was not possible to gather complete participant demographic data; however, all were experienced in elite cycling and had raced throughout the 2016 NRS season. The study was granted ethics approval by the University of the Sunshine Coast's Human Ethics Committee.

Materials

The research team observed the race from within the cycling team's support vehicle. Go Pro™ cameras were used to record the races, and a Dictaphone was used to record the verbal communications occurring within the team support car. Dictaphones were also used to record team race planning meetings, the post-race Critical Decision Method (CDM) (Klein et al. 1989) interviews and post-race team debriefs. The interviewers used a pro-forma containing the CDM interview probes. The interview transcripts were transcribed using Microsoft Word™. For data representation, the task and social networks were created using Microsoft Visio™.

Procedure

The research team provided the cycling team with an overview of the study, its aims and the data collection methodology prior to the first race of each event. The research team travelled to each race with the cycling team and observed the warm up and pre-race discussions and planning activities. The Go Pro cameras and Dictaphone were placed within the support vehicle and set to record shortly before the start of each race. Once the races began, the research team observed the peloton from the support vehicle and made hand-written notes regarding the team's performance, tactics and the interactions between the DS, the Commissaire and riders. During low workload periods of the race, the researchers were also able to discuss the team's performance with the DS, including tactics, planned activities and factors influencing rider performance.

Following the race, the research team attended the riders' post-race debrief and planning session for the next race, recording both using a Dictaphone.

At the conclusion of the planning and debrief sessions, the research team conducted one-on-one CDM interviews with each rider regarding the earlier race. The interviewers used a set of CDM probes adapted for the cycling race context and recorded each interview using a Dictaphone. Following the interview, each rider was asked to complete a social network analysis questionnaire to show who and what they interacted with during the race. This involved presenting them with a social network diagram showing all agents (riders, other riders, DS, Commissaire, cycling computer) and asking them to rate their level of communication with each during the race on a scale of low (<10 communications), medium (between 10 and 20 communications) and high (>20 communications).

The video and audio footage were transcribed verbatim using Microsoft Word. The three EAST network representations were constructed as follows:

1. *Task network*. The task network was built based on the research team's observations, a review of the video data and the interview transcripts. A draft version of the task network was reviewed and refined by the cycling team during the Battle on the Border event.

2. *Social networks*. Construction of the social network involved transforming the social network diagrams obtained following the races into a social network diagram detailing the number of interactions between the cyclists and other agents.
3. *Situation awareness networks*. Situation awareness networks were constructed through content analysis of the riders' interview transcripts. Specifically, the riders' responses to the CDM questions 'What were your specific goals for the race?' 'What information were you using throughout the race?' and 'What was the most important piece of information?' and riders' initial description of the race content were analysed by one analyst. Keywords and the relationships between them were extracted and used to construct a situation awareness network comprising concepts and the relationships between them. For example, from the sentence, 'I was checking my position in the peloton', the related concepts 'Checking', 'Position' and 'Peloton' were extracted. Once the situation awareness network was constructed, concepts were shaded based on their presence in each of the rider's interview transcripts.

RESULTS

Task Network

The task network is presented in Figure 8.1. Within Figure 8.1, the circular nodes represent distinct tasks undertaken immediately prior to and during the road races. The linkages between the nodes represent relationships between tasks; for example, the 'attack' and 'support protected rider' tasks are closely linked, since the attacks are made by domestique riders in order to either help the protected rider break away from the peloton or to tire out opposing team riders.

The coloured rings around each node represent the agents in the cycling system who undertake each task. For example, the 'attack' task is undertaken by the riders only, whereas the 'monitor other riders' task is undertaken by the riders, the Commissaire and the DS.

Two features of the task network should be noted. First, there are a range of tasks involved, including both individual tasks that are unique to specific agents and shared tasks that are common across agents. For example, the task of communicating with other riders is undertaken by team riders within the peloton (e.g. asking for assistance to initiate a break), by the DS in the support vehicle (e.g. communicating tactics with team riders) and by the Commissaire (e.g. informing other riders in the peloton and the DSs in the support vehicles of the time gap between a breakaway rider and the peloton). Second, the relationships between each task can be used to identify those tasks that are integral to team success. In this case, the task network indicates that the tasks of 'race planning and tactics', 'establish and maintain appropriate position in the peloton', 'monitor other riders', 'communicate with other riders' and 'monitor speed, cadence, power and distance' are integral to team success as they are the most connected to other tasks in the network. It is notable that these tasks are also highly interrelated.

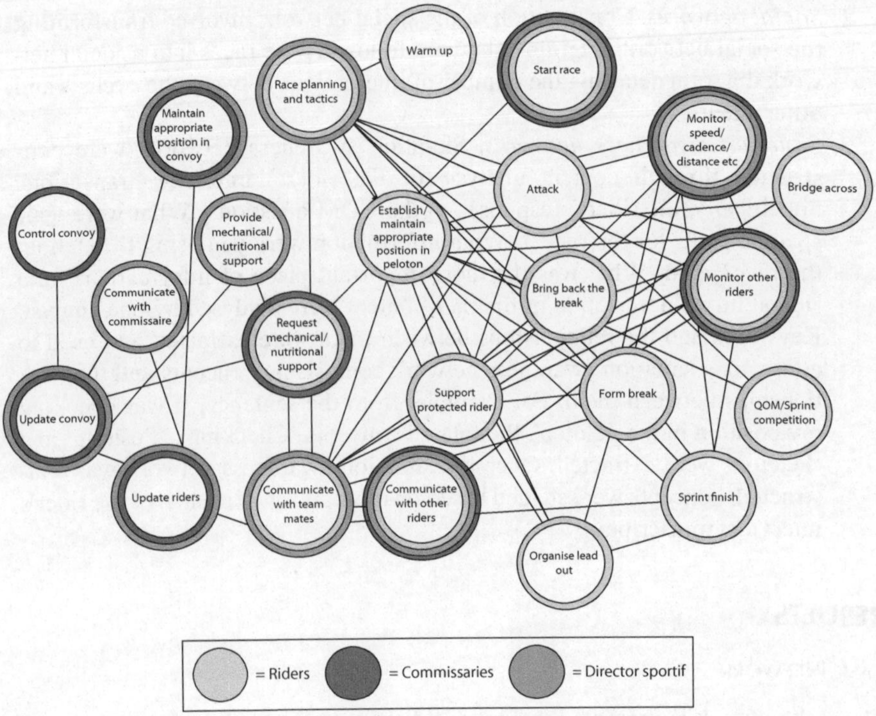

FIGURE 8.1 Generic task network for women's cycling road racing.

Social Network

The social network for the Battle on the Border stage 1 road race is presented in Figure 8.2. Within Figure 8.2, the circular nodes represent agents (e.g. riders, support vehicle, bicycle computers), and the arrows between them represent the extent to which the agents communicated with one another during the race.

The social network shows the structure of communications throughout the road race. In this case, the protected rider and domestique rider 2 are the most connected in terms of incoming and outgoing communications. In addition, both the protected rider and domestique rider 2 have a greater frequency of communications with other agents, both having five connections that were rated as 'high' on the frequency of communications scale.

Situation Awareness Network

The situation awareness network for the Battle on the Border stage 1 road race is presented in Figure 8.3. Within the network, the nodes represent pieces of information, and the lines linking the nodes represent relationships between the information (e.g. 'Computer' displays 'Speed').

The network shows that certain pieces of information were critical during the race. For example, the nodes 'Peloton', 'Position' and 'Riders' are the most connected nodes within the network, indicating that they were integral to situation awareness and decision making during the race. This included the riders self-monitoring their

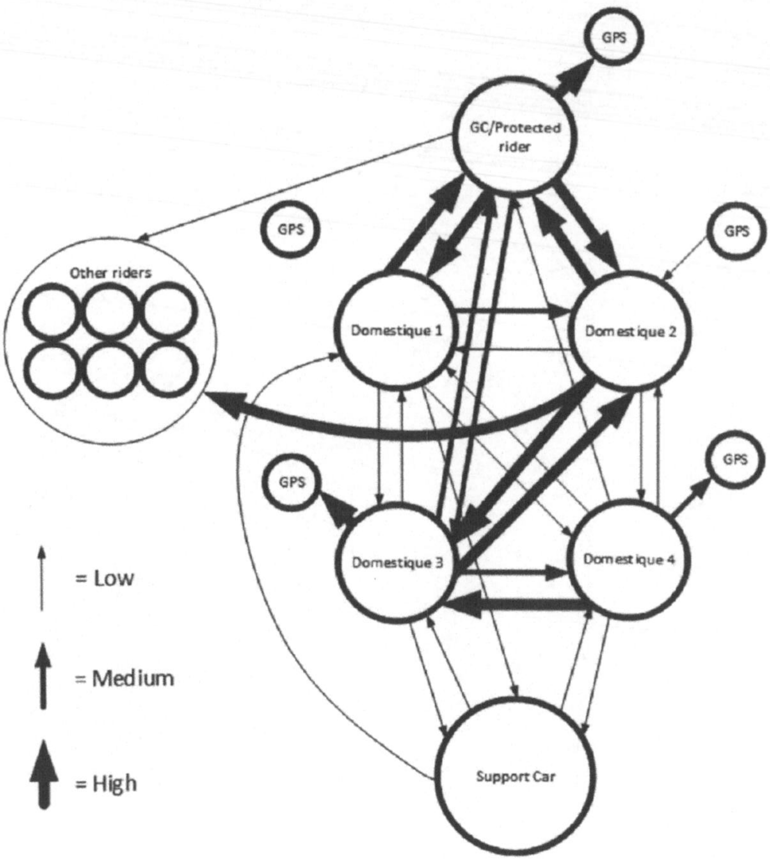

FIGURE 8.2 Social network for Battle on the Border road race stage 1.

position in the peloton, as well as the position of their teammates, and relating this information to tactics and energy levels. The network also shows a focus on the protected riders from other teams in terms of what they were doing and their position in the peloton throughout the race. The constant monitoring of own and other riders' energy levels was reported by all riders as important throughout the race and is shown through the nodes related to the 'energy' node. Nodes related to the team's race plan were also prominent. Finally, it is interesting to see safety-related nodes in the network, including the nodes 'crashes', 'weather', 'conditions' and 'pot holes'.

CONCLUSION

What Does the Analysis Tell Us About DSA and Teamwork in Elite Women's Cycling Teams?

The analysis highlights the high requirement for teamwork during elite women's racing. Salas et al.'s (2005) 'Big Five' model of teamwork identifies five key teamwork

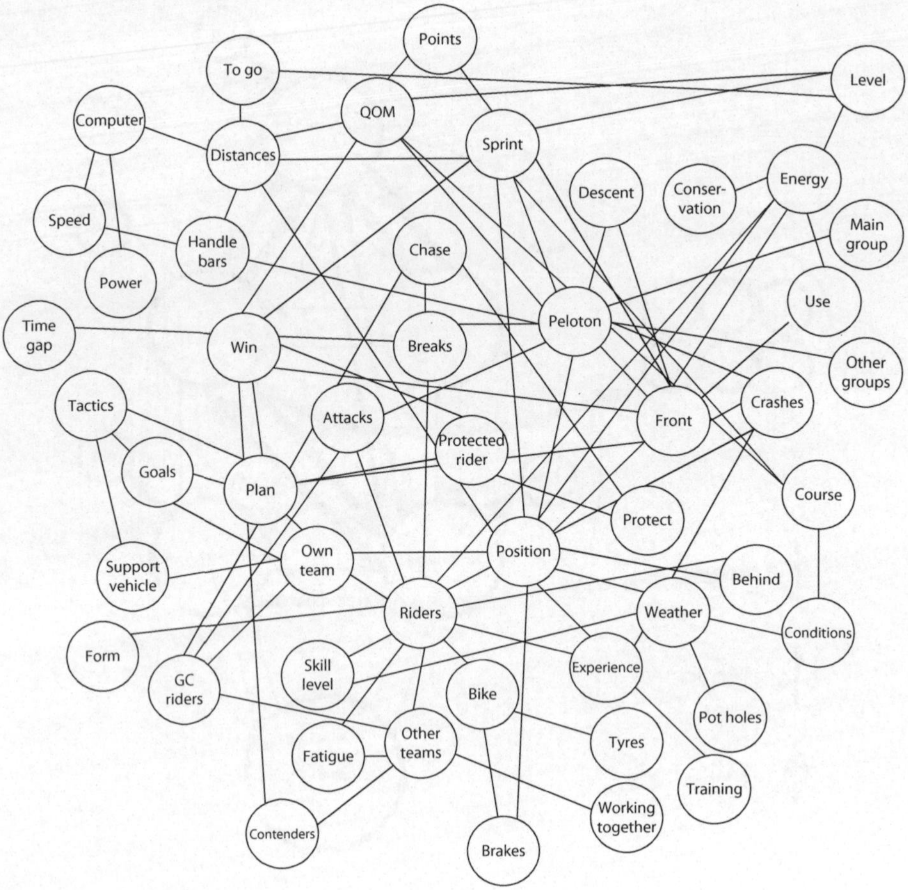

FIGURE 8.3 Situation awareness network for Battle on the Border road race stage 1.

behaviours (leadership, mutual performance monitoring, back-up behaviour, adaptability, team orientation) along with three key supporting mechanisms (shared mental models, mutual trust, communication). Throughout the three networks, it is possible to identify clear examples of key teamwork behaviours as outlined in the 'Big Five' model (see Table 8.1).

In terms of DSA, the networks demonstrate that the overall peloton possesses DSA that comprises the situation awareness of individual riders, riding teams, the peloton as a collective and the convoy and its elements (e.g. the Commissaire, race officials, medical support, neutral spares, the media and team support vehicles follow behind the peloton). In addition, the analysis shows that situation awareness is held by both human and non-human agents. For example, the situation awareness network shows the importance of non-human agents, such as the bike-mounted computer and annotated handlebars. Here, the DS annotates the bicycle handlebars to provide critical race information to the riders, such as when the sprint section of the race will occur. Riders then used their race computers in conjunction with

TABLE 8.1
Examples of 'Big Five' Teamwork Behaviours

'Big Five' Model Behaviour	Example	Source
Leadership	The tactical direction provided by both riders within the team and the DS	Task and situation awareness network
Mutual performance monitoring	Riders need to monitor their own teammates in terms of position, fatigue, etc.	Task and situation awareness network
Back-up behaviour	The ability of riders to respond to the needs of other riders in the team, e.g. chasing down a breakaway for the protected rider	Task and situation awareness network
Adaptability	The need for the riders to respond to other teams' behaviour, e.g. breaks from the peloton, attacks	Task and situation awareness network
Team orientation	The fact that all riders in the team are working together to ensure that the protected rider wins the race (team goal overriding individual goals)	Task and situation awareness network
Shared mental models	The use of race planning to ensure that the teams have a shared understanding of the plan and of the team's goals	Task and situation awareness network
Mutual trust	The belief among the riders that team members will perform their roles, e.g. the protected rider's knowledge that her teammates will sacrifice their own overall position and protect her until the sprint finish	Task and situation awareness network
Communication	Multiple communications occurring continuously between all riders	Social and task network

the information written on the handlebars to be aware of when to attack or initiate a sprint.

The importance of continual transactions of situation awareness within the peloton was also emphasised, as is the fact that the transactions can be both verbal (e.g. talking to a teammate) and non-verbal (monitoring another rider's behaviour to assess their energy level). Notably, these transactions include intra-team transactions (exchanges of awareness between teammates) and inter-team transactions (exchanges of awareness between riders from different teams).

What Are the Implications for Optimising Performance in Elite Women's Cycling Teams?

Although the analysis presented was exploratory in nature, it is possible to identify avenues that could be used to optimise race performance. The importance of pre-race planning and the implementation of the plan itself during the race was emphasised throughout the present analysis, with race planning identified as a key node within the task network and information related to the race plan identified as a key node within the situation awareness network. This suggests that robust, in-depth and detailed planning processes are likely to yield improvements in team performance, and that communication of the plan is critical, both before and throughout the race. Further, the importance of devising a race plan that is agile and adaptable and contains a series of well-defined contingency plans is apparent. This will enable teams to respond appropriately and rapidly in the face of variable tactics from opposing teams.

The importance of transactions in awareness was also highlighted; it may be that educating cyclists on forms of non-verbal transaction may provide ways in which teammates can communicate without giving away tactics to other teams. Further, behaviours designed to mask key information relating to opposition team's situation awareness may be useful to create situation awareness decrements, for example, disguising high levels of fatigue and sending out spurious information regarding race tactics. Finally, teaching cyclists ways in which to identify other teams' tactics and energy levels will be useful and serve to optimise team situation awareness.

A final interesting feature of the analysis is the importance of communication within the team, even when team riders may be spread throughout the peloton without the ability to communicate directly (radio communications are not used in Australian elite women's cycling). This suggests that future work may be required to investigate strategies for rapid and effective communications during races.

What Are the Implications for Future EAST Applications in Elite Sport?

The analysis presented provides some evidence of the utility of applying EAST in the elite sports context. EAST provided new perspectives on elite cycling, outlining the key tasks, interactions and information required during elite women's cycling road races. Further applications of the EAST framework are encouraged, both in cycling and in other forms of elite team sports such as football, rugby league and union, Australian Rules football, American football, Formula 1 racing and baseball. In addition, formal comparisons between EAST and other existing analysis approaches are recommended.

REFERENCES

De Keukelaere, C., Kermarrec, G., Bossard, C., Pasco, D. and De Loor, P. (2013). Formes, contenus et évolution du partage au sein d'une équipe de sport de haut niveau. *Le travail humain*, 76 (3), 227–255.

Klein, G. A., Calderwood, R., and MacGregor, D. (1989). Critical decision method for eliciting knowledge. *IEEE Transactions on Systems, Man and Cybernetics*, 19 (3), 462–472.

Macquet, A. C. and Stanton, N. A. (2014). Do the coach and athlete have the same 'Picture' of the situation? Distributed situation awareness in an elite sport context. *Applied Ergonomics*, 45 (3), 724–733.

Neville, T. and Salmon, P. M. (2016). Never blame the umpire? A review of situation awareness models and methods for examining the performance of officials in sport. *Ergonomics*, 59 (7), 962–975.

Neville, T., Salmon, P. M., Read, G. J. M. and Kollinitas, A. (2016). Play on or call a foul? Testing and extending DSA theory through sports officiating. *TIES*, 17 (1), 80–103.

Salas, E., Sims, D. E. and Burke, C. S. (2005). Is there a 'Big Five' in teamwork? *Small Group Research*, 36 (5), 555–599.

Salmon, P. M., Stanton, N. A., Gibbon, A. C., Jenkins, D. P. and Walker, G. H. (2010). *Human Factors Methods and Sports Science: A Practical Guide*. Taylor & Francis: Boca Raton, FL.

Salmon, P. M, Stanton, N. A., Walker, G. H., Baber, C., Jenkins, D. P. and McMaster, R. (2008). What really is going on? Review of situation awareness models for individuals and teams. *Theoretical Issues in Ergonomics Science*, 9 (4), 297–323.

Stanton, N. A., Salmon, P. M., Rafferty, L., Walker, G. H., Jenkins, D. P. and Baber, C. (2013). *Human Factors Methods: A Practical Guide for Engineering and Design*. 2nd Edition, Ashgate: Aldershot, UK.

Stanton, N. A., Salmon, P. M., Walker, G. H. (2015). Let the reader decide. A paradigm shift for situation awareness in socio-technical systems. *Journal of Cognitive Engineering and Decision making*, 9:1, 45–50.

Stanton, N. A., Salmon, P. M., Walker, G. H., Salas, E., Hancock, P. (2017). State-of-the-science: Situation Awareness Models. *Ergonomics*. 60(4): 449–466.

Trenchard, H. (2015). The peloton superorganism and protocooperative behavior. *Applied Mathematics and Computation*, 270, 179–192.

Trenchard, H., Richardson, A., Ratamero, E. and Perc, M. (2014). Collective behavior and the identification of phases in bicycle pelotons. *Physica A: Statistical Mechanics and its Applications*, 405, 92–103.

9 EAST in Automated Driving Systems

With Victoria A. Banks

INTRODUCTION

There have been rapid developments in automotive automation over the past decade whereby partial and fully automated vehicles are reaching a point where widespread deployment is feasible (National Highway Traffic Safety Administration; NHTSA 2016). It is expected that by the end of 2016, NHTSA will publish best-practice guidelines on establishing principles of safe operation for vehicles operating at Level 4 'full' automation. Fully autonomous cars promise to deliver abundant socioeconomic advantages (Casner et al. 2016), including improvements to traffic flow and mobility and significant improvements to road safety. With 90% of accidents being attributable to driver error (Smiley and Brookhuis 1987; Stanton and Salmon 2009), vehicle manufacturers may have good reason to remove drivers from active control. There are, of course, other benefits of 'driverless' vehicles, one being to give vehicle occupants time to engage in other tasks not related to driving (Fagnant and Kockelman 2015). This is seen as one of the main drivers for market implementation to improve comfort and convenience. According to the Department for Transport (2015), UK drivers spend, on average, 235 hours a year behind the wheel. This equates to approximately six working weeks where the driver has no spare capacity to engage in other tasks. The advent of fully automated vehicles could therefore completely transform our experience of driving and provide the driver with additional productive time (making the journey similar to taking public transport – without the drawbacks).

The human factors issues pertaining to vehicle automation have been speculated about since the 1970s (Sheridan 1970). Simply removing the driver from the control-feedback loop and eliminating their responsibility over safe vehicle operation does not, however, render human factors completely redundant. Instead, vehicles operating at increased levels of autonomy with 'self-driving' capabilities open up new avenues of investigation. Some of these avenues are explored within this chapter.

Levels of Automation and the Role of the Driver

Automation of the driving task brings with it a shift in the role and responsibilities of the human driver. Within the literature, Kaber and Endsley (2004) captured this idea eloquently by explaining that the role of the driver shifts from one of active operation to more of a passive monitor as the level of automation increases. Within industrial practice, the role and responsibilities of the driver are alluded to, but not explicitly described, in automation taxonomies. The Society for Automotive Engineers (SAE 2015), NHTSA (2013) and BASt Expert Group (Gasser 2014) have all developed their own versions of automated driving taxonomies. These are often used interchangeably within the literature, which can lead to confusion about what the driver can and cannot do under different levels of automation within the driving system. SAE went some way in trying to standardise these descriptions (SAE J3016). Even so, in any instance whereby a take-over request is issued or indeed possible, the driver is expected to be able to regain control of the vehicle and resume their traditional driving role. As long as the driver remains in the control-feedback loop, which would be the case for any system whereby control transitions may be made between the automated system and driver, an acknowledgement of their role and how they can be supported back into their traditional driving role remains an important area of investigation (Eriksson and Stanton 2017). Thus, we cannot assume that at higher levels of automation, the driver or human element will no longer be required. This is because any potential for a take-over request to be issued implies that (1) the vehicle does not offer full 'self-driving' functionality and (2) the driver should be able to regain control of the vehicle as may be required by the situation and/or operational limits of the system.

We also cannot overlook the fact that individuals may even *want* to resume control from the vehicle at some point. Thus, increasing levels of autonomy in driving do not eliminate all of the human factors issues that are typically associated with lower levels, for example, Levels 2 and 3. We already know from the literature that automation within the driving task can lead to decreased situation awareness (Stanton and Young 2005; Stanton et al. 2011), erratic changes to driver workload (Stanton et al. 1997; Young and Stanton 2002, 2004; de Winter et al. 2014, 2016), skill degradation (Stanton and Marsden 1996) and issues relating to trust (Walker et al. 2016), overreliance and complacency (Stanton 2015). It seems likely that some, if not all, of these will remain enduring challenges for systems designers as long as the driver remains within the control-feedback loop to some extent. This means that strategies for transferring control back to the driver after prolonged periods of autonomous driving must be carefully designed to ensure that they have appropriateness levels of situation awareness and are deemed capable of regaining control of the vehicle. The entire spectrum of driver responsibilities and workload should be considered in delivering an effective 'hand-over' between the driver and autonomous vehicle.

With this in mind, it is becoming increasingly important to acknowledge the role of the driver within an automated driving system. There are many lessons that can be taken from the field of aviation, as we increasingly see the role of the driver becoming analogous to the role of a pilot (Stanton and Marsden 1996). In aviation, Hutchins (1995) described two roles in which the pilot can serve; Pilot Flying (PF) and Pilot Not Flying (PNF). Whilst the PF is responsible for overall control of the plane, the PNF is responsible for communicating with Air Traffic Control and aircraft systems as well as completing all of the checklists that are required during each phase of flight. Thus, the burden of responsibility simply changes rather than being reduced. In recognition of the changing responsibilities of the PNF, the Federal

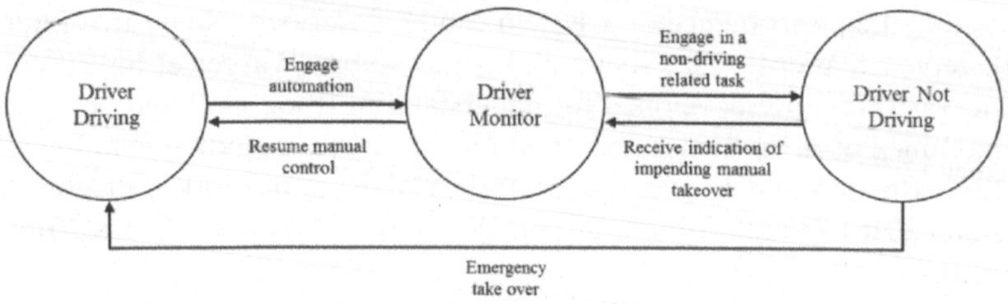

FIGURE 9.1 Driver mode transition network.

Aviation Administration (FAA 2003) altered the terminology of PNF to that of Pilot Monitor (PM). In the same way, a Driver Driving (DD) would be responsible for overall control of the vehicle and represents the traditional role of the driver, whilst a Driver Monitor (DM) would assume a similar role to that of PM. The latter role is to monitor the behaviour of the vehicle and automated sub-systems to ensure that safe and normal driving practice is maintained. However, the introduction of automation into the driving task does not necessarily mean that the driver will assume the role of DM. We already know that the perception of increased reliability can leave drivers vulnerable to becoming complacent and over-reliant on automated functionality (Parasuraman et al. 1993; Lee and See 2004). Without active vehicle control, a DM, for example, could become vulnerable to the onset of boredom or fatigue (e.g. Molloy and Parasuraman 1996; Stanton et al. 1997; Young and Stanton 2002). Thus, DMs may unintentionally drift in and out of the Driver Not Driving (DND) role. This becomes particularly problematic in instances whereby the automated systems are not able to adequately resolve a scenario without human intervention. If the driver *should* be in the role of DM (i.e. during partial automation) but is in fact behaving in a manner more akin to the role of DND, active control of the vehicle may be transferred to a DND (rather than a DM who is prepared to resume the role of DD) who may fail to respond appropriately due to a sudden increase in driver workload, reduced situation awareness (Dozza 2012) or startle (Sarter et al. 1997) (Figure 9.1).

We therefore need to better understand the impact that automation implementation has upon the role of the driver. One way to do this is to adopt the theoretical underpinnings of Distributed Cognition (DCOG), which acknowledges that both human and non-human agents can work together in pursuit of a common goal (Hutchins 1995). In driving, this essentially reflects the relationship between the human driver and automated sub-systems that become increasingly capable of performing the traditional roles of the driver (Banks et al. 2014). The DCOG paradigm provides the necessary foundations and methods to explore how the role of the driver changes within the complex sociotechnical systems in which they are involved (Walker et al. 2010; Walker et al. 2015).

METHOD

This chapter explores the changing role of the driver within automated driving systems using the Event Analysis of Systemic Teamwork (EAST) Stanton et al. 2006, 2013; Walker et al. 2006, 2010) framework. EAST is a descriptive method and proposes that the performance of a system can be described using three interlinked network representations: task, social and information (Walker et al. 2006, 2010). Whilst it was originally developed for understanding command and control activities, it has since been applied to transportation domains including aviation (Walker et al. 2010; Stanton and Harvey 2017), rail (Walker et al. 2006), driving (Banks and Stanton 2016) and maritime transport (Stanton et al. 2006; Stanton 2014; Baber et al. 2013).

Task networks are used to provide a summary of system goals (Salmon et al. 2014) and offer a description of the sequences and interdependencies that exist between individual sub-tasks that must be completed to attain these goals. Social networks are used to analyse the structure of the system in terms of the communications that take place between different system 'agents'. Finally, information networks show the information that is used and communicated by system agents during a task (Walker et al. 2010). They therefore detail aspects of communication that underpin the completion of a task and the relationships between informational nodes. This chapter uses the representations afforded by EAST to explore the differences between the DD and DM roles using network metrics. These networks can then be subjected to quantitative analysis using the Applied Graphic and Network Analyses (AGNA) tool (AGNA version 2.1; Benta 2005). AGNA is a platform-independent freeware application that can be used to analyse social networks. Nodes within the network can be analysed individually to assess agent centrality/prominence or as a whole. In driving research, network metrics can be used to identify key agents, key tasks and key informational elements required to complete the task. The following metrics have been chosen due to their previous application to the driving task (see Banks and Stanton 2016; Salmon et al. 2009; Walker et al. 2011).

Density represents the level of interconnectivity between system agents. It is expressed as a value between 0 and 1, where 0 represents a network that has no connections between system agents and 1 indicates that the network is fully connected (Kakimoto et al. 2006). It is calculated using the formula

$$\text{Network density} = 2e/n(n-1)$$

where:

e is the number of links in the network

n is the number of information elements within the network (Walker et al. 2012)

Diameter is used to analyse the connections and pathways between different system agents within the network (Walker et al. 2011). Denser networks (i.e. the route through the network is shorter and more direct) have smaller values. It is calculated using the following formula:

$$\text{Diameter} = \max_{uy} d(n_i, n_j)$$

Cohesion represents the number of reciprocal connections divided by the total number of possible connections (Stanton 2014).

Sociometric status is also of interest because it gives an indication of agent prominence as a communicator. Agents with high sociometric values are highly connected (Salmon et al. 2012). It is calculated using the formula

$$\text{Sociometric Status} = \frac{1}{(g-1)} \sum_{j=1}^{g} \left(x_{ji}, x_{ij}\right)$$

where:

g is the total number of nodes in the network

i and j are individual nodes and are the edge values from node i to node j (Salmon et al. 2012)

RESULTS

Task Networks

The driving task itself is said to be made up of approximately 2500 subtasks (see Hierarchical Task Analysis of Driving, HTAoD; Walker et al. 2015). These are divided into pre-drive, basic, operational, tactical, strategic and post-drive tasks. Whilst a DD would be involved in all of these tasks, the implementation of automation into the driving task would see the role of the human operator reduced. For this reason, the analysis focuses upon *some* of the basic, and strategic sub-tasks of driving, as it is in these categories that the changing role of the driver is most visible. Table 9.1 shows the terminologies, taken directly from the HTAoD (Walker et al. 2015) and categorised into task type, that were chosen to construct the task networks presented in this chapter. Notably, both networks should be considered as continuous processes that are performed as long as the vehicle is in operation. Figure 9.2a reflects the linkages between these sub-tasks of driving from the perspective of a DD. A DD would be physically be responsible for completing all of the basic sub-tasks of driving as well as being solely responsible for completing the strategic sub-tasks. However, as automation is introduced into the driving task, the task network evolves. Figure 9.2b represents the task network from the perspective of a DM. Whilst the tasks within the network remain largely the same, the emphasis upon

TABLE 9.1
Basic and Strategic Sub-Tasks of Driving

Basic Tasks	Strategic Tasks
Perform steering actions	Perform surveillance
Control vehicle speed	Perform navigation
Undertake directional control	Comply with rules
Negotiate bends	Respond to environmental conditions
Negotiate gradients	

Source: Walker, G. H., et al., *Human Factors in Automotive Engineering and Technology*, Ashgate, Aldershot, UK, 2015.

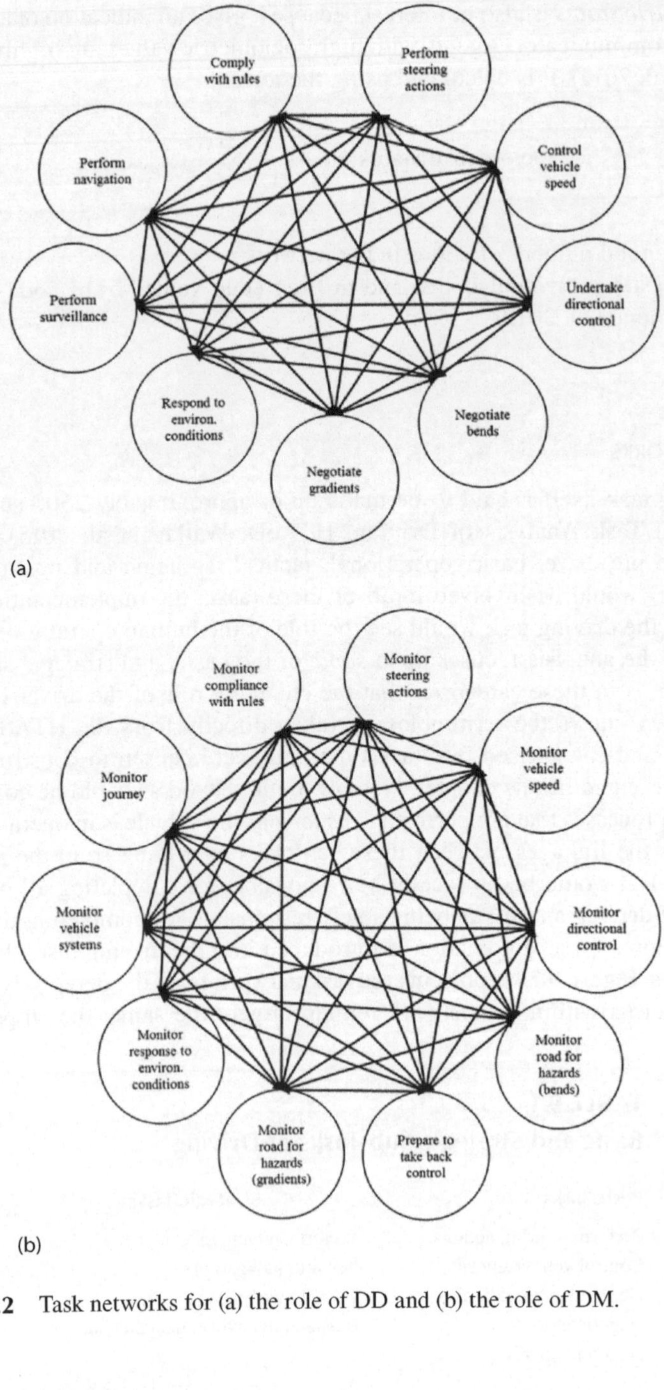

FIGURE 9.2 Task networks for (a) the role of DD and (b) the role of DM.

TABLE 9.2
Description of DD and DM Task Networks

	DD	DM
Number of nodes	9	10
Number of edges	144	160

physical input is removed. Instead, a DM would be responsible for monitoring the completion of basic tasks as well as strategic tasks. There is therefore no removal of tasks within the DM task network; the burden of responsibility simply changes from a physical input to a monitoring input.

Analysis of the whole networks, given their all-connected nature (Leavitt 1951), render more sophisticated network metrics redundant as they would elicit the same score. For this reason, only a basic description of the networks is presented in Table 9.2. It shows that whilst one additional task is added to the DM task network ('Prepare to take back control'), in comparison to the DD network, the number of edges increases by 26. This suggests that the level of task demand actually increases in a system whereby the driver adopts the role of DM. This is supported by empirical literature on mental workload, showing greater demand for semi-automated driving (Stanton et al. 2007; de Winter et al. 2014) Therefore, we must acknowledge that the DM role is likely to be more mentally demanding than the DD role – by way of an analogy, monitoring another driver (particularly a learner driver where there is a real possibility of having to intervene) can be more demanding than driving oneself. This is essentially what the DM role consists of – monitoring the behaviour of the automated system, monitoring the road environment and anticipating the behaviour of other road users as well as being prepared to regain control of the vehicle if necessary.

Social Networks

In order to further our understanding of how the role of the driver affects network dynamism, high-level social networks were constructed. The authors identified five system agents relating to the DD network (Figure 9.3a), whilst eight system agents were identified for a DM social network (Figure 9.3b). The three additional agents within the DM network represent nodes associated with automated functionality. Banks and Stanton (2016) identified the Longitudinal and Lateral Controllers as separate system agents, given their differing capabilities. The same viewpoint has been taken in the social network representation shown in Figure 9.3b.

Taken at face value, it is clear that the social network associated with the role of DM is more complex than that of DD. This is likely to be attributed to the increased communication and coordination that is required within the system network to maintain the goal of safe and normal driving practice following automation implementation. Thus, the transition between DD and DM appears to bring with it a change in overall system dynamism.

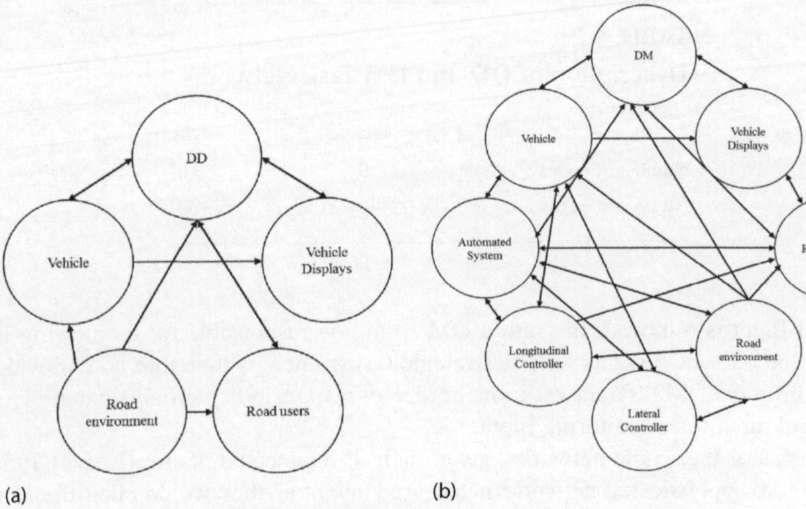

FIGURE 9.3 Social network diagrams for (a) the role of DD and (b) the role of DM.

These social network representations were transformed into association matrices to enable further analysis using network metrics. Both networks can be defined as binary (i.e. they can be represented by a zero-one matrix) and non-symmetric (i.e. directed). The results of the social network analysis using network density, diameter and cohesion calculation are shown in Table 9.3. The results show that the DM social network is denser than the DD network. This is unsurprising, considering that the DM role is only made possible by the introduction of automation into the driving system. This brings with it the creation of new links and interactions between system agents. The analysis also revealed that the DM social network is more cohesive than the DD social network. No difference was found in network diameter between the DD and DM social networks.

In addition, sociometric status was calculated for both DD and DM social networks (see Table 9.4). The identification of key agents (denoted by asterisks) is based upon the rule that any value above the mean sociometric status value reflects network dominance (Salmon et al. 2009). Unsurprisingly, the DD role is considered to be the most dominant

TABLE 9.3
Basic Description of the DD and DM Social Networks

	DD	DM
Number of nodes	5	8
Number of edges	10	37
Network density	0.50	0.70
Network diameter	2.00	2.00
Network cohesion	0.30	0.54

TABLE 9.4
Sociometric Status for the Roles of DD and DM

Node	DD	DM
Driver	1.75[a]	1.57[a]
Vehicle	1.00	1.00
Vehicle displays	0.75	1.00
Road users	0.75	1.14
Road environment	0.75	1.29
Automated system	–	2.00[a]
Longitudinal controller	–	1.14
Lateral controller	–	1.00
Mean	*1.00*	*1.32*

[a] Key system agents based upon the rule that any value above the mean sociometric status value reflects dominance (From Salmon, P. M. et al., *Distributed Situation Awareness: Advances in Theory, Measurement and Application to Teamwork,* Ashgate, Aldershot, UK, 2009.)

agent in a system whereby full responsibility of the driving task lies with the human operator. In an automated system where the driver adopts the role of DM, both the DM and automated system become the most prominent system agents (see Table 9.4). Notably, the automated system ranks highest in sociometric status, which is unsurprising given its role within the system of vehicle operation. Even so, this representation demonstrates that as long as the driver adheres to their responsibilities of monitoring system behaviour, they maintain a central role within the control-feedback loop.

Of course, ensuring that drivers actually adhere to their changing responsibilities is important, because if a driver transitions to a DND, problems may arise. It is difficult to predict and understand the behaviour of a DND because they could be engaged in any task of their choosing. For example, let's assume that a DM becomes engaged in a non-driving-related secondary task that keeps their eyes averted from the road environment for approximately 26 seconds (Eriksson and Stanton 2017). This essentially would signal the breaking of links between the driver, vehicle displays, road users and road environment within Figure 9.3b. A 'broken-links' analysis can be used to demonstrate how momentary lapses in efficient monitoring could impact upon overall network dynamism. This approach essentially represents communication breakdowns between system agents (Stanton and Harvey 2017) similar to that of removing nodes from the network altogether (e.g. Baber et al. 2013). 'Broken links' have typically been used by EAST analysts to analyse accidents post event to explore vulnerabilities within the system (e.g. Griffin et al. 2010; Rafferty et al. 2012). In this instance, reciprocal relationships occurring between the DM and other social agents, shown in Figure 9.3b, are rescinded. The consequences of this are clearly shown in the results of a social network analysis (see Tables 9.5 and 9.6). As is to be expected, network density and cohesion is rapidly reduced. The

TABLE 9.5
Results of a 'Broken-Links' Analysis Whereby Links Are Broken Between the DM, Vehicle Displays, Road Users and Environment

	DM (Complete Network)	DM (Broken Links Indicative of DND Mode Transition)
Number of nodes	8	8
Number of edges	37	30
Network density	0.70	0.54
Network diameter	2.00	2.00
Network cohesion	0.54	0.39

TABLE 9.6
Sociometric Status in a 'Broken Links' Network

Agent	DM (Complete Network)	DM (Broken Links Indicative of DND Mode Transition)
DM	1.57*	0.86
Vehicle	1.00	1.43*
Vehicle displays	1.00	0.71
Road users	1.14	0.86
Road environment	1.29	1.14
Automated system	2.00*	2.00*
Longitudinal controller	1.14	1.14
Lateral controller	1.00	1.00

prominence of the driver within this network also reduces rapidly (see Table 9.6). This places further pressure on systems designers to ensure that their systems are safe, reliable and failsafe, especially given the prominence of the automated system and vehicle under these circumstances. Providing that remaining system agents have the information required to function effectively and are able to use this information to perform appropriately in any given context, safety should be maintained (Salmon et al. 2012) even if the DM transitions to a DND for short periods. There is evidence that individual agents are able to compensate for each other, enabling the system to maintain safe operation (Stanton et al. 2006; Stanton 2016). For example, if a driver fails to respond to system warnings, some autonomous systems can bring the vehicle to a 'safe stop' without any human input. Similarly, an Autonomous Emergency Brake system operates in the background and will only activate if activation thresholds are met. However, this is only possible if system agents have all of the information required. This implicates the 'information network'.

INFORMATION NETWORKS

According to Walker et al. (2010), information networks show the information that is used by and communicated by agents during a task. Their work specifically explored systemic situation awareness of air traffic control using observational notes, interview and live audio feeds. The following information networks were created by analysing driver verbalisations collected as part of another study by Banks and Stanton (2015), which utilised the Critical Decision Method (Klein and Armstrong 2005) to elicit information relating to driver decision making. In total, 48 verbal transcripts were analysed to construct the DD and DM information networks. The DD information network was constructed using evidence from verbal transcripts collected during manual driving (i.e. no automation), whilst the DM information network was constructed using evidence from verbal transcripts collected during a drive in which the basic and strategic sub-tasks of driving (Table 9.1) were automated.

Each 'informational node' within the transcripts was identified and paired with its closest relation (i.e. nodes from the same sentence). This strategy was used to build up a network of information concepts relating to the driving task. The network representations were presented to subject matter experts for verification. Figure 9.4 presents a schematic overview of the DD information network, including 42 nodes and 46 connections. Sociometric status was calculated to identify key concepts. The results indicate that for a DD, the following information nodes were most prominent:

1. *Traffic type*: Includes vehicles, pedestrians and public transport or services.
2. *Traffic*: Considers the properties of road users, such as speed and route.
3. *Infrastructure*: Considers physical aspects such as road type, capacity and lane markings.

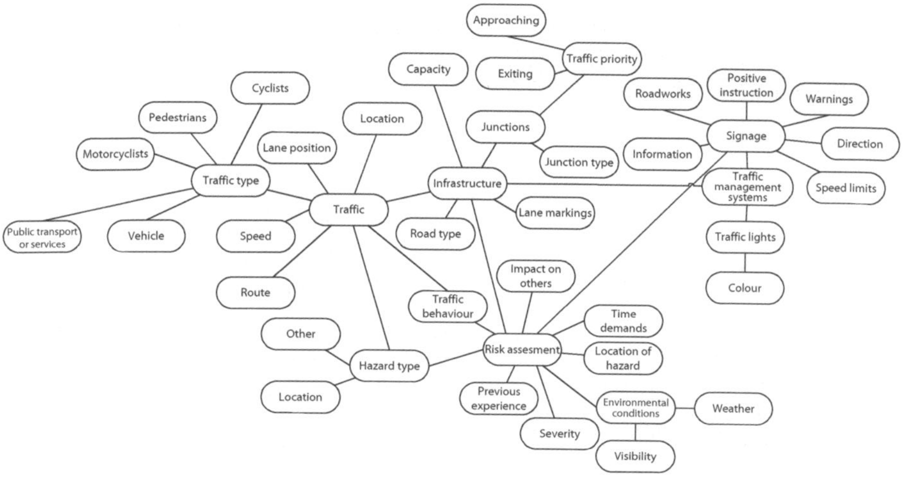

FIGURE 9.4 Schematic overview of the DD information network.

4. *Risk assessment*: Includes hazard type, previous experience and environmental conditions.
5. *Signage*: Considers the meaning and information presented on road signs including speed, warnings and instruction.

The informational nodes coming off these key nodes within Figure 9.4 provide greater detail pertaining to specific situations and contexts within driving (Stanton 2013).

In contrast, the DM information network, presented in Figure 9.5, includes 53 nodes and 61 connections. Sociometric status was once again calculated and revealed that in addition to the key nodes outlined above for a DD, a DM must also consider the following:

1. *Autonomous driving features*: Consider system limits and system-controlled responses.
2. *System mode*: Considers the state of the system, either active or inactive.
3. *Feedback*: Includes the modalities in which information is fed back to the driver.

Thus, despite the intention of autonomous driving features being to reduce the burden of responsibility placed upon the driver, it actually increases the amount of information that a driver must consider in order to maintain safe driving practice. In order to further understand how this additional information requirement impacts upon network dynamism, the information networks as presented in Figures 9.4 and 9.5 were further analysed using the AGNA software tool. The results of this analysis are shown in Table 9.7.

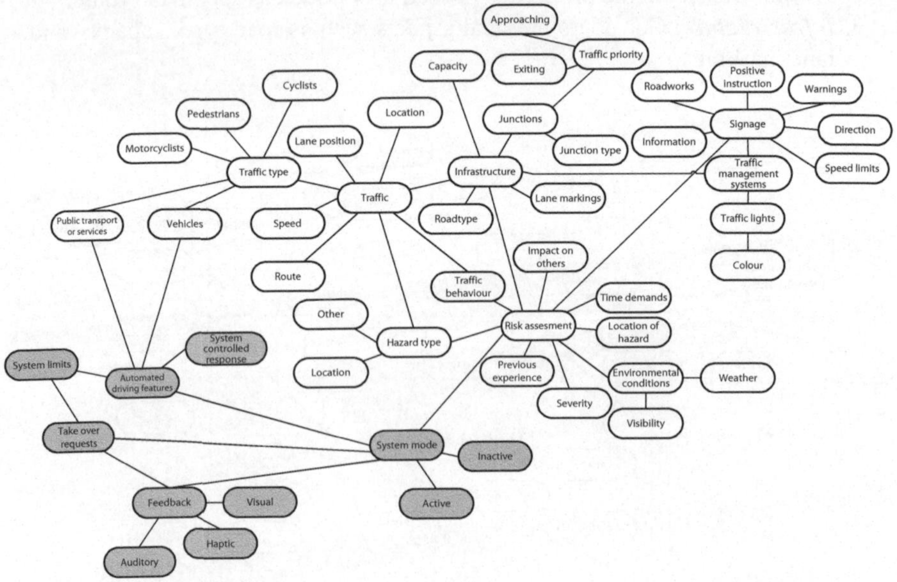

FIGURE 9.5 Schematic overview of the DM information network (grey boxes indicate 'new' nodes within the information network as a result of automation implementation).

TABLE 9.7
Information Network Metrics for the Roles of DD and DM

Metric	DD	DM
Number of nodes	42	53
Number of edges	84	112
Network density	0.05	0.04
Network diameter	6.00	7.00
Network cohesion	0.04	0.03

DISCUSSION

This chapter shows how the application of EAST, and the representations afforded it, can be used to analyse the changing role of the driver. The application of quantitative network metrics to analyse these networks enables us to see how the changing role of the driver impacts upon overall system structure. Importantly, these representations are intended to provide a foundation to the discussion surrounding the role of the driver within automated driving systems. The findings do go a long way to explain why the empirical research shows high levels of mental workload with driving automation that requires constant monitoring (Stanton et al. 1997; Young and Stanton 2004; de Winter et al. 2014). It is a paradox that this level of automation actually results in more, rather than less, work for the driver (Stanton 2015).

From a theoretical modelling perspective, the DND role is particularly problematic. Driver disengagement from the primary driving task is a serious concern. The literature is cluttered with instances whereby engagement in secondary tasks can lead to performance decrements (Eriksson and Stanton 2017). For example, Merat et al. (2014) found that if an automated system disengaged at the point where the driver's attention was diverted away from the road centre, resumption of manual control was erratic for up to 40 seconds after the transfer of control. The main concern for the DND role is that it is not possible to construct a task, social or information network for the DND role. This is because drivers are free to participate in any task of their choosing. This makes it difficult to predict and understand the behaviour of a DND. A 'broken-links' analysis, however, points to a dramatic shift in system structure when a DM transitions to a DND role. Whilst some may argue that the DM role becomes redundant at higher levels of autonomy (e.g. Gasser 2014), the authors caution that this is not strictly the case. The Department for Transport's report (2015), for example, recognizes that some Level 4 systems may still offer a full set of controls that enable manual driving. This means that at some point, the DND could be required to regain control of the vehicle, whether this be due to a 'forced' transfer of control due to some form of mechanical failure (e.g. sensor failure), through choice (e.g. the driver may want to abort or change the destination of travel or they may simply want to be in control) or simply because autonomous driving features only operate in *some* driving modes at Level 4. For this reason, the DND will need to adopt the role of DM during the exchange of control between them and the autonomous vehicle. The success of this transition of control will, however, be based upon a number of interacting psychological constructs including situation awareness, workload, trust and skill (Stanton and Young 2000; Heikoop et al. 2016). If vehicle manufacturers are to handle this transition effectively, a greater understanding of how drivers appraise and make use of higher-level autonomy is needed (Richards and Stedmon 2016). Thus, whilst less emphasis is placed upon the driver as the level of automation increases, vehicle manufacturers still need to think about ways in which the driver can be supported if and when they choose to regain control of the vehicle, especially during early versions of highly automated driving systems. This is because the DD, DM and DND are closely related and likely to be adopted interchangeably throughout the duration of a drive, especially during the intermediate phases of automation.

Whilst the DND role represents the aspiration of many OEMs, there are no such systems that exist on the market today that allow this to happen. If a driver does find themselves in the role of DND, they should be supported back into the role of DM to ensure that the overall goal of the system network can be appropriately maintained. For contemporary vehicle automation systems, the role of DND could be seen as a form of automation misuse (Parasuraman and Riley 1997) given the functional limits of automated architecture. Much more research is needed to further populate these network representations for the varying role of the driver to validate the networks proposed in this chapter. The hypothetical representations presented in this chapter do, however, provide an avenue for discussion as they represent how an 'ideal' network may function. Realistically, however, the authors acknowledge that prolonged exposure to high levels of automation (and driver inactivity) could result in issues surrounding boredom or fatigue (e.g. Stanton et al. 1997; Young and Stanton 2002). Strategies to support the role and maintenance of the DM role are therefore important avenues for further research. We already know from the literature that the level and type of automation can have a direct effect on driver engagement (Stanton et al. 1997; Merat et al. 2014), changes to driver workload (Stanton et al. 1997; Young and Stanton 2002; de Winter et al. 2014), situation awareness (Stanton et al. 2011; Dozza 2012) and decision making (Banks and Stanton 2015).

CONCLUSIONS

The qualitative models of DD and DM support and explain the findings from empirical studies on driver workload changes with automated systems. The increase in workload associated with monitoring automated systems may be explained by the fact that there are simply more tasks to undertake. These qualitative insights are supplemented with the quantitative network statistics, showing the importance of the role change when driving with an automated system (i.e. from the human driver to the computer driver). Modelling future systems is an important step for Ergonomics as a discipline, as it allows us to anticipate the likely behaviour of future technologies and the role for humans. In this way, we can compare alternative designs to identify which are likely to have the better outcomes. Future research should seek to extend and validate the models presented in this chapter using empirical data generated from observational studies exploring driver behaviour at varying levels of automation. In particular, prolonged exposure to high levels of autonomy may elicit DND behaviours that enable us to properly understand the complexities associated with bringing a driver in such a role back into the control-feedback loop.

ACKNOWLEDGEMENTS

Professor Neville A. Stanton was funded by the European Marie Curie International Training Network project on the Human Factors of Automated Driving (PITN-GA-2013-605817) and the EPSRC/JLR TASCC project (Human Interaction: Designing Autonomy in Vehicles – HI:DAVe – Grant number: EP/N011899/1).

REFERENCES

Baber, C., Stanton, N. A., Atkinson, J., McMaster, R. and Houghton, R. J. (2013). Using social network analysis and agent-based modelling to explore information flow using common operational pictures for maritime search and rescue operations. *Ergonomics*, 56 (6), 889–905.

Banks, V. A. and Stanton, N. A. (2015). Contrasting models of driver behaviour in emergencies using retrospective verbalisations and network analysis. *Ergonomics*, 58 (8), 1337–1346.

Banks, V.A. and Stanton, N.A. (2016). Driver-centred vehicle automation: Using network analysis for agent-based modelling of the driver in highly automated driving systems. *Ergonomics*, DOI:10.1080/00140139.2016.1146344.

Banks, V. A., Stanton, N. A. and Harvey, C. (2014). Sub-systems on the road to vehicle automation: Hands and feet free but not 'mind' free driving. *Safety Science*, 62, 505–514.

Benta, M. (2005) Studying communication networks with AGNA 2.1. *Cognition Brain Behaviour*, 9, 567–574.

Casner, S. M., Hutchins, E. L. and Norman, D. (2016). The challenges of partially automated driving. *Communications of the ACM*, 59 (5), 70–77.

Department for Transport. (2015). The pathway to driverless cars: Summary report and action plan. Available at: https://www.gov.uk/government/uploads/system/uploads/attachment_data/file/401562/pathway-driverless-cars-summary.pdf [Accessed 19.09.2016].

de Winter, J. C. F., Happee, R., Martens, M. H. and Stanton, N. A. (2014). Effects of adaptive cruise control and highly automated driving on workload and situation awareness: A review of the empirical evidence. *Transportation Research Part F: Traffic Psychology and Behaviour*, 27, 196–217.

de Winter, J. C. F., Stanton, N. A., Price, J. S. and Mistry, H. (2016). The effects of driving with different levels of unreliable automation on self-reported workload and secondary task performance. *International Journal of Vehicle Design*, 70 (4), 297–324.

Dozza, M. (2012). What factors influence drivers' response time for evasive maneuvers in real traffic? *Accident Analysis & Prevention*, 58, 299–308.

Endsley, M. R. and Kaber, D. B. (1999). Level of automation effects on performance, situation awareness and workload in a dynamic control task. *Ergonomics*, 42 (3), 462–492.

Eriksson, A. and Stanton, N. A. (2017). Take-over time in highly automated vehicles: Transitions to and from manual control. *Human Factors*, DOI:10.1177/0018720816685832.

Fagnant, D. J. and Kockelman, K. (2015). Preparing a nation for autonomous vehicles: Opportunities, barriers and policy recommendations for capitalising on self-driven vehicles. *Transportation Research Part A*, 77, 167–181.

Federal Aviation Administration. (2003). Standard operating procedures for flight deck crew members. AC No: 120-71A, 27.2.03.

Gasser, T. (2014). Vehicle automation: Definitions, legal aspects, research needs. UNECE Workshop: German Federal Highway Research Institute: Towards a new transportation culture: Technology innovations for safe, efficient and sustainable mobility. Brussels, 17–18 November 2014.

Heikoop, D. D., de Winter, J. C. F., van Arem, B. and Stanton, N. A. (2016). Psychological constructs in driving automation: A consensus model and critical comment on construct proliferation. *Theoretical Issues in Ergonomics Science*, 17 (3), 284–303.

Hutchins, E. (1995). How a cockpit remembers its speed. *Cognitive Science*, 19 (3), 265–288.

Kaber, D. B. and Endsley, M. R. (2004). The effects of level of automation and adaptive automation on human performance, situation awareness and workload in a dynamic control task. *Theoretical Issues in Ergonomics Science*, 5 (2), 113–153.

Kakimoto, T., Kamei, Y., Ohira, M. and Matsumoto, K. (2006). Social network analysis on communications for knowledge collaboration in OSS communities. In: *Proceedings of The 2nd International Workshop on Supporting Knowledge Collaboration in Software Development* (KCSD'06), pp. 35–41 Tokyo, Japan.

Leavitt, H. J. (1951). Some effects of certain communication patterns on group performance. *Journal of Abnormal and Social Psychology*, 46, 38–50.

Lee, J. D. and See, K. A. (2004). Trust in automation: Designing for appropriate reliance. *Human Factors*, 46, 50–80.

Merat, N., Jamson, A. H., Lai, F. C., Daly, M. and Carsten, O. M. (2014). Transition to manual: Driver behaviour when resuming control from a highly automated vehicle. *Transportation Research Part F: Traffic Psychology and Behaviour*, 27 (B), 274–282.

National Highway Traffic Safety Administration. (2013). *Preliminary Statement of Policy Concerning Automated Vehicles System*. Washington, DC.

National Highway Traffic Safety Administration. (2016). Update to 'Preliminary Statement of Policy Concerning Automated Vehicles System'. Available at: http://www.nhtsa.gov/Research/Crash+Avoidance/Automated+Vehicles [Accessed 19.09.2016].

Nowakowski, C., Shladover, S. E. and Tan, H. S. (2015). Heavy vehicle automation: Human factors lessons learned. *Procedia Manufacturing*, 3, 2945–2952.

Parasuraman, R., Molloy, R. and Singh, I. L. (1993). Performance consequences of automation-induced 'complacency'. *The International Journal of Aviation Psychology*, 3, 1–23.

Parasuraman, R. and Riley, V. (1997). Humans and automation: Use, misuse, disuse, abuse. *Human Factors: The Journal of the Human Factors and Ergonomics Society*, 39 (2), 230–253.

Richards, D. and Stedmon, A. (2016). To delegate or not to delegate: A review of control frameworks for autonomous cars. *Applied Ergonomics*, 53 (B), 383–388.

Salmon, P. M., Stanton, N. A. and Young, K. L. (2012). Situation awareness on the road: Review, theoretical and methodological issues, and future directions. *Theoretical Issues in Ergonomics Science*, 13 (4), 472–492.

Salmon, P. M., Stanton, N. A., Walker, G. H. and Jenkins, D. P. (2009). *Distributed Situation Awareness: Advances in Theory, Measurement and Application to Teamwork*. Ashgate: Aldershot, UK.

Salmon, P. M., Walker, G. H. and Stanton, N. A. (2016). Pilot error versus sociotechnical systems failure: A distributed situation awareness analysis of Air France 447. *Theoretical Issues in Ergonomics Science*, 17 (1), 64–79.

Sarter, N. B., Woods, D. D. and Billings, C. E. (1997). Automation surprises. *Handbook of Human Factors and Ergonomics*, 2, 1926–1943.

Sheridan, T. B. (1970). Big brother as driver: New demands and problems for the man at the wheel. *Human Factors: The Journal of the Human Factors and Ergonomics Society*, 12, 95–101.

Smiley, A. and Brookhuis, K. A. (1987). Alcohol, drugs and traffic safety. In: J.A. Rothengatter and R. A. de Bruin (Eds.), *Road Users and Traffic Safety* (pp. 83–105).Van Gorcum: Assen, The Netherlands.

Smith, B. W. (2013). SAE levels of driving automation. Available at: http://cyberlaw.stanford.edu.loda [Accessed 11.04.2016].

Society for Automotive Engineers (2015). Taxonomy and definitions for terms related to on-road motor vehicle automated driving systems. Available at: http://standards.sae.org/j3016_201401/. [Accessed 12.10.2015].

Stanton, N. A. (2014). Representing distributed cognition in complex systems: How a submarine returns to periscope depth. *Ergonomics*, 57 (3), 403–418.

Stanton, N. A. (2015). Responses to autonomous vehicles. *Ingenia*, 62, 9.

Stanton, N. A. (2016). Distributed situation awareness. *Theoretical Issues in Ergonomics Science*, 17 (1), 1–7.

Stanton, N. A., Dunoyer, A. and Leatherland, A. (2011). Detection of new in-path targets by drivers using stop and go adaptive cruise control. *Applied Ergonomics*, 42 (4), 592–601.

Stanton, N. A. and Marsden, P. (1996). From fly-by-wire to drive-by-wire: Safety implications of automation in vehicles. *Safety Science*, 24 (1), 35–49.

Stanton, N. A. and Harvey, C. (2017). Beyond human error taxonomies in assessment of risk in sociotechnical systems: A new paradigm with the EAST 'broken-links' approach. *Ergonomics*, 60 (2) 221–233.

Stanton, N. A. and Salmon, P. M. (2009). Human error taxonomies applied to driving: A generic driver error taxonomy and its implications for intelligent transport systems. *Safety Science*, 47 (2), 227–237.

Stanton, N. A., Salmon, P. M, Rafferty, L., Walker, G. H., Baber, C. and Jenkins, D. P. (2013). *Human Factors Methods: A Practical Guide for Engineering and Design*. Ashgate: Aldershot, UK.

Stanton, N. A., Stewart, R., Harris, D., Houghton, R. J., Baber, C., McMaster, R., Salmon, P., Hoyle, G., Walker, G., Young, M. S., Linsell, M., Dymott, R. and Green, D. (2006). Distributed situation awareness in dynamic systems: Theoretical development and application of an ergonomics methodology. *Ergonomics*, 49 (12–13), 1288–1311.

Stanton, N. A. and Young, M. (2000). A proposed psychological model of driving automation. *Theoretical Issues in Ergonomics Science*, 1 (4), 315–331.

Stanton, N. A. and Young, M. (2005). Driver behaviour with adaptive cruise control. *Ergonomics*, 48, 1294–1313.

Stanton, N. A., Young, M. and McCaulder, B. (1997). Drive-by-wire: The case of mental workload and the ability of the driver to reclaim control. *Safety Science*, 27 (2–3), 149–159.

Walker, G. H., Gibson, H., Stanton, N. A., Baber, C., Salmon, P. and Green, D. (2006). Event analysis of systematic teamwork (EAST): A novel integration of ergonomic methods to analyse C4i activity. *Ergonomics*, 49 (12–13), 1345–1369.

Walker, G. H., Stanton, N. A., Baber, C., Wells, L., Gibson, H., Salmon, P. and Jenkins, D. (2010). From ethnography to the EAST method: A tractable approach for representing distributed cognition in Air Traffic Control. *Ergonomics*, 53 (2), 184–197.

Walker, G. H., Stanton, N. A. and Salmon, P. M. (2011). Cognitive compatibility of motorcyclists and car drivers. *Accident Analysis & Prevention*, 43, 878–888.

Walker, G. H., Stanton, N. A. and Salmon, P. M. (2015). *Human Factors in Automotive Engineering and Technology*. Ashgate: Aldershot, UK.

Walker, G. H., Stanton, N. A. and Salmon, P. M. (2016). Trust in vehicle technology. *International Journal of Vehicle Design*, 70 (2), 157–182.

Walker, G. H., Stanton, N. A., Salmon, P. M., Jenkins, D. P., Monnan, S. and Handy, S. (2012). Communications and cohesion: a comparison between two command and control paradigms, *Theoretical Issues in Ergonomics Science*, 13(5), 508–527.

Walker, G. H., Stanton, N. A., Salmon, P. M., Jenkins, D. P. and Rafferty, L. A. (2010). Translating concepts of complexity to the field of ergonomics. *Ergonomics*, 53 (10), 1175–1186.

Young, M. S. and Stanton, N. A. (2002). Malleable attentional resources theory: A new explanation for the effects of mental underload on performance. *Human Factors*, 44 (3), 365–375.

Young, M. S. and Stanton, N. A. (2004). Taking the load off: Investigations of how adaptive cruise control affects mental workload. *Ergonomics*, 47 (8), 1014–1035.

10 EAST in Future Road Transportation Systems

With Victoria A. Banks, Gary Burnett and Setia Hermawati

INTRODUCTION

Over the past two decades, there have been major developments in the integration of wireless and autonomous technologies in the road transportation network (Talebpour and Mahmassani 2016). Automated vehicles in particular are quickly becoming an engineering reality (Stanton 2015), and whilst much research has primarily focused upon driver–automation interaction (e.g. Banks et al. 2014; Zeeb et al. 2015; Louw and Merat 2017), many issues remain. Some of these issues relate to how automation can be regulated, legislated and standardised, but more importantly, we do not fully understand how automation will impact overall road system behaviour. For example, Atkins Mobility (2016) speculate that if automation brings about improvements to road safety, we may see a future wherein crash barriers are no longer necessary and roadway signs become redundant as information can be shared using Vehicle-to-Vehicle (V2V) and Vehicle-to-Infrastructure (V2I) communication streams and other location services (e.g. Global Positioning Systems [GPS]). V2V and V2I are made possible through both the use of Dedicated Short Range Communication (DSRC) sensors and wireless network architectures such as 5G. In vehicles, DSRC represents on-board sensor units whilst external roadside units represent the means to achieve V2I communication. This wireless connectivity would enable intra-vehicle communication as well as real-time communication with traffic management systems. Essentially, camera and radar-based technologies enable the vehicle to 'see' (e.g. vision systems that process video data and fuse with microwave radar data), whilst technologies such as DSRC enable the car to 'talk' (i.e. transmit data to other vehicles and infrastructure) and 'listen' (i.e. receive data from other vehicles and infrastructure).

The concept of Connected and Autonomous Vehicles (CAV) is not new, with research and innovation dating back to the early 1990s (De La Fortelle 2005). The Science and Technology Select Committee (2017) cite numerous benefits associated with CAV, including increased accessibility and mobility and improvements to road safety and congestion. KPMG (2015) hypothesise that by 2030, all new vehicles sold within the United Kingdom will be 'fully connected'. It is clear, then, that future transportation systems will be reliant upon the exchange of information between both human and non-human entities to ensure effective system functioning. This type of communicative behaviour is the essence of Distributed Cognition (DCOG) Hutchins 1995), whereby interactions take place between humans, resources and materials across space and time(Hollan et al. 2000; Hutchins 1995). DCOG is related to the theory of 'transactional memory', whereby individuals (both human and non-human entities) tend to rely upon others to remember things for them (Stanton et al. 2015). Thus, DCOG is characterised by multiple system 'agents' that work together in order to achieve a common goal (Hutchins 1995). 'Agents' in this sense can receive, hold and share information with one another in order to pursue a common goal (Clark 1995). This implies the need for communication and coordination to exist between them (Christoffersen and Woods 2002; Stanton 2014; Eriksson and Stanton 2017). Of course, there are many challenges associated with Vehicle-to-Anything (V2X) communication, including, but not limited to, issues relating to the size of the network (both in terms of geography and availability). This means that in some instances, the exchange of information between system agents (both human and non-human) may be delayed, inaccurate or incomplete. DCOG provides a theoretical framework that can be used in the effective design of new communication and interactive technologies to support the relationship that exists between human and non-human agents by highlighting areas of potential weakness (Hollan et al. 2000). To date, DCOG has only been used to consider small sociotechnical systems in isolation, such as an airline cockpit (Hutchins 1995) and command teams (Stanton 2014). In this chapter, we argue that DCOG provides not only the theoretical foundation but also the methods that can be used to explore complex sociotechnical systems at a macro level.

The road transportation system is a good example of a macro-level sociotechnical system. This is because it can dynamically configure itself to ensure that multiple sub-systems acting within it (e.g. Traffic Management Centres [TMC] work alongside External Agencies [EA]) can operate simultaneously to achieve various functions. Thus, the network is based upon a large number of complex interactions and interdependencies between multiple system agents at a number of levels (Salmon et al. 2014). These include system agents within the road environment (RE) (e.g. drivers, pedestrians and vehicles), TMC (e.g. traffic management and CCTV operators) and EA (e.g. radio stations and emergency services). Whilst these categories of system agent are typically analysed independently, this chapter applies the principles underpinning DCOG to all of the agents and agencies involved within the road transportation network. Given the uncertainty as to how the transportation network will be affected by CAV functionality, this chapter provides a comparison between non-CAV and CAV networks to explore how network dynamism may change as a result of increased connectivity. This comparison is important because it provides initial insights into how system agents will react to, and interact with, intelligent transportation systems.

METHOD

DCOG in complex sociotechnical systems can be further explored and understood using the Event Analysis of Systemic Teamwork framework (EAST) Stanton et al. 2008). EAST is a descriptive method that proposes that a system can be described using three interlinked network representations: task, social and information (Walker et al. 2006, 2010). Task networks provide analysts with a means to show the processes involved in attaining network goals (Salmon et al. 2014). They can provide a description of the sequences and interdependencies that exist between individual sub-tasks that must be completed to attain these goals. Social networks are used to analyse the structure of the system in terms of the communications that take place between different system 'agents'. Finally, information networks show the information that is used by, and communicated by, system agents during a task (Stanton et al. 2008). Information networks detail aspects of communication that underpin the completion of a task as well as the relationships that exist between these different informational nodes. EAST has been used to focus upon specific tasks within varied domains, including aviation (Sorensen et al. 2011), rail (Walker et al. 2006), driving (Banks and Stanton 2016) and maritime transport, (Stanton et al. 2006, 2017a, 2017b; Stanton 2014; Baber et al. 2013), providing meso-level representations of DCOG (Grote et al. 2014). However, this chapter goes further by using the representations afforded by EAST (Stanton et al. 2008) to explore DCOG at a macro level (Grote et al. 2014). EAST makes it possible to provide an overview of how different agents and agencies within the road transportation network can function simultaneously within a shared space (i.e. the road network). The networks can then be subjected to quantitative analysis using the Applied Graphic and Network Analysis (AGNA) tool (AGNA, version 2.1; Benta 2005). AGNA is a platform-independent freeware application that can be used to analyse task, social and information networks. Nodes within each network can either be analysed individually to assess agent centrality/prominence or as a whole to give an overall impression of system complexity. Network metrics can be used to identify key agents, tasks and informational elements within system operation. Within driving research, the following network metrics have been applied to analyse EAST representations:

Density represents the level of interconnectivity between system agents. It is expressed as a value between 0 and 1, where 0 represents a network that has no connections and 1 indicates that the network is fully connected (Kakimoto et al. 2006). It is calculated using the formula

$$\text{Network density} = 2e / n(n-1)$$

where:

e is the total number of links within the network

n is the number of nodes within the network (Walker et al. 2012)

Diameter is used to analyse the connections and pathways that exist between nodes within the network (Walker et al. 2011). Denser networks (i.e. the route through the network is shorter and more direct) have smaller values. It is calculated using the formula

$$\text{Diameter} = \max_{\text{uy}} d\left(n_i, n_j\right)$$

where $d(n_i,n_j)$ is the 'largest number of [agents] which must be traversed in order to travel from one [agent] to another when paths which backtrack, detour, or loop are excluded from consideration' (i.e. max_{uy},) (Weisstein 2008; Harary 1994). *Cohesion* represents the number of reciprocal connections divided by the total number of possible connections (Stanton 2014).

Finally, *sociometric status* provides an indication of agent prominence (Houghton et al. 2006; Salmon et al. 2014). Key agents (i.e. those most prominent within the network) have higher sociometric values (Salmon et al. 2012). It is calculated using the formula

$$\text{Sociometric Status} = \frac{1}{(g-1)} \sum_{j=1}^{g} (x_{ji}, x_{ij})$$

where:

g is the total number of nodes in the network
i and j are individual nodes
x_{ji} are the number of communications between node j and node i
x_{ij} are the number of communications between node i and node j (Salmon et al. 2012; Houghton et al. 2006)

RESULTS

Identification of System Agents

For the purposes of this analysis, a total of 21 system agents were identified from previous work conducted by Price (2016) and Banks and Stanton (2016). Their work specifically sought to identify system agents involved in Traffic Management operations (e.g. Price 2016) and within automated driving environments (e.g. Banks and Stanton 2016). The 21 agents broadly span three operational categories: RE, TMC and EA (see Table 10.1 for complete list and descriptions). These agents represent the main human and non-human entities that can be found within the road transportation network.

Task Networks

On the basis of the list presented in Table 10.1, it is possible to consider the types of tasks in which these system agents engage and how they may be related. This makes it possible to construct a high-level task network for the entire road transportation system involving all 21 agents. Walker et al. (2006) suggest that task networks can show how sub-tasks may relate to other sub-tasks based upon their functional or temporal properties.

The task network for the road transportation system, shown in Figure 10.1, should be viewed as a continuous process to reflect the notion that the system always remains active (i.e. 24 hours a day, 7 days a week). This network representation does not consider any CAV or intelligent functionality. The task network portrays the relationships that exist between tasks in a non-sequential manner. Whilst not exhaustive, this representation includes some of the main tasks (n=19) associated with RE, TMC and EA agents (and the interconnections that may exist between them).

TABLE 10.1
System Agents Involved in the Road Transportation System

Operational Category	Subcategory	Agent	Description
Road Environment (RE) – this category represents all agents that are present within the road environment	Drivers (DD / DM)*	Host Driver Host Passenger Other Drivers Other Passengers	The categories of individuals occupying vehicles Role of driver dependent upon level of automation (DD, DM)
	Vehicles (Connected and Autonomous Vehicles; CAV)	Host Vehicle Other Vehicles Services/Goods Vehicles Emergency Vehicles	The categories of traffic using (or potentially using) the road network Addition of intelligent sensors enable CAV function
	External Roadside Equipment (ERE)	Traffic Monitoring Equipment	E.g. CCTV cameras and induction loops
		Traffic Management Equipment	E.g. traffic lights and Variable Message Signs (VMSs)
		Vulnerable Road Users (VRU)	E.g. cyclists and pedestrians
Traffic Management Centre (TMC) – this category represents all agents that have direct access to information relating to the overall traffic situation		TMC Operator	Responsible for managing traffic
		Closed-Circuit Television (CCTV) Applications	Controls the TMC's CCTV cameras
		Urban Traffic Management Control (UTMC) Applications	Collects data relating to road environment (e.g. vehicle counts)
		Police Closed-Circuit Television (CCTV) Personnel	Monitor CCTV for crime, assisting police operations
External Agencies (EA) – this category represents all agents that both share and receive information relating to traffic situations		Radio Stations	Distribute information to traffic and other agents
		Information Providers	Provide additional information (e.g. Met Office, Highways England)
		Other Transport Control Centres	Includes other road TMCs as well as public transport control centres (e.g. Bus)
		Emergency Services Control Centres	Manage emergency service operations
		Traffic Data Distribution Services	Dissemination of information to traffic and third parties

Note: DD and DM reflect the 'Driver Driving' and 'Driver Monitoring' roles identified by Banks, V. A. and Stanton, N. A., *Theoretical Issues in Ergonomics Science*, in press.

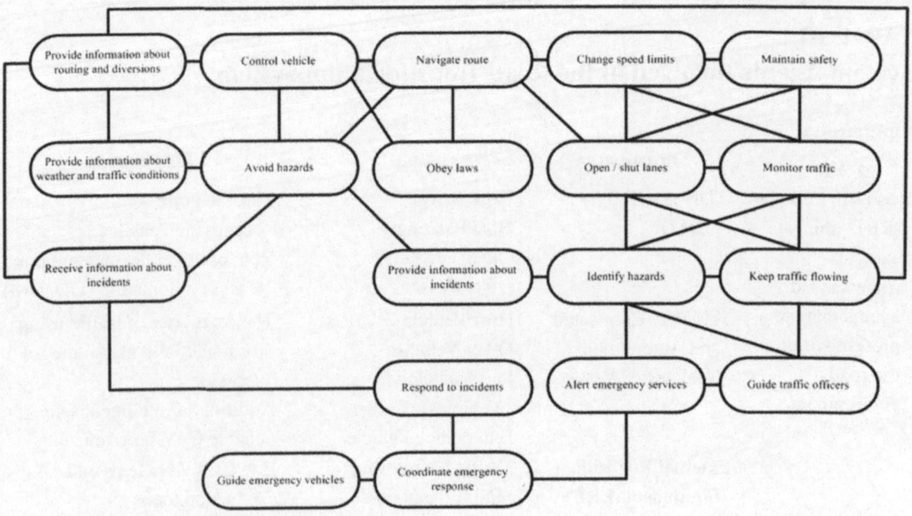

FIGURE 10.1 Task network showing high-level tasks essential for effective functioning of the road transportation network.

Analysis of the task network presented above was conducted using AGNA (Benta 2005). The AGNA analysis indicates that the network can be described as binary (i.e. it can be represented by a zero-one matrix) and symmetric (i.e. it is non-directed). In total, the network has 68 connections (i.e. the number of in-out connections between nodes). The network has a diameter of 6 (i.e. six hops from one side of the network to the other) and a density score of 0.2 (i.e. a low distribution network).

However, in order to compare and contrast how non-CAV and CAV task networks may differ, a combined networks approach was adopted (e.g. Stanton 2014). A combined task and social network approach enables us to see 'who' is involved in completing tasks within the transportation network in parallel to keep the task network functioning. In terms of DCOG, this approach provides a means to explore the allocation of system function among social agents spanning the RE, TMC and EA categories (outlined in Table 10.1). For the purposes of discussion, the authors assume that the same basic task network exists regardless of whether intelligent features exist within it. Thus, Figure 10.2 represents the task network (shown in Figure 10.1) coded by social agents for a non-CAV network, whilst Figure 10.3 represents the task network for a CAV network – note that subcategories of RE have been added.

Figure 10.2 essentially demonstrates that 14 out of 20 tasks are completed in isolation by a single agent category (i.e. RE, TMC or EA). In practical terms, this means that there is a clear division between tasks associated with different agent categories. Importantly, the role of the driver reflected within Figure 10.2 represents that of 'Driver Driving' (DD) (Banks and Stanton 2018). The DD role essentially represents the idea that the driver is responsible for completing all of the physical and cognitive tasks associated with driving (Walker et al. 2015) and is thus responsible for controlling the vehicle, avoiding hazards, navigating routes, obeying laws, monitoring traffic and maintaining safety in the non-CAV network portrayed in Figure 10.2.

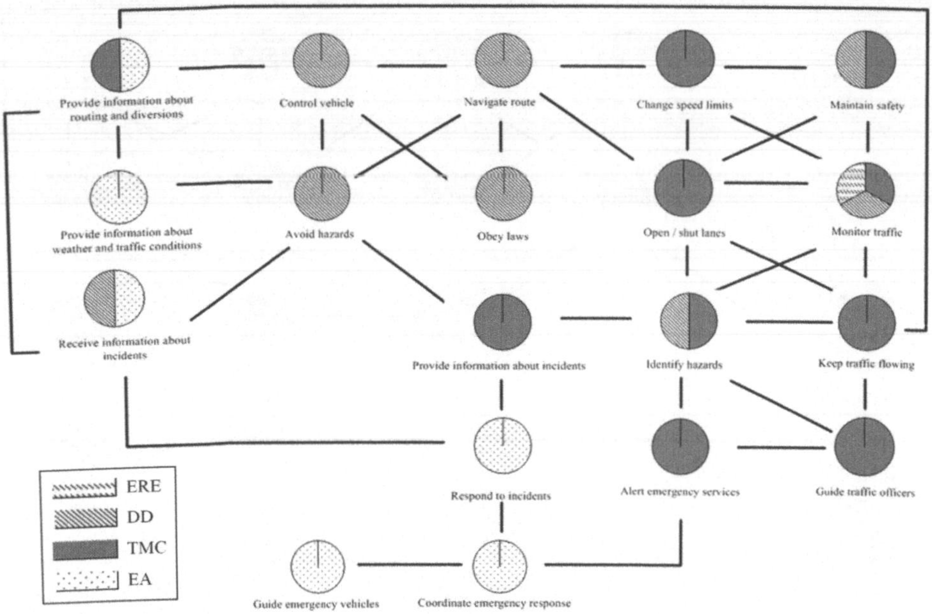

FIGURE 10.2 Task network for a non-CAV transportation network coded by social agents.

In contrast, Figure 10.3 represents the task network (coded by social agents) for a CAV network. Here we see many more tasks becoming shared (i.e. completed in parallel by two or more agents) in comparison to the non-CAV network. Tasks completed in isolation appear to have halved as a result of CAV implementation. This indicates that the divisions between the RE, TMC and EA categories are less rigid, opening up possibilities to create new communication channels in an effort to improve information sharing and network cohesiveness. 'Task sharing' is particularly more pronounced within the RE category, as CAV vehicles are capable of completing the same tasks as a traditional DD. To acknowledge the changing role of the driver within a CAV network, an additional driver role has been added to this network representation. A Driver Monitor (DM) reflects the 'intended' supervisory role of the driver during the intermediate phases of automation (Banks and Stanton 2018). A DM would need to continue monitoring the behaviour of the vehicle and automated sub-systems despite the transfer of physical control and aspects of decision making to CAV vehicles. This is to ensure that safe and normal driving practice is maintained. Note that the role of DD remains in the representation to reflect the potential for mixed traffic driving scenarios (e.g. both automated and manually driven vehicles may occupy the road), but it also demonstrates further the utility of CAV to perform all of the tasks typically associated with a DD.

Social Networks

To better understand the structure of communications that may occur between agents within RE, TMC and EA, social networks were constructed to demonstrate

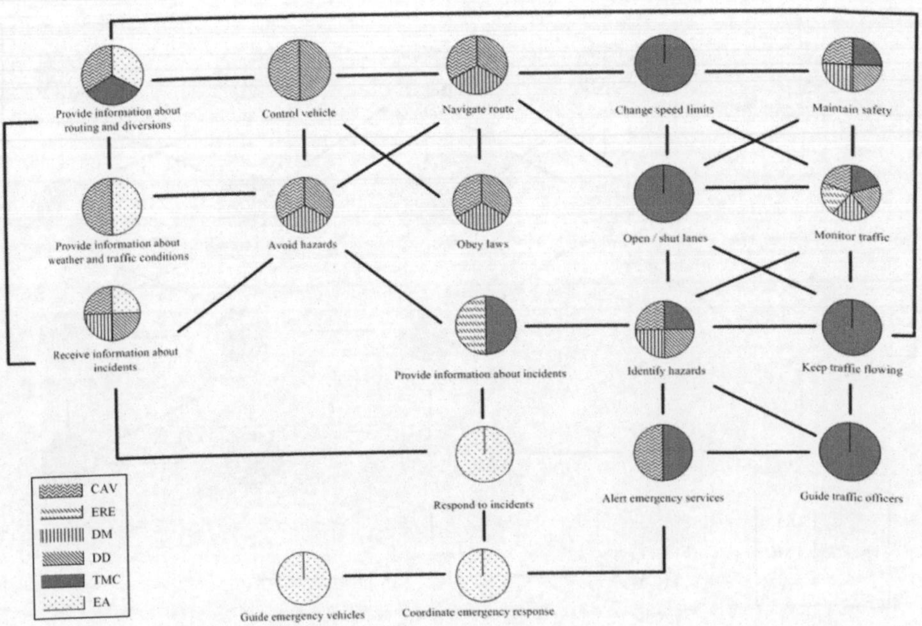

FIGURE 10.3 Task network for CAV coded by social agents.

the directional flow of communication between system agents for both non-CAV and CAV networks. Social networks essentially provide a means to assess the relationship (i.e. in terms of communication and activity) between different system agents (Stanton 2014). 'Communication' between system agents can range from being obvious (e.g. gestures between drivers such as flashing lights, radio broadcasts sharing traffic updates, vehicle displaying information on internal interface to alert driver to changes in system status, driver manipulating controls within vehicle to alter its behaviour) to being 'invisible' (e.g. Urban Traffic Management Control [UTMC] data can be used to change behaviour of traffic management equipment). Simplified versions of these networks were then constructed and validated by Subject Matter Experts (SMEs) for ease of interpretation and are presented in Figures 10.4 and 10.5. Four SMEs from a leading automotive manufacturer and members of the Human Factors Research Group at the University of Nottingham were involved in this process, with a collective 18 years' experience within the driving automation field.

Figure 10.4 presents a simplified social network of a non-CAV transportation network. Here, we see that whilst links exist between the RE, TMC and EA, communications typically remain within operational categories. In contrast, Figure 10.5 presents a social network for future CAV networks and shows that the links that exist between different operational categories become stronger as new links are formed. This is because wireless connectivity between vehicles and infrastructure enables both intra-vehicle, inter-vehicle and real-time communication with traffic management systems. CAV functionality therefore binds the social network more

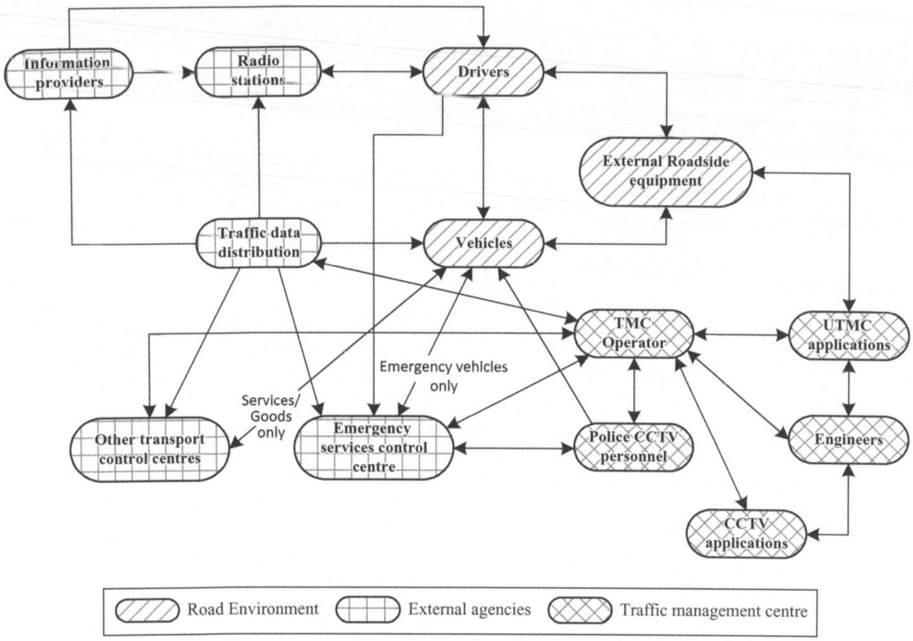

FIGURE 10.4 Simplified social network representation of macro-level communications within a non-CAV network.

closely together, bridging the 'communication gap' between different operational categories.

Further analysis of these networks was performed using AGNA (Benta 2005) to demonstrate that even simplified social network models of macro-level systems can be used to identify differences between agent communications in non-CAV and CAV transportation systems. Table 10.2 presents the findings of this analysis and shows that whilst the networks have the same number of nodes (i.e. number of system agents), the CAV network incorporates additional connections (i.e. demonstrating the increased communication that exists within the network). The CAV network yields a smaller diameter, which is perhaps attributable to the more direct nature of interactions within it. With fewer hops required to get from one side of the network to the other, together with greater network density, CAV implementation does indeed appear to improve network efficiency.

The calculation of sociometric status was used to assess agent prominence within the networks. To enable us to see more clearly the shift in agent prominence occurring as a result of CAV implementation, only agents holding different sociometric status values between the social networks are highlighted in Table 10.3. Agents within the RE had the largest changes in prominence. Both 'Vehicle' and 'External Roadside Equipment' categories also gain prominence, which is attributable to the introduction of V2V and V2I sensor units. These changes also influence dynamic shifts in other areas of the transportation network, as, for example, UTMC becomes even more central to TMC operation. In addition, CAV implementation also strengthens the link between RE and EA agents.

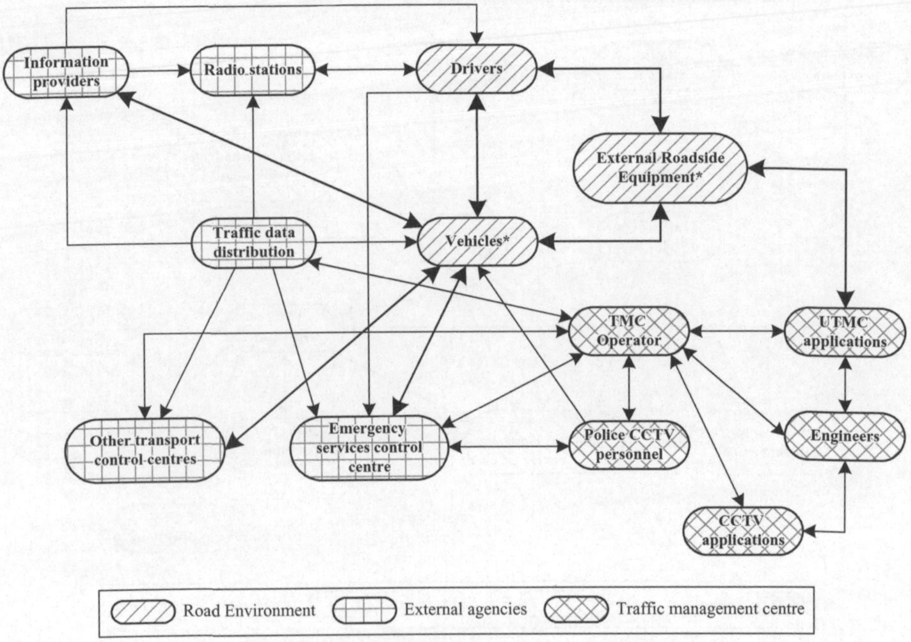

FIGURE 10.5 Simplified social network representation of macro-level communications within future CAV networks. Bold lines represent new and strengthened links between agents as a result of CAV implementation.

TABLE 10.2
Contrasting Network Metrics for Non-CAV and CAV Social Networks

Social Network Analysis	Non-CAV	CAV
Number of nodes (i.e. system agents)	13	13
Number of connections	43	45
Diameter	5.0	4.0
Density	0.28	0.37

Information Networks

The final representational medium within the EAST framework is information networks (Stanton et al. 2008, 2013; Walker et al. 2006, 2010). Information networks detail aspects of communication that underpin the foundations of the system and how they are linked. The process of constructing information networks typically follows the interrogation of operator verbalisations. These may be captured using observational notes, interviews and live audio feeds recorded during task completion (e.g. Walker et al. 2010). Information nodes are identified and then paired with their closest relation. For example, nodes within the same sentence will be linked

TABLE 10.3
Contrasting Sociometric Status for Agents Involved in Non-CAV and CAV Networks

Category	Agent	Non-CAV Network	CAV Network
Road environment (RE)	Drivers	0.67*	1.00*
	Vehicle	0.83*	1.67*
	External roadside equipment	0.50	1.00*
Traffic management centres (TMC)	TMC operator	1.17*	1.17*
	CCTV applications	0.33	0.33
	UTMC applications	0.50	0.67*
	Police CCTV personnel	0.42	0.42
	Engineers	0.50	0.50
External agencies (EA)	Radio stations	0.33	0.33
	Information providers	0.25	0.42
	Other transport control centres	0.42	0.58
	Emergency services control centres	0.67*	0.83*
	Traffic data distribution	0.58*	0.58
	Mean	*0.55*	*0.73*

Note: Highlighted Agents Yield Different Sociometric Status Values as a Result of CAV Implementation. Asterisks Identify 'Key' Agents.

(e.g. Stanton 2014; Walker et al. 2010). This strategy quickly builds a network of information concepts (Banks and Stanton 2018). According to Stanton (2014), information may be related temporally or spatially to different agents or tasks. The information network presented in Figure 10.6 was constructed based upon this strategy and combines driver-orientated knowledge (see Banks and Stanton 2018) with TMC-orientated knowledge (see Price 2016) to represent a non-CAV transportation network. This representation contains 53 nodes and 119 connections and demonstrates that in order to ensure the effective functioning of the road transportation network, a plethora of information must be captured, processed and distributed between all of the agents involved. In contrast, the information network representing the operation of a CAV network has 18 additional nodes and a further 66 connections (see Figure 10.7). These additional nodes essentially represent the information that can be generated from the implementation of CAV into the network.

The information networks presented in Figures 10.6 and 10.7 were subjected to further analysis using AGNA (Benta 2005), which showed that whilst network dynamism does not significantly change at a global level (see Table 10.4), the CAV network grows rapidly, even at a high level of analysis.

Underlying differences between non-CAV and CAV information networks is important because we need to ensure that agents within the network are equipped

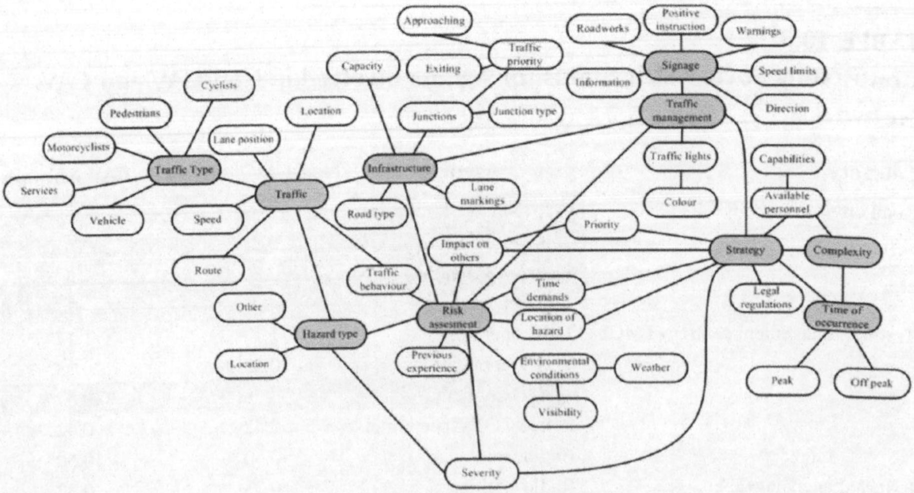

FIGURE 10.6 Information network for non-CAV networks (note that 'key' nodes highlighted in grey are based upon the calculation of sociometric status – see Table 10.5).

with 'key' knowledge about its functioning. Sociometric status was calculated to assist in the identification of key informational nodes within the non-CAV and CAV networks above. Data were ranked in descending order to enable the authors to highlight key informational nodes within both non-CAV and CAV information networks. The results are shown in Table 10.5 and appear to confirm that any node with four

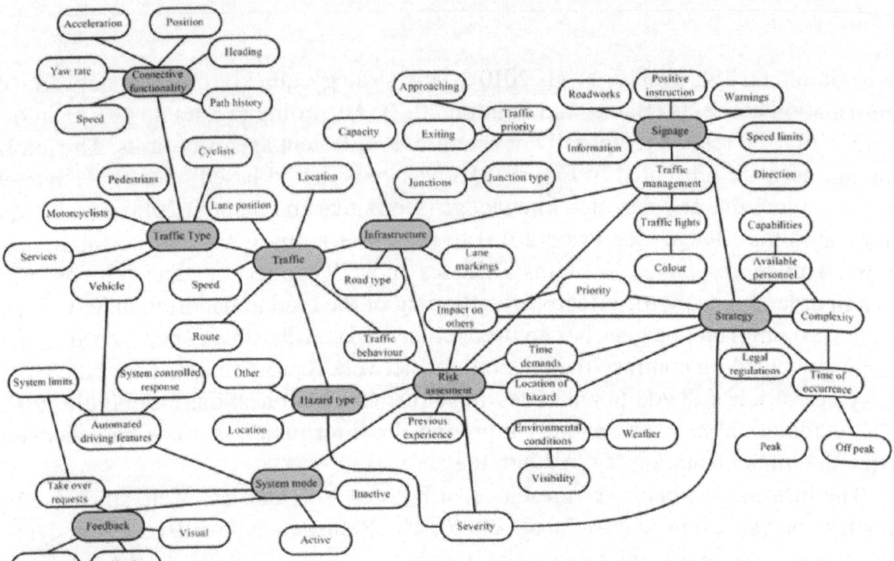

FIGURE 10.7 Hypothetical information network for CAV networks (note that 'key' nodes highlighted in grey are based upon the calculation of sociometric status – see Table 10.5).

TABLE 10.4
Comparison of Network Metrics Relating to the Information Networks for Non-CAV and CAV Networks

Information Network Analysis	Non-CAV	CAV
Number of nodes	53	71
Number of connections	119	185
Diameter	8	8
Density	0.04	0.04

TABLE 10.5
Key Informational Nodes Based Upon Rank Sociometric Status Scores

	Non-CAV	CAV
1	Strategy	Connective functionality
2	Risk assessment	Strategy
3	Traffic	Risk assessment
4	Signage	Traffic
5	Infrastructure	Signage
6	Traffic type	Infrastructure
7	Hazard type	Traffic type
8	Time of occurrence	System mode
9	Traffic management	Feedback
10	Complexity	Hazard type

or more connections should be considered to play a central role in the knowledge base that underpins effective functioning of the system (Banks and Stanton 2018). Table 10.5 shows that key terminologies within non-CAV and CAV networks remain largely the same, although there is greater emphasis upon 'connectivity', particularly in relation to system mode and feedback in CAV networks.

DISCUSSION

This chapter presents a practical method for analysing and evaluating DCOG at a macro level. Unlike previous applications that have tended to explore DCOG in meso-level systems, the authors of this chapter have taken a much 'broader' systems view. Rather than isolating a sub-system operating within a much larger sociotechnical system, the authors have taken the road transportation system as their 'unit of analysis' (Grote et al. 2014). This in turn provides a macro-level account of DCOG that has enabled us to explore the dependencies and interdependencies that exist between human and non-human agents across different operational domains for both non-CAV and CAV networks.

Inspired by Hutchins (1995), the authors have shown how cognitive processes can be distributed socially across system agents and internal and external tasks and through time (acknowledging that earlier processes may transform the nature of later processing) using EAST. Although EAST has been primarily used to focus on meso-level activities within varied domains (e.g. Baber et al. 2013; Banks and Stanton 2016; Stanton et al. 2006; Stanton 2014; Walker et al. 2006), this chapter shows that network representations can also provide a means to explore macro-level DCOG within much larger sociotechnical systems. Rather than focusing upon an individual component of a system (for example, the RE), the authors have considered other dependencies and interdependencies spanning multiple operational environments (i.e. the inclusion of TMC and EA).

EAST was used to describe non-CAV and CAV networks in terms of task, social and information networks to help improve our knowledge of how these systems may work in practice. Whilst the network representations presented in this chapter are by no means exhaustive, they do provide a basis for discussion into (a) the agents that play a role in our road transportation system, (b) how these agents are connected to one another, (c) the types of communication or information that are shared within the network and (d) how this may change as a result of CAV implementation. Regardless, any new conceptualisation based upon future transportation systems will need to consider the changes in these sociotechnical networks.

The combined task and social network approach using the EAST framework can provide insight into DCOG within the transportation network by identifying 'who' is doing 'what'. It demonstrates that even though the underlying task network for the transportation system remains the same, a connected and autonomous system can lead to multiple system agents completing the same tasks in parallel. As long as CAV architecture complements existing infrastructures and frameworks that uphold non-CAV networks, tasks spanning the RE, TMC and EA categories could all benefit from CAV implementation. Specifically, CAV should increase the efficiency and reliability of network operations for all social agents involved (Talebpour and Mahmassani 2016). However, there are a number of issues that must be considered in order to make CAV networks a success. For example, there remain issues relating to cybersecurity, compatibility of data sharing between different cloud computing platforms and 'car hacking' (e.g. Checkoway et al. 2011; Frost and Sullivan 2014). With non-human agents set to become the most prominent agents within a CAV network, designers must ensure that cloud computing platforms are secure, reliable and efficient. They must be capable of holding, processing, selecting and re-transmitting relevant information to relevant social agents. A failure within the network could, after all, have far-reaching implications in terms of RE, TMC and EA operations. For example, a loss of network coverage could mean information is shared intermittently, which will affect the reliability and efficiency of the network. It is therefore important to consider the differences between CAV and non-CAV networks, because until the network is fully connected (across space and time), transitions between CAV and non-CAV operation are still likely to occur.

Quantitative analysis of networks using network metrics enables systematic assessment of existing systems (i.e. non-CAV) and future systems (i.e. CAV). The network analysis presented in this chapter is rudimentary but arguably provides insight into the distributed nature of system function, which has not previously been explored. Network metrics have been used to quantitatively analyse the underlying differences between non-CAV and CAV networks. Sociometric status, in particular, has been useful in identifying key agents or nodes within the networks as well as highlighting how the implementation of CAV can change the prominence of individual agents or nodes. These methods and tools provide a means to explore possible strengths and weaknesses within system function. In the future, EAST may be used to provide further insight into potential network resilience issues via a 'broken-links' approach (Stanton and Harvey 2017). After all, it is likely that failure in one part of the CAV network is likely to have a substantial knock-on effect in others, especially if human agents become accustomed to CAV functionality. Perceptions of competence and expectancy can all impact upon the way in which agents interact on the road (e.g. Rudin-Brown and Parker 2004).

CONCLUSIONS

Vehicle automation and the connected services that they offer appear set to revolutionise the way in which people interact with and behave within the road transportation network. Essentially, CAV points to a future whereby most, if not all, of the potential communication links between vehicles and infrastructure within the RE may be connected. Thus, rather than CAV eliminating ergonomics and human factors issues from the transportation network entirely, it could actually increase the importance of them as it highlights the types of activities human and non-human agents will engage in. Whilst networked simulators exist, enabling multiple drivers to interact in a single virtual environment, research outputs are limited to a single operational environment (e.g. Sawyer and Hancock 2012). Thus, it has only been possible to explore how intelligent transport systems, such as CAV, will impact individual agents involved within the road environment. EAST goes further than this and provides a framework that can enable researchers to visualise the impact of CAV on a much larger scale. Adopting sociotechnical systems design in this way could offer a means to improve the performance of the whole system rather than individual components in isolation (Stanton 2014). Modelling system interaction between system agents can provide an extensive qualitative overview of DCOG within large-scale network operations.

REFERENCES

Atkins Mobility. (2016). Connected & autonomous vehicles: Introducing the future of mobility. Available at http://www.atkinsglobal.co.uk. [Accessed 10.01.2017].

Baber, C., Stanton, N. A., Atkinson, J., McMaster, R. and Houghton, R. J. (2013). Using social network analysis and agent-based modelling to explore information flow using common operational pictures for maritime search and rescue operations. *Ergonomics*, 56(6), 889–905.

Banks, V. A. and Stanton, N. A. (2016). Driver-centred vehicle automation: Using network analysis for agent-based modelling of the driver in highly automated driving systems. *Ergonomics*, 59(11), 1442–1452.

Banks, V. A. and Stanton, N. A. (2018). Analysis of driver roles: Modelling the changing role of the driver in automated driving systems using EAST. *Theoretical Issues in Ergonomics Science*, (In press).

Banks, V. A., Stanton, N. A. and Harvey, C. (2014). Sub-systems on the road to vehicle automation: Hands and feet free but not 'mind' free driving. *Safety Science*, 62, 505–514.

Benta, M. (2005). Studying communication networks with AGNA 2.1. *Cognition Brain Behaviour*, 9, 567–574.

Checkoway, S., McCoy, D., Kantor, B., Anderson, D., Shacham, H., Savage, S., Koscher, K., Czeskis, A., Roesner, F. and Kohno, T. (2011). Comprehensive experimental analyses of automotive attack surfaces. *Proceedings of USENIX Security Symposium*. San Francisco, CA.

Christoffersen, K. and Woods, D. D. (2002). How to make automated systems team players. *Advances in Human Performance and Cognitive Engineering Research*, 2, 1–12.

De La Fortelle, A. (2005). Cybercars: Past, present and future of the technology. 12th Word Congress on Intelligent Transport Systems. San Francisco, CA.

Eriksson, A. and Stanton, N. A. (2017). Takeover time in highly automated vehicles: Noncritical transitions to and from manual control. *Human Factors*, 59(4), 689–705. DOI:10.1177/0018720816685832.

European Road Transport Research Advisory Council (ERTRAC). (2015). ERTRAC automated driving roadmap, July 2015. [Accessed 12.13.2016].

Frost and Sullivan. (2014). Cybersecurity in the automotive industry. 30 October. https://www.ertrac.org/uploads/documentsearch/id38/ERTRAC_Automated-Driving-2015.pdf [Accessed 01.09.2017].

Grote, G., Stanton, N. A. and Weyer, J. (2014). Beyond human-centred automation. *Ergonomics*, 57(3), 289–294.

Harary, F. (1994). *Graph Theory*. Addison-Wesley: Reading, MA.

Hollan, J., Hutchins, E. and Kirsh, D. (2000). Distributed cognition: Toward a new foundation for human–computer interaction research. *ACM Transactions on Computer Human Interaction*, 7, 174–196.

Houghton, R. J., Baber, C., McMaster, R., Stanton, N. A., Salmon, P. M., Stewart, R. and Walker, G. H. (2006). Command and control in emergency services operations: A social network analysis. *Ergonomics*, 49, 1204–1225.

Hutchins, E. (1995). How a cockpit remembers its speeds. *Cognitive Science*, 19(3), 265–288.

KPMG. (2015). Connected and autonomous vehicles – The UK economic opportunity. Available at: https://www.smmt.co.uk/wp-content/uploads/sites/2/CRT036586F-Connected-and-Autonomous-Vehicles-%E2%80%93-The-UK-Economic-Opportu.1.pdf.

Lavrinc, D. (2014). This is how bad self-driving cars suck in the rain. Available at: http://jalopnik.com/this-is-how-bad-self-driving-cars-suck-in-the-rain-.

Louw, T. and Merat, N. (2017). Are you in the loop? Using gaze dispersion to understand driver visual attention during vehicle automation. *Transportation Research Part C: Emerging Technologies*, 76, 35–50.

Price, J. (2016). Human factors in the design of traffic management systems. University of Southampton, Faculty of Engineering and the Environment, London. EngD thesis.

Rudin-Brown, C. M. and Parker, H. A. (2004). Behavioural adaptation to adaptive cruise control (ACC): Implications for preventative strategies. *Transportation Research Part F*, 7(2), 59–76.

Salmon, P. M., Lenné, M. G., Walker, G. H., Stanton, N. A. and Filtness, A. (2014). Using the event analysis of systemic teamwork (EAST) to explore the conflicts between different road user groups when making right hand turns at urban intersections. *Ergonomics*, 57 (11), 1628–1642.

Salmon, P. M., Stanton, N. A. and Young K. L. (2012). Situation awareness on the road: review, theoretical and methodological issues, and future directions. *Theoretical Issues in Ergonomics Science*, 13(4), 472–492.

Sawyer, B. D. and Hancock, P. A. (2012). Development of a linked simulation network to evaluate intelligent transportation system vehicle to vehicle solutions. *Proceedings of the Human Factors and Ergonomics Society Annual Meeting*, 56(1), 2316–2320.

Science and Technology Select Committee. (2017). Connected and autonomous vehicles: The future? Available at: https://www.publications.parliament.uk/pa/ld201617/ldselect/ldsctech/115/115.pdf.

Society for Automotive Engineers. (2016). Taxonomy and definitions for terms related to on-road motor vehicle automated driving systems. http://standards.sae.org/j3016_201401/ [Accessed 12.12.2016].

Sorensen, L. J., Stanton, N. A. and Banks, A. P. (2011). Back to SA school: Contrasting three approaches to situation awareness in the cockpit. *Theoretical Issues in Ergonomics Science*, 12(6), 451–471.

Stanton, N. A. (2014). Representing distributed cognition in complex systems: How a submarine returns to periscope depth. *Ergonomics*, 57(3), 403–418.

Stanton, N. A. (2015). Responses to autonomous vehicles. *Ingenia*, 62, 9.

Stanton, N. A., Baber, C. and Harris, D. (2008). *Modelling Command and Control: Event Analysis of Systemic Teamwork*. Ashgate: Aldershot, UK.

Stanton, N. A. and Harvey, C. (2017). Beyond human error taxonomies in assessment of risk in sociotechnical systems: A new paradigm with the EAST 'broken-links' approach. *Ergonomics,* 60(2), 221–233.

Stanton, N. A. and Roberts, A. P. J. (2017a). Examining task, social and information networks in submarine command and control. *IEEE Transactions on Human-Machine Systems*, DOI: 10.1109/THMS.2017.2720659.

Stanton, N. A., Roberts, A. P. J. and Fay, D. T. (2017b). Up periscope: Understanding submarine command and control teamwork during a simulated return to periscope depth. *Cognition, Technology and Work*, 19 (2–3), 399–417.

Stanton, N. A., Salmon, P. M., Rafferty, L., Walker, G. H., Baber, C. and Jenkins, D. P. (2013). *Human Factors Methods: A Practical Guide for Engineering and Design*. Ashgate: Aldershot, UK.

Stanton, N. A., Salmon, P. M. and Walker, G. H. (2015). Let the reader decide: A paradigm shift for situation awareness in sociotechnical systems. *Journal of Cognitive Engineering and Decision Making*, 9 (1), 44–50.

Stanton, N. A., Stewart, R., Harris, D., Houghton, R. J., Baber, C., McMaster, R., Salmon, P. M. et al. (2006). Distributed situation awareness in dynamic systems: Theoretical development and application of an ergonomics methodology. *Ergonomics*, 49 (12–13), 1288–1311.

Talebpour, A. and Mahmassani, H. S. (2016). Influence of connected and autonomous vehicles on traffic flow stability and throughput. *Transportation Research Part C: Emerging Technologies*, 71, 143–163.

Walker, G. H., Gibson, H., Stanton, N. A., Baber, C., Salmon, P. and Green, D. (2006). Event analysis of systematic teamwork (EAST): A novel integration of ergonomic methods to analyse C4i activity. *Ergonomics*, 49 (12–13), 1345–1369.

Walker, G. H., Stanton, N. A., Baber, C., Wells, L., Gibson, H., Salmon, P. and Jenkins, D. (2010). From ethnography to the EAST method: A tractable approach for representing distributed cognition in Air Traffic Control. *Ergonomics*, 53 (2): 184–197.

Walker, G. H., Stanton, N. A. and Salmon, P. M. (2015). *Human Factors in Automotive Engineering and Technology*. Ashgate: Aldershot, UK.

Walker, G. H. Stanton, N. A. and Salmon, P. M. (2011). Cognitive compatibility of motorcyclists and car drivers. Accident Analysis & Prevention, 43(3), 878–888.

Walker, G. H., Stanton, N. A., Salmon, P. M., Jenkins, D. P., Monnan, S. and Handy, S. (2012). Communications and cohesion: a comparison between two command and control paradigms, Theoretical Issues in Ergonomics Science, 13(5), 508–527.

Weisstein, E. W. (2008). Graph diameter. From MathWorld – A Wolfram Web Resource. Available at: http://mathworld.wolfram.com/GraphDiameter.html. [Accessed 6 October 2017].

Zeeb, K., Buchner, A. and Schrauf, M. (2015). What determines the take-over time? An integrated model approach of driver take-over after automated driving. *Accident Analysis & Prevention*, 78, 212–221.

Section III

Future Developments in EAST

11 STAMPING on EAST

Adding a Control Network to EAST to Examine the Safety Controls in the Railway Level Crossing System Lifecycle

With Gemma J. M. Read, Natassia Goode, Eryn Grant, Clare Dallat, Tony Carden and Anjum Naweed

INTRODUCTION

The popularity of systems thinking is now such that various systems ergonomics methods are being applied across the transport domains, including rail (Salmon et al. 2016b; Stefanova et al. 2015), road (Parnell et al. 2017; Salmon et al. 2016a), aviation (Allison et al. 2017) and maritime transport (Lee et al. 2017). In the case of rail transport specifically, these methods have been applied to a diverse set of issues, including railway level crossings (RLXs) (Salmon et al. 2016b; Stefanova et al. 2015), ticketing system design (Read et al. 2015), accident analysis (Chen et al. 2015; Ouyang et al. 2010; Salmon et al. 2013), signals passed at danger (Madigan et al. 2016) and train cabin design (Jansson et al. 2006). Given the increasing complexity of rail transport systems, it is likely that systems ergonomics methods will increasingly be applied for analysis and design in rail systems worldwide.

Although all of these methods share a common alignment with systems theory, most have differing theoretical underpinnings, giving rise to their own unique analysis approach. Systems Theoretic Accident Model and Process (STAMP), for example, is based on control theory and seeks to identify what control structure is in place in a given system and, in the event of an accident, how the adaptive control function within the system failed. Event Analysis of Systemic Teamwork (EAST), on the other hand, focuses on the relationship between tasks, agents and information and uses networks to identify issues that impact performance (e.g. information bottlenecks, overloaded agents). Further differences can be found when considering the focus of other systems ergonomics methods such as Hierarchical Task Analysis (goals), Cognitive Work Analysis (CWA) (constraints) and Functional Resonance Analysis Method (FRAM) (performance variability and resonance). As a result, each systems ergonomics method brings with it a unique set of strengths and weaknesses, and there has been much discussion on the potential benefits of converging different methods together (Dallat et al. in press; Houghton et al. 2015; Naweed 2014; Salmon et al. 2010, 2011; Stanton et al. 2013, 2016; Patriarca et al. 2017).

Researchers and practitioners are therefore beginning to explore the utility of integrating systems ergonomics methods. Patriarca et al. (2017) combined the abstraction hierarchy from CWA with FRAM, suggesting that using the two in an integrated manner enables additional analyses of intra-agent and intra-system-level interactions. Houghton et al. (2015) combined social network analysis (SNA) and CWA to examine military planning tasks, concluding that using the two approaches together enables a richer understanding of system structure. Stanton et al. (2016) integrated CWA and SNA to compare different cockpit crewing configurations and their likely impact on safety on future flight decks. CWA was used to define system constraints, whereas SNA metrics were used as a measure of system resilience. Given that modern-day systems are stretching the capabilities of our methodological toolkit (Salmon et al. 2017; Walker et al. 2017), and the fact that many methods do not fully align with systems thinking, it is our opinion that the integration of systems ergonomics methods represents an important area of investigation for our discipline. This builds on a long history and tradition of methods integration in ergonomics research and practice (e.g. Kirwan 1992; Stanton et al. 2005, 2013).

In the case of EAST, it is apparent that adding further forms of network may enhance the utility of the framework. Candidate additional EAST networks that the authors have discussed include a network of safety controls, a network of unexpected emergent behaviours and a risk network. This chapter investigates the utility of adding a safety controls network based on principles from the STAMP method (Leveson 2004). To test this, we undertook a case study in which EAST and an additional control network were used to examine safety management during the design and operation of RLXs systems in Victoria, Australia. The four networks were used to examine the system and its safety controls with a view to identifying recommendations for strengthening safety management at RLXs.

RESEARCH CONTEXT: RAILWAY LEVEL CROSSINGS

Collisions between trains and vehicles at RLXs remain a persistent road and rail safety issue worldwide. In Australia, there were 97 fatalities resulting from collisions between trains and vehicles at RLXs between 2000 and 2009 and 92 collisions between trains and pedestrians at RLXs between 2002 and 2012 (ATSB 2012; Independent Transport Safety Regulator 2011). Whilst Australia is used here as an example, the problem is of a similar magnitude in the United States of America and Europe (Evans 2011; Federal Railroad Administration 2013).

On top of the personal and social impacts of RLX collisions, the financial burden is also significant. In 2010, the annual cost of RLX incidents in Australia was estimated at just over one hundred million dollars, taking into account human and property damage costs and other costs such as emergency service attendance, delays, investigation and insurance (Tooth and Balmford 2010). Given the combined personal, societal and economic burden created by RLX collisions, they continue to represent an area of strategic importance for road and rail safety authorities (Read et al. 2017).

Analyses of fatal collisions provide evidence that RLX collisions have a complex web of interrelated causes spanning overall road and rail systems. In short, they are a systems problem (Read et al. 2017; Salmon et al. 2016b; Stefanova et al. 2015). This has important implications for the methods used to understand and improve RLX safety. Describing and understanding RLX systems requires analysis methods that consider the behaviour of overall road and rail systems (Read et al. 2013; Wilson and Norris 2014). Key elements requiring analysis include

- the tasks that are undertaken across the system as part of RLX design and operation
- the diverse set of agents that reside within the RLX system and undertake design and operation tasks
- the interactions between agents (human and non-human) when undertaking these tasks
- the information required for successful completion of these tasks
- the factors influencing behaviour across the system, including safety controls, environmental factors and financial and production pressures

It is apparent that no one method can completely fulfil these requirements. EAST, for example, considers tasks, agents and information but does not explicitly consider safety controls and their impact on behaviour. STAMP, on the other hand, considers controls but does not detail the tasks and information required when enacting controls. It is our contention, then, that a richer analysis of sociotechnical systems behaviour can be achieved by integrating different systems ergonomics methods.

The study was undertaken to identify the range of factors influencing behaviour and safety in RLX systems and to identify opportunities for introducing new interventions design to improve safety. To achieve this, the EAST framework and STAMP control structure methods were integrated and used together to examine a RLX system 'lifecycle', including all activities involved in the design, implementation, operation and removal (for grade separation) of RLXs in Victoria, Australia. A short description of how EAST and the STAMP control structure principles were integrated is given below.

PART 1: INTEGRATING EAST AND STAMP

STAMP and Control Theory

STAMP (Leveson 2004), originally developed as an accident analysis methodology, is underpinned by systems and control theory. The method takes the view that accidents result from the inadequate control or enforcement of safety constraints – when disturbances, failures or dysfunctional interactions between components are not handled by existing control mechanisms (Leveson 2004). In the RLX context, for example, the model might suggest that collisions involving vehicles and trains at RLXs occur when controls such as flashing lights and warning bells, education campaigns and road rules and regulations fail to prevent drivers from attempting to traverse the crossing when a train is present. STAMP therefore views safety as an issue of control and one that is managed through a control structure that has the goal of enforcing constraints on actors across the system.

It is worth noting that a broad view of control is adopted when using STAMP. Leveson (2004) describes various forms of control, including managerial, organisational, physical, operational and manufacturing-based controls; that is, system behaviour is controlled not only by engineered systems and direct intervention but also by policies, procedures, shared values and other aspects of the surrounding organisational and social culture. The first phase of STAMP involves building a control structure to describe the system under analysis and the control relationships that exist between actors and organisations related to both system design and system operation. The control structure model views systems as comprising interrelated components that maintain a state of dynamic equilibrium through feedback loops of control and information (Leveson 2004). Accordingly, control structure models incorporate a series of hierarchical system levels and describe the actors and organisations that reside at each level. Control and feedback loops are included to show what control mechanisms are enacted down the hierarchy and what information about the status of the system is sent back up the hierarchy.

A generic control structure model is presented in Figure 11.1 (Leveson 2004). The left-hand side of Figure 11.2 shows a generic control structure for system development, whereas the right-hand side shows a generic control structure for system

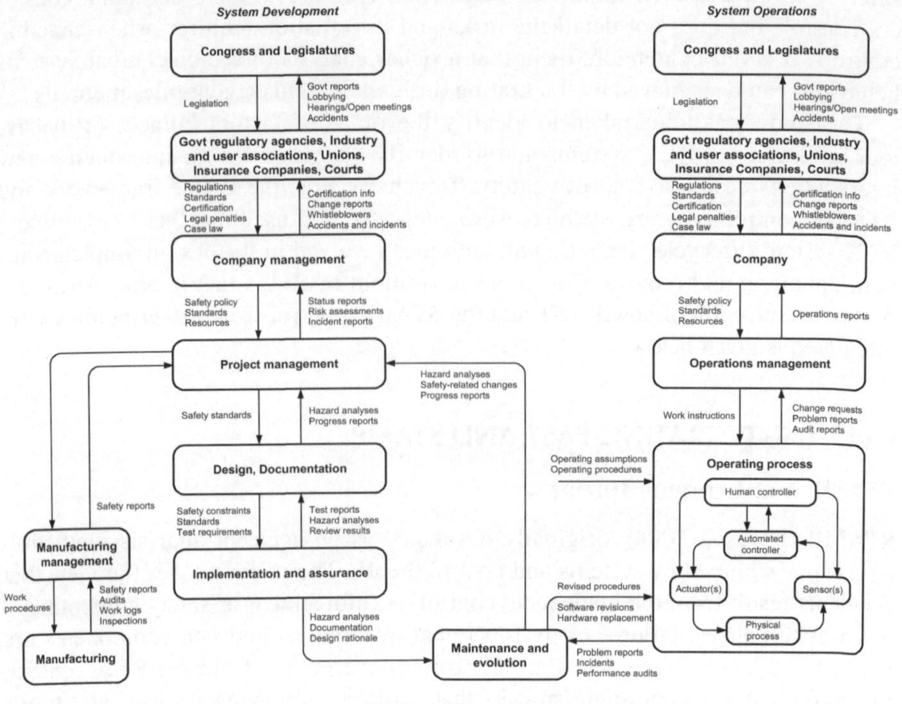

FIGURE 11.1 STAMP generic control structure model.

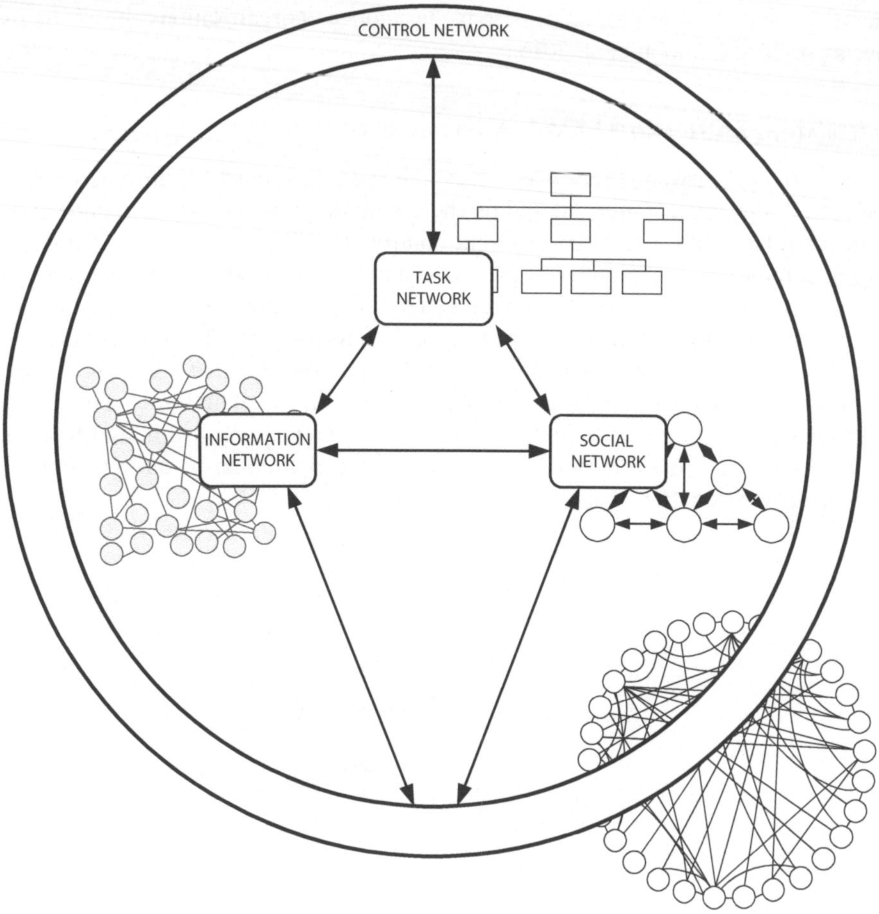

FIGURE 11.2 EAST network of network approach with added control network.

operation. The arrows flowing down the hierarchy represent control relationships (or reference channels, Leveson 2004), and the arrows flowing up the hierarchy represent feedback loops (or measuring channels, Leveson 2004). In relation to the example described above, the rail level crossing infrastructure would enact the control of 'warnings' on road users. In turn, information regarding collisions, near misses, violations of the warnings and road rules and maintenance requirements for the warning infrastructure represent forms of feedback that would be passed back up the system. These feedback loops enable road and rail safety authorities and policy makers to assess the status of controls and whether they are having the desired impact.

The control structure has been employed as a modelling tool to describe systems and the control and feedback loops involved in safety management (Salmon et al. 2016a). Whilst the usefulness of the control structure has been reported, there are some notable limitations: (1) it describes control and feedback loops only and does not examine the relationships between controls; (2) it is difficult to represent control and feedback loops that exist between actors and/or organisations *within* levels of the hierarchy; and (3) it is often difficult to place actors and organisations across the five levels provided (Salmon et al. 2016a).

Integrating EAST and STAMP: A Network of Controls

Integration of the two methods was achieved by adding a fourth 'control network' to the EAST framework. This is based on the notion that it is important to understand both the relationships between different controls (i.e. design standards dictate the nature of the warnings provided at the RLX) and the relationships between controls, tasks, agents and information (i.e. enacting a particular control requires certain tasks to be undertaken by certain agents using specific information). The new four-network EAST approach is represented in Figure 11.2. It can be seen that conceptually in a given system, the control network sits around the task, social and information triad of networks, with controls acting to constrain the interactions within and between these original networks. Figure 11.3 provides a representation of the relationships between controls, tasks, agents and information.

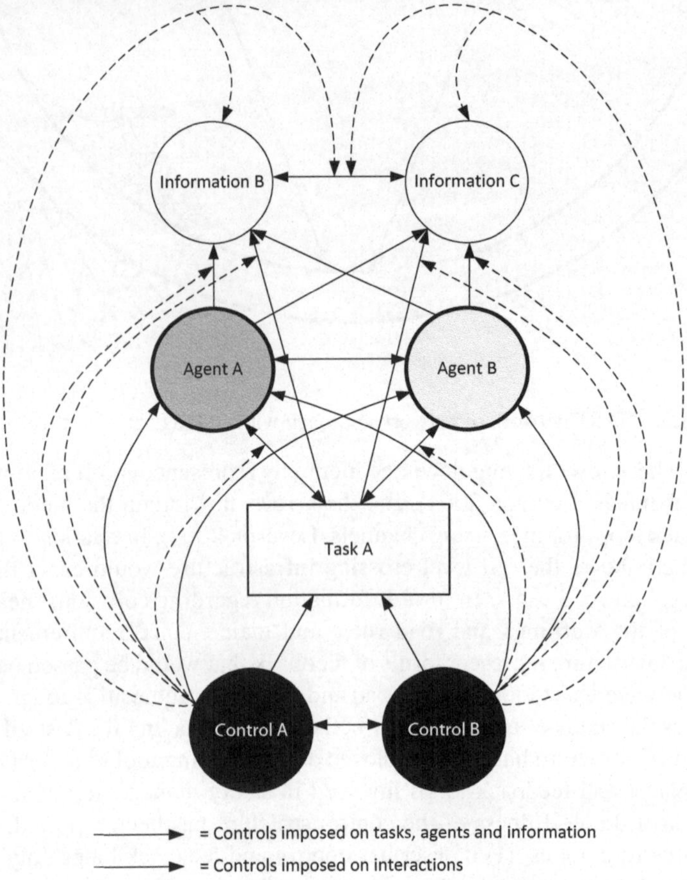

FIGURE 11.3 Interaction of controls, tasks, agents and information.

For example, the task of negotiating a RLX safely is a task undertaken by both road vehicle drivers (Agent A) and train drivers (Agent B). When negotiating the RLX, both use various pieces of information such as the status of the warning devices, the speed and location of the train and the behaviour of other users such as pedestrians. Road vehicle drivers and train drivers are bound by controls such as the warnings themselves, rules and regulations and standard operating procedures. Likewise, the nature of information they are using is influenced by controls such as design standards (e.g. timing of warnings). It is important to note that controls are enacted on tasks, agents and information and also on the interactions between them. For example, various controls at the RLX (e.g. boom gates, flashing lights, train horn, road rules and regulations) are designed to manage the interactions between drivers, their vehicle and train drivers and the train.

PART 2: APPLYING THE INTEGRATED METHOD TO RLX SAFETY MANAGEMENT

METHODOLOGY

Data Inputs

Seven analysts with significant experience in applying either EAST or STAMP in a range of domains (e.g. defence, road and rail transport, aviation, maritime transport) were involved in conducting the analysis. The analysts had extensive experience with and background information on rail safety (see Naweed et al. 2015; Salmon et al. 2016b; Stanton and Walker 2011) and, more specifically, the Victorian RLX system, as they had recently completed a four-year research programme involving various analyses of different RLXs as well as the wider RLX sociotechnical system (see Read et al. 2017; Salmon et al. 2013, 2016b). This background knowledge and information was also supplemented by an additional documentation review (e.g. road and rail safety strategy and policy documents, crash investigation reports, relevant academic literature).

EAST Network Development

Initially task, social and information networks were developed in a workshop setting involving four analysts. Two of the analysts had extensive experience in applying EAST in a range of transportation settings (Salmon et al. 2014; Stanton 2014; Stanton et al. 2016; Walker et al. 2006, 2010). The remaining two analysts were human factors researchers with significant experience of applying systems ergonomics methods in a range of safety critical domains (e.g. Newnam and Goode 2015; Read et al. 2016; Salmon et al. 2014). One of these analysts also had significant experience in the area of rail safety (having worked for 9 years in rail safety regulation and on various rail safety research projects).

Prior to the workshop, participants were provided with the following materials to support network construction:

1. *Hierarchical Task Analysis (HTA) of the RLX system lifecycle.* A HTA (Stanton 2006) of the RLX system lifecycle was developed based on the data described above. The HTA describes the goals, sub-goals and operations

required when designing, implementing, operating and removing RLXs in Victoria, Australia.

2. *ActorMap of rail level crossing stakeholders*. An ActorMap was developed to identify the stakeholders involved in the RLX system lifecycle (Read et al. 2017). The resulting ActorMap showed which stakeholders currently share the responsibility for RLX system lifecycle of RLXs in Victoria, Australia. The stakeholders were placed across a hierarchy of systems levels in line with Rasmussen's risk management framework (Rasmussen 1997).

The task, social and information networks were developed based on a process of first identifying nodes, then identifying relationships between the nodes and finally by reviewing the nodes and relationships for internal consistency. The analysis rules surrounding what constituted a node and a relationship for each network representation are presented in Table 11.1.

Following the workshop, a draft control network was constructed by two analysts based on a review of Salmon et al.'s (2016a) in-depth analysis of rail level crossing systems; Salmon et al.'s (2016b) road safety control structure; and a STAMP control structure model of rail safety under development by the authors as part of a separate project. Development of the control network followed the process above whereby nodes (controls) were first identified followed by the relationships between them (see Table 11.1 for examples). The draft control structure was subsequently reviewed by the other analysts from the original workshop.

Network Analysis

Each network was analysed using quantitative metrics that have previously been used to interrogate EAST networks (e.g. Salmon et al. 2014; Stanton 2014). In the present study, the following metrics were applied to each network:

1. *Network density (overall network)* – Network density represents the level of interconnectivity of the network in terms of relations between nodes. Density is expressed as a value between 0 and 1, with 0 representing a network with no connections between nodes and 1 representing a network in which every node is connected to every other concept (Kakimoto et al. 2006; cited in Walker et al. 2011). Higher density values are indicative of a well-connected network in which tasks, agents, information and controls are tightly coupled.
2. *Sociometric status (individual nodes)*. Sociometric status provides a measure of how 'busy' a node is relative to the total number of nodes within the network under analysis (Houghton et al. 2006). In the present analysis, nodes with sociometric status values greater than the mean sociometric status value plus one standard deviation are taken to be 'key' (i.e. the most connected) nodes within each network. These nodes represent either key tasks, agents, pieces of information or controls. For example, in the case of the control network, the node with the highest sociometric status is the control that is the most interrelated with other controls.
3. *Centrality (individual nodes)*. Centrality is used to examine the standing of a node within a network based on its geodesic distance from all other

TABLE 11.1
Analysis Rules Regarding Relationships between Nodes in EAST Networks in the Context of RLX Design, Implementation, Operation or Removal

Network	Nodes	Relationships	Examples
Task network	Represent high-level tasks that are required during the design, implementation, operation and removal of RLXs. High-level tasks were extracted from the sub-ordinate goals level of the HTA	Represent instances where the conduct of one high-level grouping of tasks (i.e. task network node) influences, is undertaken in combination with, or is dependent on, another group of tasks	The nodes 'Select site for upgrade' and 'Announce upgrade programme' are linked because the upgrade programme announcement cannot be made until sites have been selected
Social network	Represent human, technological or organisational agents who undertake one or more of the tasks involved in the design, implementation, operation and removal of RLXs (as identified in the HTA and task network)	Represent instances where agents within the social network interact with one another during the design, implementation, operation or removal of RLXs	The nodes 'RLX warnings and barriers' and 'Road users' are linked as the warnings and barriers communicate with road users to inform them that a train is approaching
Information network	Represent grouped categories of information that is required by agents when undertaking RLX the design, implementation, operation and removal tasks (as identified in the task and social network)	Represent instances where information influences other information or is used in combination with other information in the network during the design, implementation, operation or removal of RLXs	The nodes 'Traffic' and 'Separation' are linked as road vehicles and trains require separation from one another (i.e. Traffic 'requires' Separation)
Control network	Represent controls that are enacted on the tasks, agents, information and interactions during RLX design, implementation, operation and removal. A broad view of controls is adopted to incorporate both engineering, managerial and social controls	Represents instances where controls are either used together or where one control influences the nature and/or enactment of another control	The nodes 'Design standards', 'Boom gates' and 'Flashing lights' are linked as the design standard dictates the characteristics of both warnings in terms of form, presentation and timing

nodes in the network (Houghton et al. 2006). Central nodes represent those that are closer to the other nodes in the network as, for example, information passed from one to another node in the network would travel through less nodes. Houghton et al. (2006) point out that well-connected nodes can still achieve low centrality values as they may be on the periphery of the network. For example, in the case of the control network nodes, higher centrality status values are those that are closest to all other controls in the network as they have direct rather than indirect links with them.

RESULTS

Task Network

The RLX system lifecycle task network is presented in Figure 11.4. The outcomes for the task network analysis are presented in Table 11.2. For the sociometric status and centrality analysis, Table 11.2 includes the values for the nodes that scored above the mean sociometric status and centrality values for the network. Nodes that achieved values above the mean + standard deviation for sociometric status and centrality are shaded grey.

As shown in Figure 11.4, the RLX system lifecycle is underpinned by a network of 15 high-level tasks. Analysis of the task network in Table 11.2 reveals a network density of 0.2, which is indicative of a relatively loosely connected network (i.e. tasks are not tightly coupled). The sociometric status and centrality analyses reveal that 'Risk management' and 'Monitor performance' are the key tasks within the network.

Social Network

The RLX system lifecycle social network is presented in Figure 11.5. The outcomes for the SNA are presented in Table 11.3. Table 11.3 follows the same convention as Table 11.2 for the sociometric status and centrality values.

As shown in Figure 11.5, a diverse set of 27 agents undertake tasks throughout the RLX system lifecycle and operation. These agents range from RLX components (e.g. barriers, flashing lights), RLX users (e.g. drivers, pedestrians, road vehicles, trains), infrastructure owners and rail operators to the regulator, government departments, unions, courts and the media. Analysis of the social network shown in Table 11.3 reveals a network density of 0.18, which is again indicative of a relatively loosely connected network in which agents are not tightly coupled. The sociometric status and centrality analysis reveals that 'Government departments', 'Rail operators', 'Rail infrastructure managers', 'Regulators' (sociometric status) and 'Train drivers' (Centrality only) are key agents in the RLX system lifecycle.

Information Network

The RLX system lifecycle information network is presented in Figure 11.6. The outcomes for the information network analysis are presented in Table 11.4. Table 11.4 follows the same convention as Table 11.2 for the sociometric status and centrality values.

As shown in Figure 11.6, 28 information nodes are required throughout the RLX system lifecycle. Analysis of the information network in Table 11.5 reveals a network density of 0.09, which is indicative of a very loosely connected network. The sociometric status analysis reveals that 'Risk level' and 'Design concepts' are the key nodes

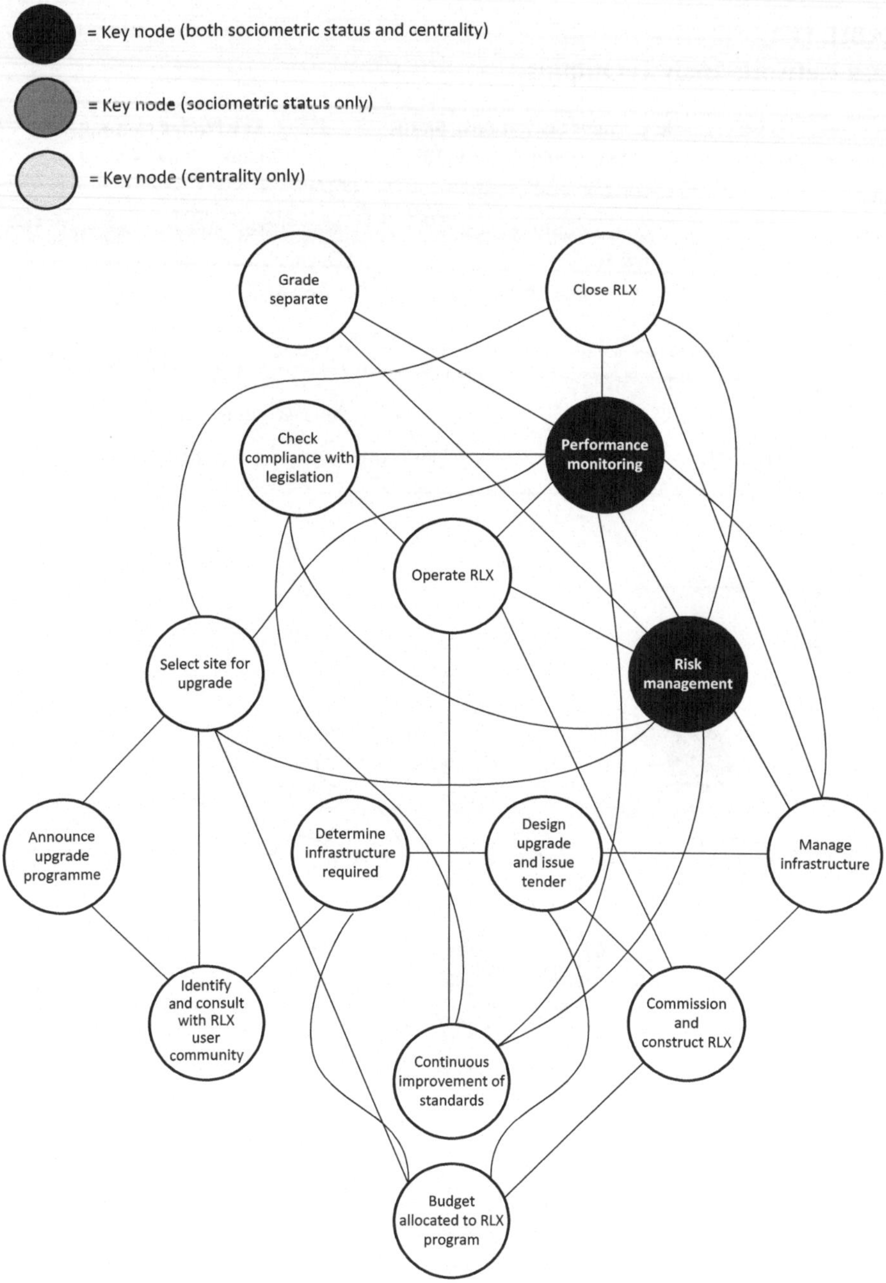

FIGURE 11.4 RLX system lifecycle task network.

TABLE 11.2
Task Network Analysis Outputs

Network Metrics	Key Nodes Sociometric Status (Mean = 0.4, SD = 0.22)	Key Nodes Centrality (Mean = 7.81, SD = 1.56)
Nodes = 15	Risk management = 1.0	Risk management = 10.88
Edges = 42	Monitor performance = 0.79	Monitor performance = 9.98
Density = 0.2	Operate RLX = 0.5	Operate RLX = 9.23
	Check compliance with legislation = 0.43	Close RLX = 8.96
	Select site for upgrade = 0.43	Check compliance with legislation = 8.6
	Manage infrastructure = 0.43	Select site for upgrade = 8.83
		Manage infrastructure = 8.12

Note: Cells Shaded Grey Represent Nodes with Values above the Mean + Standard Deviation.

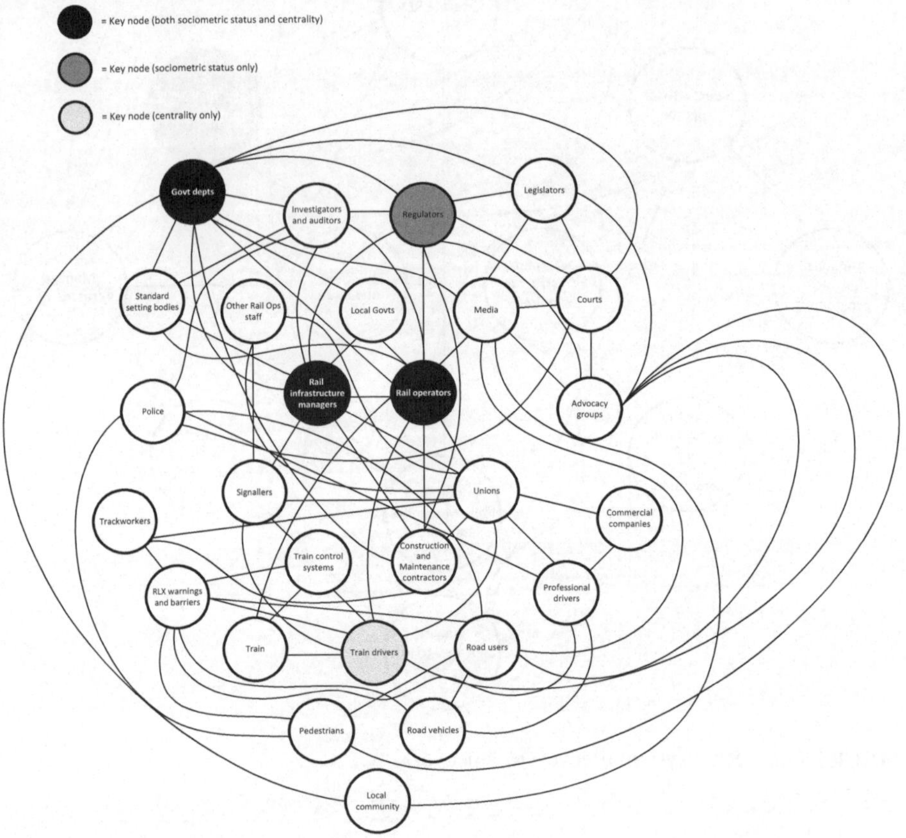

FIGURE 11.5 Social network for RLX system lifecycle.

TABLE 11.3
Social Network Analysis Outputs

Network Metrics	Key Nodes Sociometric Status (Mean = 0.37, SD = 0.17)	Key Nodes Centrality (Mean = 14.36, SD = 2.48)
Nodes = 27	Government depts. = 0.81	Government depts. = 17.39
Edges = 138	Rail operators = 0.74	Rail operators = 17.39
Density = 0.18	Rail infrastructure managers = 0.66	Rail infrastructure managers = 16.36
	Regulators = 0.52	Train drivers = 16.21
	Advocacy groups = 0.48	Investigators and auditors = 15.6
	Road users = 0.48	Advocacy groups = 15.45
	Investigators and auditors = 0.44	Regulators = 15.2
	RLX warnings and barriers = 0.44	Media = 14.39
	Train drivers = 0.44	Unions = 14.39
	Media = 0.44	Road users = 14.39
	Unions = 0.41	Police = 14.39
	Professional drivers = 0.41	Signallers = 14.39
	Legislators = 0.37	
	Courts = 0.37	

Note: Cells Shaded Grey Represent Nodes with Values above the Mean + Standard Deviation.

within the information network. The key nodes according to the centrality analysis are 'Budget', 'Demographics', 'Mitigations', 'Speed' and 'Surrounding land use'.

Control Network

The RLX system lifecycle control network is presented in Figure 11.7. The outcomes for the control network analysis are presented in Table 11.5. Table 11.5 follows the same convention as Table 11.2 for the sociometric status and centrality values.

As shown in Figure 11.7, 30 controls are present that constrain activities across the RLX system lifecycle. Analysis of the control network in Table 11.5 reveals a network density of 0.11, which is indicative of a relatively loosely connected network. The sociometric status analysis reveals that 'Audits and Inspections', 'Design standards' and 'Road rules' are the key controls within the control network. Key controls according to the centrality analysis are 'Insurance premiums', 'Timetables' and 'Train protection devices'. Other nodes that achieved sociometric status and centrality scores above the mean include controls relating to the RLX itself (e.g. 'Train', 'Vehicle', 'Signals'); safety strategy (e.g. 'Policy and strategy', 'Targets'); enforcement (e.g. 'Enforcement', 'Legal Penalties'); and education (e.g. 'Education', 'Initiatives').

DISCUSSION

This chapter presents the findings from a case study in which a control network was added to EAST to enable EAST to incorporate a control network analysis.

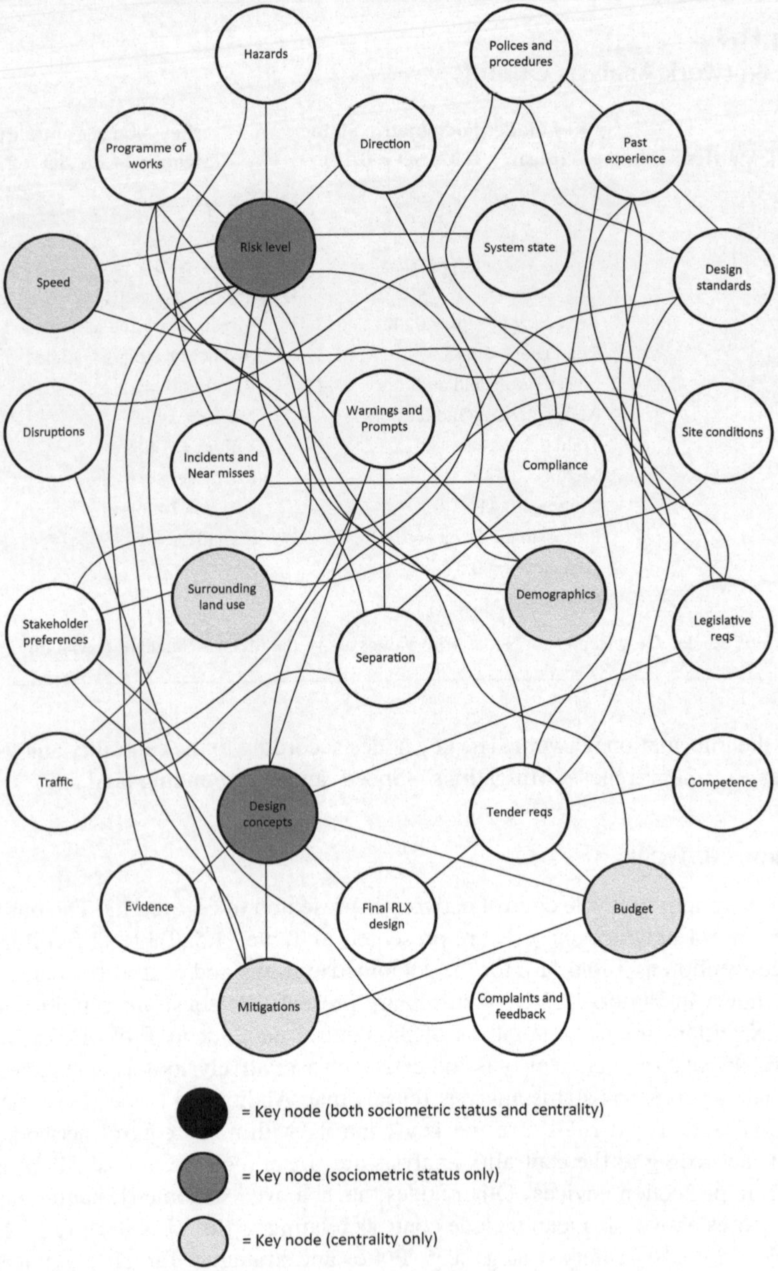

FIGURE 11.6 Information network for RLX system lifecycle.

TABLE 11.4
Information Network Analysis Outputs

Network Metrics	Key Nodes Sociometric Status (Mean = 0.17, SD = 0.1)	Key Nodes Centrality (Mean = 16.05, SD = 6.04)
Nodes = 28	Risk level = 0.55	Budget = 27.36
Edges = 65	Design concepts = 0.37	Demographics = 26.85
Density = 0.09	Tender requirements = 0.3	Mitigations = 26.35
	Past experience = 0.26	Speed = 23.72
	Incidents and near misses = 0.22	Surrounding land use = 22.23
	Mitigations = 0.22	Final RLX design = 21.89
	Separation = 0.19	Stakeholder preferences = 21.89
	Compliance = 0.19	System state = 21.56
	Warnings and prompts = 0.19	Site conditions = 20.62
	Policies and procedures = 0.19	Direction = 20.62
	Design standards = 0.19	Disruptions = 19.49
	Site conditions = 0.19	Complaints and feedback = 17.35
	Stakeholder preferences = 0.19	
	Legislative requirements = 0.19	

Note: Cells Shaded Grey Represent Nodes with Values above the Mean + Standard Deviation.

TABLE 11.5
Control Network Analysis Outputs

Network Metrics	Key Nodes Sociometric Status (Mean = 0.22, SD = 0.13)	Key Nodes Centrality (Mean = 18.70, SD = 14.74)
Nodes = 30	Audits and inspections = 0.66	Insurance premiums = 89.8
Edges = 97	Design standards = 0.59	Timetables = 34.53
Density = 0.11	Road rules = 0.37	Train protection devices = 37.41
	Train = 0.34	Signals = 29.93
	Initiatives = 0.31	Education = 23.63
	Vehicle = 0.28	
	Policy and strategy = 0.28	
	Targets = 0.28	
	Legal penalties = 0.24	
	Enforcement = 0.24	
	Education = 0.24	

Note: Cells Shaded Grey Represent Nodes with Values above the Mean + Standard Deviation.

The intention was to extend EAST's utility for analysing the behaviour of complex sociotechnical systems. The addition of a control network analysis component to EAST enables analysts to identify the system of controls present within a particular system and to examine how different controls are related to one another.

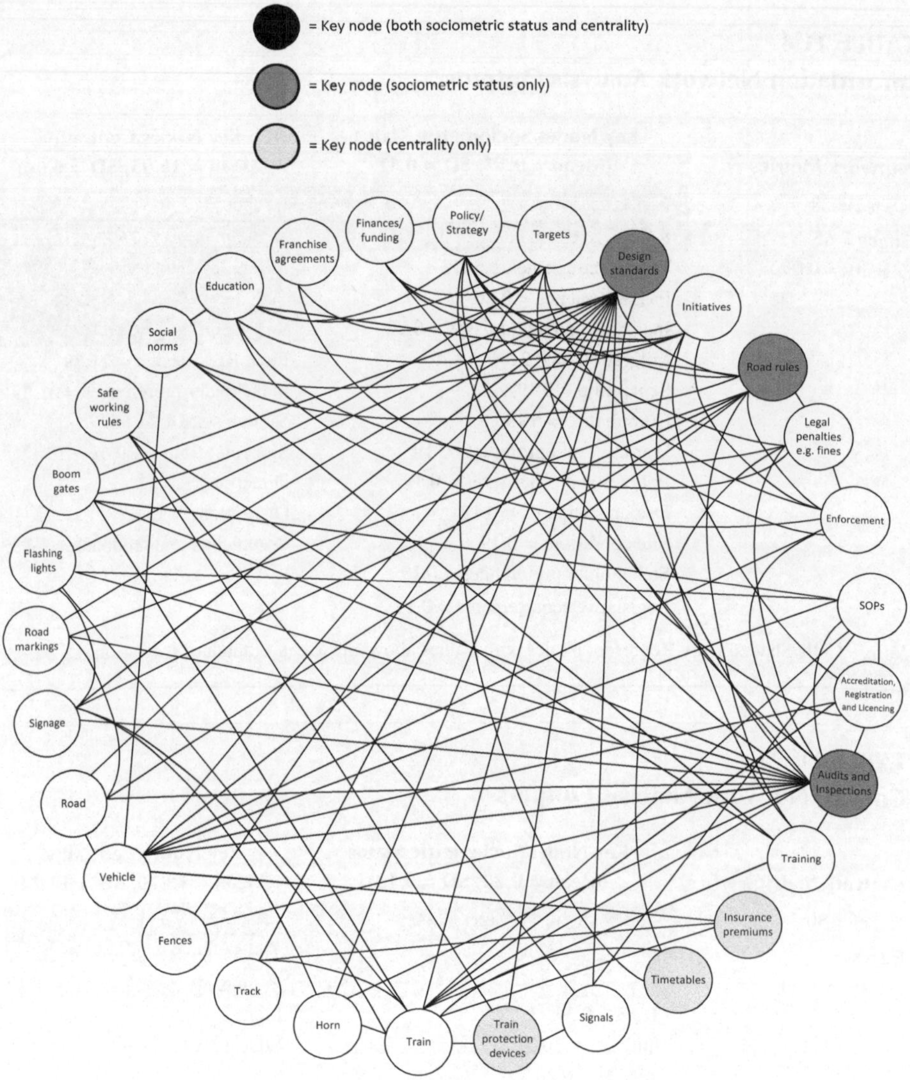

FIGURE 11.7 Control network for RLX system lifecycle.

Integration with the task, social and information networks also allows analysts to examine the relationship between controls and tasks, agents and information as well as the extent to which controls are likely to be effective. Finally, network metrics can be used to compare the relevant importance of controls. This provides an additional capability over and above the STAMP control structure, and it is anticipated that it will useful for safety critical systems that rely heavily on the use of controls to manage risk.

An important contribution of the analysis presented was the use of EAST to describe and analyse the overall sociotechnical system – in this case, the RLX system design lifecycle. This is a key capability of the framework; however, few EAST analyses have sought to examine overall sociotechnical systems. The analysis presented in this chapter demonstrates that it is possible to use EAST to analyse the behaviour of overall systems and that a richer analysis is produced in doing so. It is hoped that readers of this book will be encouraged to apply EAST to analyse entire sociotechnical systems in other areas.

Implications for Railway Level Crossing Safety Management

Examination of the four networks reveals some important implications for RLX safety management. The task network identified risk management and performance monitoring as critical tasks. Further, the information network identified 'risk level' as one of the key information requirements, whereas the control network identified audits and inspections as one of the key safety control measures. Assessing the risks at different RLXs and subsequently implementing appropriate risk management strategies therefore ostensibly represent key activities within the RLX system lifecycle. Despite this, recent Australian studies have suggested that elements of these tasks may be sub-optimal, including a lack of incident and near miss reporting and analysis mechanisms (Salmon et al. 2016) and inadequate risk assessment tools (Salmon et al. 2013). The findings suggest that road and rail stakeholders should attempt to strengthen their risk management and performance monitoring activities through the development of more comprehensive risk management processes (e.g. incorporating human factors data) and performance monitoring controls (e.g. incident and near miss reporting systems, regular behavioural assessments and increasing data collection mechanisms at RLXs such as closed-circuit television cameras).

All four of the networks had relatively low density scores, indicating that the RLX system lifecycle as a whole is loosely coupled. This intuitively makes sense, as the analysis describes activities, decisions and actions that occur across various temporal and spatial planes. However, it may be that attempting to increase the coupling of tasks, agents, information and controls in some cases will prove beneficial for safety management. For example, in the social network, standards-setting bodies and local governments were not well connected with other agents. Ostensibly, there would be benefits in better integrating both in the RLX system lifecycle. This would ensure that design standards are flexible and appropriate and are updated based on feedback from other agents regarding performance throughout the RLX system lifecycle. This suggestion is confirmed through the control network in which design standards was identified as one of the key controls in terms of impact on other controls.

CONCLUSION

This chapter has demonstrated an extended version of the EAST framework that enables it to analyse risk controls in addition to task, social and information networks. This extension enhances the analytical and explanatory power of the EAST framework, and the analysis presented provides a rich understanding of the RLX system lifecycle and RLX safety management activities. Potential improvements around design standards, the coupling of agents and organisations, risk management and performance monitoring and the clarification of roles and responsibilities across the wider system were identified. It is hoped that researchers and practitioners conduct further applications of the modified EAST framework across the safety critical domains.

REFERENCES

Allison, C., Revell, K. M. A., Sears, R. and Stanton, N. A. (2017). Systems Theoretic Accident Model and Process (STAMP) safety modelling applied to an aircraft rapid decompression event. *Safety Science*, vol. 98, October 2017, pp. 159–166.

Australian Transport Safety Bureau. (2012). *Australian Rail Safety Occurrence Data: 1 July 2002 to 30 June 2012 (ATSB Transport Safety Report RR-2012-010).* Canberra, Australia: ATSB.

Chen, L., Zhao, Y. and Zhao, T. (2015). An AcciMap analysis on the China-Yong-Wen railway accident. In: P. W. Tse et al. (Eds.), *Engineering Asset Management-Systems, Professional Practices and Certification.* New York City, NY: Springer International Publishing, 1247–1253.

Dallat, C., Salmon, P. M. and Goode, N. (in press). Risky systems versus risky people: To what extent do risk assessment methods consider the systems approach to accident causation? A review of the literature. *Safety Science*, accepted for publication March 2017.

Evans, A. W. (2011). Fatal accidents at railway level crossings in Great Britain 1946–2009. *Accident Analysis & Prevention*, 43(5), 1837–1845.

Federal Railroad Administration. (2013). *Railroad Safety Statistics 2012 Preliminary Annual Report.* Department of Transportation, Washington, D.C.

Houghton, R. J., Baber, C., McMaster, R., Stanton, N. A., Salmon, P. M., Stewart, R. and Walker, G. H. (2006). Command and control in emergency services operations: A social network analysis. *Ergonomics*, 49, 1204–1225.

Houghton, R. J., Baber, C., Stanton, N. A., Jenkins, D. P., Revell, K. (2015). Combining network analysis with Cognitive Work Analysis: insights into social organisational and cooperation analysis. *Ergonomics*, 58:3, 434–449.

Independent Transport Safety Regulator. (2011). Level crossing accidents in Australia. ITSR, Sydney. Available at: http://www.onrsr.com.au/__data/assets/pdf_file/0020/2963/Transport-safety-bulletin-Issue-2-Level-crossing-accidents-in-Australia-August-20112.PDF. Accessed 10/02/15.

Jansson, A., Olsson, E. and Erlandsson, M. (2006). Bridging the gap between analysis and design: Improving existing driver interfaces with tools from the framework of cognitive work analysis. *Cognition, Technology & Work*, 8, 41–49.

Karsh, B. T., Waterson, P. and Holden, R. J. (2014). Crossing levels in systems ergonomics: A framework to support 'mesoergonomic' inquiry. *Applied Ergonomics*, 45(1), 45–54.

Kirwan, B. (1992). Human error identification in human reliability assessment: Part 2: detailed comparison of techniques. *Applied Ergonomics*, 23, 371–381.

Lee, S., Bo Moh, Y., Tabibzadeh, M. and Meshkati, N. (2017). Applying the AcciMap methodology to investigate the tragic Sewol Ferry accident in South Korea. *Applied Ergonomics*, 59(B), 517–525.

Leveson, N. G. (2004). A new accident model for engineering safer systems. *Safety Science*, 42(4), 237–270.

Madigan, R., Golightly, D. and Madders, R. (2016). Application of Human Factors Analysis and Classification System (HFACS) to UK rail safety of the line incidents. *Accident Analysis & Prevention*, 97, 122–131.

Naweed, A. (2014). Investigations into the skills of modern and traditional train driving. *Applied Ergonomics*, 45(3), 462–470.

Naweed, A., Rainbird, S. and Dance, C. (2015). Are you fit to continue? Approaching rail systems thinking at the cusp of safety and the apex of performance. *Safety Science*, 76, 101–110.

Newnam, S., and Goode, N. (2015). Do not blame the driver: A systems analysis of the causes of road freight crashes. *Accident Analysis & Prevention*, 76, 141–151.

Office of the National Rail Safety Regulator. (2016). *Rail Safety Report 2015–2016*. Office of the National Rail Safety Regulator: Adelaide, Australia.

Ouyang, M., Hong, L., Yu, M-H. and Fei, Q. (2010). STAMP-based analysis on the railway accident and accident spreading: Taking the China–Jiaoji railway accident for example. *Safety Science*, 48(5), 544–555.

Parnell, K., Stanton, N. A. and Plant, K. L. (2017). What's the law got to do with it? Legislation regarding in-vehicle technology use and its impact on driver distraction. *Accident Analysis and Prevention*, 100, 1–14.

Patriarca, R., Bergström, J. and Di Gravio, G. (2017). Defining the functional resonance analysis space: Combining abstraction hierarchy and FRAM. *Reliability Engineering & System Safety*, 165, 34–46.

Rasmussen, J. (1997). Risk management in a dynamic society: A modelling problem. *Safety Science*, 27(2/3), 183–213.

Read, G., Salmon, P. M. and Lenné, M. G. (2013). Sounding the warning bells: The need for a systems approach to rail level crossing safety. *Applied Ergonomics*, 44, 764–774.

Read, G., Salmon, P. M. and Lenné, M. G. (2016). When paradigms collide at road and rail interface. *Ergonomics*, 59(9), 1135–1157.

Read, G., Salmon, P. M, Lenné, M. G. and Jenkins, D. P. (2015). Designing a ticket to ride with the Cognitive Work Analysis Design Toolkit. *Ergonomics*, 58(8), 1266–1286. doi:10.1080/00140139.2015.1013576.

Read., G. J. M., Beanland, V., Lenné, M. G., Stanton, N. A. and Salmon, P. M. (2017). *Integrating Human Factors Methods and Systems Thinking for Transport Analysis and Design*. CRC Press: Boca Raton, FL.

Salmon, P. M., Lenné, M. G., Mulvihill, C., Young, K., Cornelissen, M., Walker, G. H. and Stanton, N. A. (2016b). More than meets the eye: Using cognitive work analysis to identify design requirements for safer rail level crossing systems. *Applied Ergonomics*, 53(Part B), 312–322.

Salmon, P. M., Lenné, M. G., Walker, G. H., Stanton, N. A. and Filtness, A. (2014). Using the Event Analysis of Systemic Teamwork (EAST) to explore conflicts between different road user groups when making right hand turns at urban intersections. *Ergonomics*, 57(11), 1628–1642.

Salmon, P. M., Read, G., Stanton, N.A . and Lenné, M. G. (2013). The Crash at Kerang: Investigating systemic and psychological factors leading to unintentional non-compliance at rail level crossings. *Accident Analysis and Prevention*, 50, 1278–1288.

Salmon, P. M., Read, G. J. M. and Stevens, N. (2016). Who is in control of road safety? A STAMP control structure analysis of the road transport system in Queensland, Australia. *Accident Analysis and Prevention*, 96, 140–151.

Salmon, P. M., Stanton, N. A., Jenkins, D. P. and Walker, G. H. (2010). Hierarchical task analysis versus cognitive work analysis: Comparison of theory, methodology, and contribution to system design. *Theoretical Issues in Ergonomics Science*, 11(6), 504–531.

Salmon, P. M., Walker, G. H., Read, G. J. M., Goode, N. and Stanton, N. A. (2017). Fitting methods to paradigms: Are ergonomics methods fit for systems thinking? *Ergonomics*, 60(2), 194–205.

Stanton, N. A. (2006). Hierarchical task analysis: Developments, applications, and extensions. *Applied Ergonomics*, 37(1), 55–79.

Stanton, N. A. (2014). Representing distributed cognition in complex systems: How a submarine returns to periscope depth. *Ergonomics*, 57(3), 403–418.

Stanton, N. A., Harris, D. and Starr, A. (2016). The future flight deck: Modelling dual, single and distributed crewing options. *Applied Ergonomics*, 53, 331–342.

Stanton, N. A., Salmon, P. M., Rafferty, L., Walker, G., Baber, C. and Jenkins, D. P. (2013). *Human Factors Methods: A Practical Guide for Engineering and Design*. Second Edition. Ashgate: Aldershot, UK.

Stanton, N. A., Salmon, P. M., Walker, G. H., Baber, C. and Jenkins, D. (2005) *Human Factors Methods: A Practical Guide for Engineering and Design*. First Edition. Ashgate: Aldershot, UK.

Stanton, N. A. and Walker, G. H. (2011). Exploring the psychological factors involved in the Ladbroke Grove rail accident. *Accident Analysis & Prevention*, 43(3), 1117–1127.

Stefanova, T., Burkhardt, J.-M., Filtness, A., Wullems, C., Rakotonirainy, A. and Delhomme, P. (2015). Systems-based approach to investigate unsafe pedestrian behaviour at level crossings. *Accident Analysis & Prevention*, 81, 167–186.

Tooth, R. and Balmford, M. (2010). *Railway Level Crossing Incident Costing Model*. LECG for the Rail Industry Safety and Standards Board: Canberra, Australia.

Vicente, K. J. (1999). *Cognitive Work Analysis: Toward Safe, Productive, and Healthy Computer-based Work*. Lawrence Erlbaum Associates: Mahwah, NJ.

Vivekanandan Dhukaram, A. and Baber, C. (2015). Modelling elderly cardiac patients decision making using Cognitive Work Analysis: Identifying requirements for patient decision aids. *International Journal of Medical Informatics*, 84(6), 430–443.

Walker, G. H., Gibson, H., Stanton, N. A., Baber, C., Salmon, P. M. and Green, D. (2006). Event analysis of systemic teamwork (EAST): A novel integration of ergonomics methods to analyse C4i activity. *Ergonomics*, 49, 1345–1369.

Walker, G. H., Stanton, N. A., Salmon, P. M. (2011). Cognitive compatibility of motorcyclists and car drivers. *Accident Analysis & Prevention*, 43 (3): May 2011, pp. 878–888.

Walker, G. H., Salmon, P. M., Bedinger, M., Stanton, N. A. (2017). Quantum ergonomics: Shifting the paradigm of the systems agenda. *Ergonomics*, 60(2), 157–166.

Walker, G. H., Stanton, N. A., Baber, C., Wells, L., Jenkins, D. P. and Salmon, P. M. (2010). From ethnography to the EAST method: A tractable approach for representing distributed cognition in air traffic control. *Ergonomics*, 53(2), 184–197.

Wilson, J. R., Norris, B. J. (2005). Rail human factors: Past, present and future. *Applied Ergonomics*, 36(6), 649–660.

12 The EAST 'Broken-Links' Approach
Assessing Risk in Sociotechnical Systems

With Catherine Harvey

INTRODUCTION

The term *Sociotechnical Systems* (STS) is used to refer to the interaction between humans and machines, from the small and simple to the large and highly complex (Walker et al. 2008; Walker et al. 2010b; Read et al. 2015). These sub-systems operate and are managed as independently functioning (autonomous) entities with their own goals, but they must collaborate with other sub-systems to achieve the higher goals of the STS (Dul et al. 2012; Wilson 2012). A key characteristic is that these goals can only be achieved by the STS and not by individual sub-systems functioning in isolation (Rasmussen 1997; von Bertalanffy 1950). STS present unique challenges for safety management and risk assessment (Rasmussen 1997; Alexander and Kelly 2013; Flach et al. 2015; Waterson et al. 2015). Traditional approaches to risk assessment, such as THERP (Technique for Human Error Rate Prediction) Swain and Guttmann 1983), TRACEr (Technique for the Retrospective and Predictive Analysis of Cognitive Errors) Shorrock and Kirwan 2000) and SHERPA (Systematic Human Error Reduction and Prediction Approach) Embrey 1986), are typically reductionistic in nature (Stanton et al. 2013), focusing on individual tasks and technologies rather than the system as a whole (Stanton 2006; Stanton and Stevenage 1998; Stanton et al. 2009; Waterson et al. 2015). These methods use error taxonomies to identify risk, but recent research has suggested that the term *human error* is obsolete (Dekker 2014). In its place, the term *human performance variability* has been proposed, which includes both normative and non-normative performance. This latter approach emphasises the broad spectrum of human behaviour rather than a dichotomy and, therefore, a need to build resilient systems (Hollnagel et al. 2006). The Systemic Accident Analysis (SAA) approach treats systems as whole entities with complex, non-linear networks (Underwood and Waterson 2013). A number of SAA methods were assessed for their potential for prospective risk analysis within STS in a previous study (Stanton et al. 2012). Some system methods incorporate error taxonomies, such as CREAM (Cognitive Reliability and Error Analysis Method) Hollnagel 1998), HFACS (Human Factors Analysis and Classification System) Shappell and Wiegmann 2001) and

STPA (System-Theoretic Process Analysis) (Leveson 2012), which, given the recent shift away from the term *human error* is something of a conundrum. Rather than considering risks in systems to be the result of error, the approach taken in this chapter is to propose risks as the failure to communicate information via social and task networks. This type of failure may be seen in several major incidents. For example, in the MS *Herald of Free Enterprise* accident (1987), the state of the bow doors was not communicated to the ship's bridge (Noyes and Stanton 1997), so the ship left harbour and subsequently capsized. In the Kegworth air disaster (1989), the aircraft failed to communicate which engine was on fire, leading to the pilots shutting down the wrong engine (Griffin et al. 2010; Plant and Stanton 2012). In the Ladbroke Grove rail incident (1999), the signals failed to communicate to the train driver that the section of the rail network was protected (Moray et al. 2016). Rather than stopping, the driver actually increased his speed as he passed the red signal, leading to a collision with an oncoming high-speed intercity train (Stanton and Walker 2011). So, rather than conceive these behaviours as errors, we have reconceived them as the failure to communicate information in the system. This is a new paradigm for risk assessment that incorporates the value of a holistic perspective for appreciating the relationships between the various sub-systems and the network diagrams for visualising important aspects of STS, such as the constraints on communications (see Flach et al. 2015). In order to analyse the information communications in systems, the Event Analysis of Systemic Teamwork (EAST) method was selected. EAST takes a different approach to the error taxonomic methods by modelling and analysing STS-level interactions. In a previous study, Stanton (2014) analysed communications between various actors within a submarine control room; in contrast, this case study analyses a retrospective account of actions within a Royal Navy training activity and is conducted at the macro level (Grote et al. 2014). The aim of this chapter is to extend the EAST network-level analysis to include risk prediction by 'breaking' links within networks.

The EAST method was first proposed by Stanton et al. (2005, 2013) and further elaborated by Stanton et al. (2008) for modelling distributed cognition in STS. The method represents distributed cognition in networks, which enables both qualitative and quantitative investigations to be performed (Stanton 2014). One of the main advantages of EAST is its aim to capture the whole system, as opposed to reductionist methods, which split a system into constituent parts for analysis (Walker et al. 2010a). It is therefore considered in this study to be a suitable technique for representing a STS and potential non-normative behaviours. The analysis describes a system as three different types of network:

- *Social*: Representing the agents (human, technical and organisational) within a system and communications between them
- *Task*: Representing the activities performed by the system and the relationships between them
- *Information*: Representing the information that is used and communicated within a system and links between different information types

The social, task and information networks are developed individually and then combined to create a complete social-task-information network diagram, showing all the links and information flows (i.e. distributed cognition) within a network of networks. EAST has been applied in many domains, including aviation (Stewart et al. 2008, Walker et al. 2010a, Stanton et al. 2016), the military (Stanton et al. 2006, Stanton 2014), road (Salmon et al. 2014) and rail (Walker et al. 2006) environments and the emergency services (Houghton et al. 2006, Baber et al. 2013). The aim of this work is to extend the EAST method to consider risk in systems via a case study and to provide an initial STS method evaluation criteria presented by Harvey and Stanton (2014). The premise of the risk assessment is that STS failures are predominately caused by the failure to communicate information between agents and tasks. This will be studied within the context of the following case study.

CASE STUDY OF HAWK MISSILE SIMULATION TRAINING

Operation of the Hawk jet to simulate missile attacks against surface ships by the UK Royal Navy was selected as the case study. This activity is viewed as a STS because it comprises many interconnected sub-systems, which are themselves complex. The context for this study is illustrated in an AcciMap (Rasmussen 1997; Jenkins et al. 2010) in Figure 12.1. The AcciMap places different subsystems within the STS at different levels and shows the links in communication and decision making between the sub-systems. Each node in the AcciMap is labelled (a, b, c, etc.) to correspond with the description in the text in the following paragraphs. The year in which each event occurred is also included, where applicable, to give an indication of timescale.

The Royal Navy uses the Hawk jet to simulate air attacks on ships during sea training of ships' gunners and radar operators (event 'a' in Figure 12.1). The Hawks

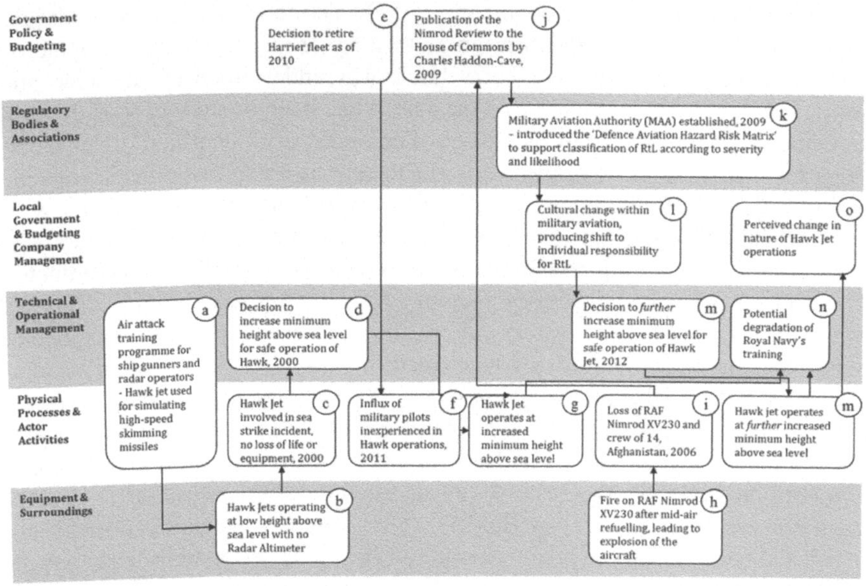

FIGURE 12.1 AcciMap showing sub-systems within the Hawk jet STS [Please note that the labels do not indicate a timeline; rather, they are added for clarity of the description below].

are used to simulate enemy aircraft attacks and high-speed skimming missiles fired against ships (Royal Navy 2012). In order to perform these simulation activities, the Hawk must be flown at a low height above sea level (b); however, the Hawk is not equipped with a Radar Altimeter (Rad-Alt), which provides a highly accurate measure of the altitude of the aircraft above the sea. This makes flying the Hawk accurately at very low levels extremely difficult, and a high level of expertise is required to perform this safely. Prompted by events over a number of years, there have been some significant changes to the method for assessing the safety of the Hawk STS.

In 2000, a Hawk jet was involved in a sea-strike incident as a consequence of very low-level flight (c). Although there was no resulting loss of life, this incident prompted a decision by the Royal Navy to increase the minimum allowable flying height above sea level for the Hawk (d).

As part of its Strategic Defence and Security Review, HM Government (2010) took the decision to retire the Harrier jet from service in October 2010 (e). As a result of this, a number of Royal Navy pilots who would have flown the Harrier were diverted into the Hawk program (f). Traditionally, the Hawk has been flown by civilian pilots under contract to the Royal Navy (2012); these pilots have extensive military experience in fast jets, which includes low-level flight supported by a Rad-Alt. This experience provided mitigation against the Risk to Life (RtL) for the Hawk air attack simulation task; however, the cohort of military pilots did not have this same level of experience, and the RtL had to be reassessed in light of this (g).

In 2006, RAF Nimrod XV230 suffered a catastrophic explosion after a mid-air refuelling procedure (h); this caused the deaths of all 12 crew members plus two mission specialists and the total loss of the aircraft (i). The government requested a comprehensive review into the airworthiness and safe operation of the Nimrod (j), which was delivered by Haddon-Cave (2009). The report described the development of the safety case for the Nimrod as 'a story of incompetence, complacency, and cynicism' (p. 161) and concluded that it was undermined by the widespread assumption that the Nimrod was safe because it had been flown successfully for the preceding 30 years (Haddon-Cave 2009). The report also identified organisational changes in the years prior to the Nimrod accident as having significant influence; these included a shift in organisational culture toward business and financial targets 'at the expense of functional values such as safety and airworthiness' (Haddon-Cave 2009, p. 355). As a consequence of the findings, Haddon-Cave (2009) recommended the establishment of an *independent* Military Aviation Authority (MAA) to properly assess RtL and shape future safety culture (k). Further recommendations included the need for strong *leadership*, a greater focus on *people* to deliver 'high standards of safety and airworthiness' (Haddon-Cave 2009, p. 355) and increased *simplicity* of rules and regulations. The tragic consequences of the Nimrod accident, along with the recommendations of the Haddon-Cave report, effected a culture change within military aviation; this resulted in a decision to assign *individual* accountability for RtL assessments to 'Duty Holders' (DH), whereas previously, responsibility for risk had been held at the organisation level (l). The newly established MAA produced guidelines for the assessment of RtL in the form of the Defence Aviation Hazard-Risk Matrix (MAA 2011), which supports the classification of single risks according to their estimated severity (catastrophic, critical, major, minor) and likelihood (frequent, occasional, remote, improbable). The resulting risk level determines at which level of DH the risk is held.

The organisational changes brought about by the events described above (i.e. influx of junior pilots) prompted reassessment of the RtL for the Hawk air attack simulation activity. The goal of safety management in the UK military is to reduce risk to a level which is As Low As Reasonably Practicable (ALARP); this is reached when 'the cost of further reduction is grossly disproportionate to the benefits of risk reduction' (Ministry of Defence 2007). The RtL for all Hawk operations is frequently reassessed, and the shift in pilot experience levels, as described above, prompted changes to the RtL for the Hawk air attack simulation activity. In order to reduce this RtL to a level that was ALARP, a decision was taken by the Royal Navy DH with subject matter expert (SME) advice to further increase the minimum height above sea level (m). A potential consequence of this decision is the degradation of Royal Navy surface fleet training against very live low-level targets, as the Hawk can no longer accurately simulate sea-skimming missile attacks on surface ships (n). These events have changed the nature of Hawk operations within the UK Ministry of Defence (MoD(2007)) (o).

Potential risks to the safe operation of the missile simulation activity are, in part, assessed according to the MAA's Regulatory Articles (MAA 2011). This assessment is based on the principle that risks can be tolerated provided they are reduced to ALARP. The MAA regulatory policy outlined its approach to the management of RtL:

> Aviation DHs [Duty Holders] are bound to reduce the RtL within their AoR [Area of Responsibility] to at least tolerable and ALARP; the application of effective and coherent risk management processes will be fundamental to achieving this.
>
> **(MAA 2011, p. 18).**

Regulatory Article (RA) 1210 – Management of Operating Risk (Risk to Life) – defined risk as:

> a measure of exposure to possible loss [combining] severity of loss (how bad) and the likelihood of suffering that loss (how often).
>
> **(MAA 2011, p. 1).**

The MAA suggested that risks can be identified via a number of different methods including previous occurrences, checklists, HAZOPS (Hazard and Operability), zonal hazard (safety) analyses and error trend monitoring. Previous work has showed that these techniques are likely to be inadequate for the analysis of STS (Stanton et al. 2012). RA 1210 specifically encourages the use of fault trees as accident models 'to assist understanding of the interrelationship between risks and to support the prioritisation of effort to maximise safety benefit' (MAA 2010, p. 6). This technique, along with other traditional error and risk prediction methods, does not account for the interactions of distributed actors within a STS (Salmon et al. 2011b). Furthermore, there is also no clear method outlined by the MAA for structuring risk identification; for example, the recommendation is that a combination of these methods should be used with the aim of identifying all credible risks, but there is no way of knowing when all possible credible risks have been defined and therefore how many methods to use and when to stop applying them.

The Hawk RtL case study was identified through interviews with a subject matter expert (SME) as part of this project. The analysts were provided with a high-level overview of the case study in an initial interview with the SME. This was followed up by a second, in-depth interview about the case study with the SME, conducted by two analysts. This resulted in a detailed account of the Hawk-Frigate STS, which was supplemented by extra information from official documentation including MAA (2010, 2011) guidelines, the official report into the Nimrod accident (Haddon-Cave 2009) and Royal Navy safety assessment guidance (Royal Navy 2012). The EAST method (Stanton 2014) was used to develop the three network diagrams, based upon the analysis of all case study information, in an iterative process that involved the SME providing feedback during development.

ANALYSIS OF NETWORKS

Social Network Analysis (SNA) metrics provide quantitative measures that represent the structures and relations between nodes in the EAST networks (Baber et al. 2013; Driskell and Mullen 2005; Walker et al. 2009). The SNA metrics describe individual nodes (including reception, emission, eccentricity, sociometric status, centrality, closeness, farness and betweenness). The SNA metrics applied in the current study, along with their descriptions, are presented in Table 12.1. Analysis software, AGNA version 2.1 (Benta 2005), was used to calculate the SNA metrics. For each EAST network, key

TABLE 12.1
SNA Metrics, along with their Descriptions

	Safety Constraint	Description
Node-level metrics	Emission	The number of edges (links to other nodes) originating at that node
	Reception	The number of edges incident to that node
	Sociometric status	Number of communications received and emitted relative to the number of nodes in the network
	Bavelas-Leavitt (B-L) centrality	Degree of connectivity to other nodes in the network
	Eccentricity	Length of the longest geodesic path originating in that node (a geodesic path is the path between two given nodes that has the shortest possible length)
	Closeness	Inverse of the sum of the geodesic distances from that node to all the other nodes, that is, the extent to which a node is close to all other nodes
	Standard closeness	Closeness multiplied by (g-1), where g is the number of nodes in the network
	Farness	Sum of the geodesic distances from that node to all other nodes
	Betweenness	Frequency with which a node falls between pairs of other nodes in the network

nodes were identified according to sociometric status. Sociometric status was selected to define key nodes because it identifies the prominence of an individual node's communications with the rest of the network, which influences the whole network's performance (Stanton 2014). In a STS, all of the nodes will have complex safety management rules and behaviours; however, as the 'key' nodes have the largest number of connections to the rest of the network, these nodes will have the highest degree of influence over the behaviour of the entire STS. Sociometric status key nodes are defined as nodes that have a higher sociometric status score than the sum of the mean sociometric status score plus the standard deviation sociometric status score for all nodes in the network. SNA metrics were calculated for the EAST networks created for this case study; key agents for sociometric status are indicated in the social, task and information network diagrams below.

RESULTS

The step-by-step application of the shortened version of EAST is described in detail in the following sections. This is accompanied by the outputs of the method along with interpretation of the results. The first stage in EAST was the identification of all social, task and information nodes within the Hawk missile simulation case study, based on the SME's account of activities, which informed the analysts' knowledge of the case study. The nodes were arranged in social, task and information networks and links drawn between related nodes. Related nodes were those between which some information was transferred. As well as providing a visual representation of a STS, the EAST network diagrams can be analysed to produce quantitative SNA metrics.

Social Network

Seven social 'agents' and their connections were identified from the Hawk RtL case study with the SME; these are shown in Figure 12.2. The social network was constructed by first identifying the main agents that are in the system, then by examining the interdependencies between those agents. The SME agreed that the social network was a reasonable representation of the main agents and their relationships. The 'edges', or links, between the agents show where information is transferred and the direction of transfer. There are 19 edges in total; in some cases, information transfer is reciprocal, but in others it only goes in one direction between two agents. The 'pilot' node was identified as the key agent according to sociometric status (1.33), with the highest number of links to and from other nodes in the network; in fact, the pilot receives and/or emits information from/to all of the other agents in the social network.

The pilot had the highest betweenness score (10.0) as it is located on the paths between a number of other agent pairs. The pilot also had the highest score for reception (6), highlighting a high degree of connectivity to other nodes in the network and indicating that the pilot's actions and communications are integral to the functioning of the STS. The high farness score for the regulator (10) indicates that this agent is located farthest from most other nodes, and this is supported by the information

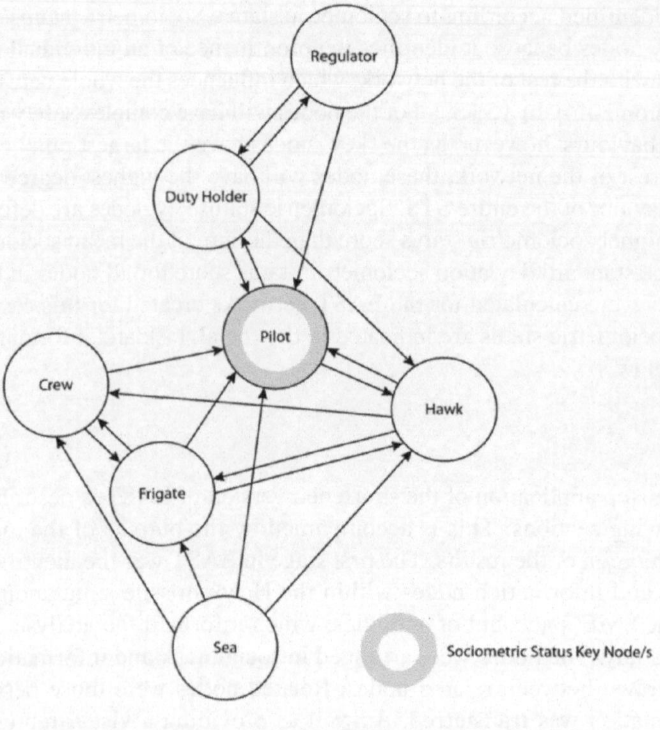

FIGURE 12.2 Social nodes and their links within the STS.

in the case study, which showed that the regulator really only communicates with the DH and possibly the pilots but has no contact with the frigate or crew. This is because the regulator in this case is the MAA, which does not have direct control over the Navy's surface ship operations. The sea scored highest for emission (4) and lowest for reception (0) as it does not receive information from any other nodes but is used for feedback only. In this sense, the sea can be regarded as a 'passive' agent, as it cannot respond to feedback; the social agents can only respond to it.

Task Network

Ten tasks nodes and their connections were identified from the Hawk RtL case study with the SME; these are shown in Figure 12.3. The task network was constructed by first identifying the main tasks that are performed by the system, then by examining the interdependencies between the tasks. The SME agreed that the task network was a reasonable representation of the system. There are 12 edges in total, and in all cases, the transfer is uni-directional.

In the task network, key nodes were identified as 'safe control of aircraft to simulate missile' and 'issuing of RtL document', which had sociometric status scores of .56 and .44 respectively. The task network contains more nodes but fewer edges than the social network, indicating that there are fewer communications between tasks.

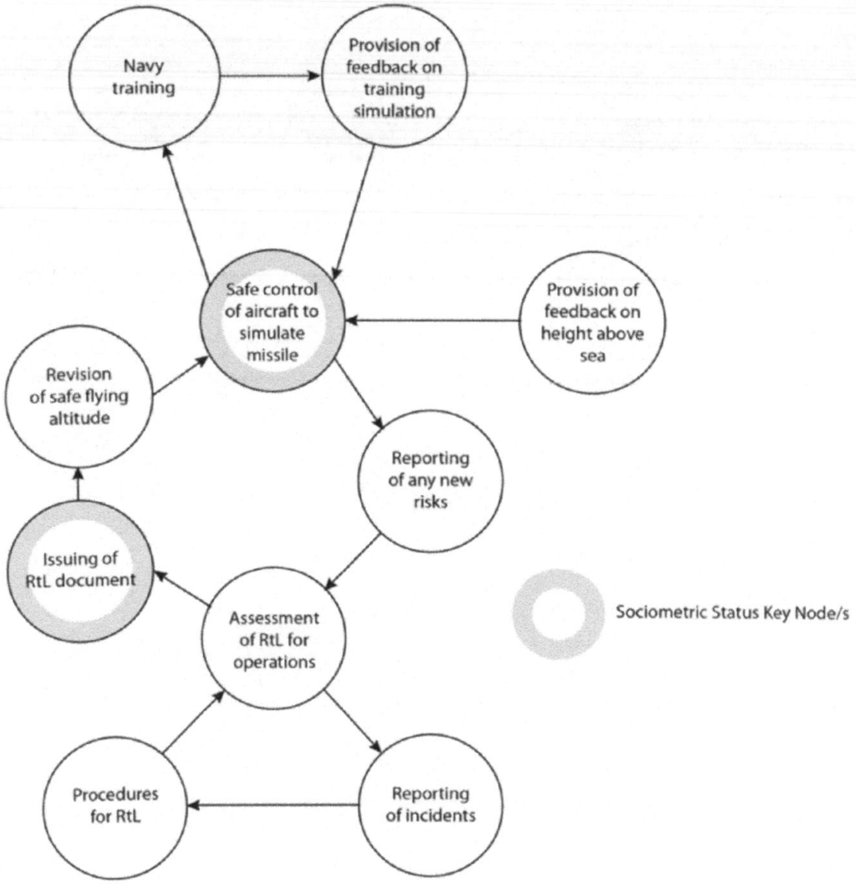

FIGURE 12.3 Task nodes and their links within the STS.

Cohesion is zero because there are no mutual, or bidirectional, links between nodes. The highest score for betweenness was for 'safe control of aircraft to simulate missile' (45), demonstrating that this task is integral in the STS as it is located between a high number of other task nodes. This is unsurprising as this task can be considered to be the main objective of the STS configuration investigated in this case study. This task also scored highest on emission (2), reception (3) and B-L centrality (6.26), as well as sociometric status (.56), showing a high level of connectivity to other nodes.

Information Network

EAST identified 25 information nodes, and their connections were based the Hawk RtL case study and further knowledge of the STS from the SME; these are shown in Figure 12.4. There are 50 edges in total; however, in this case, the links are not directional.

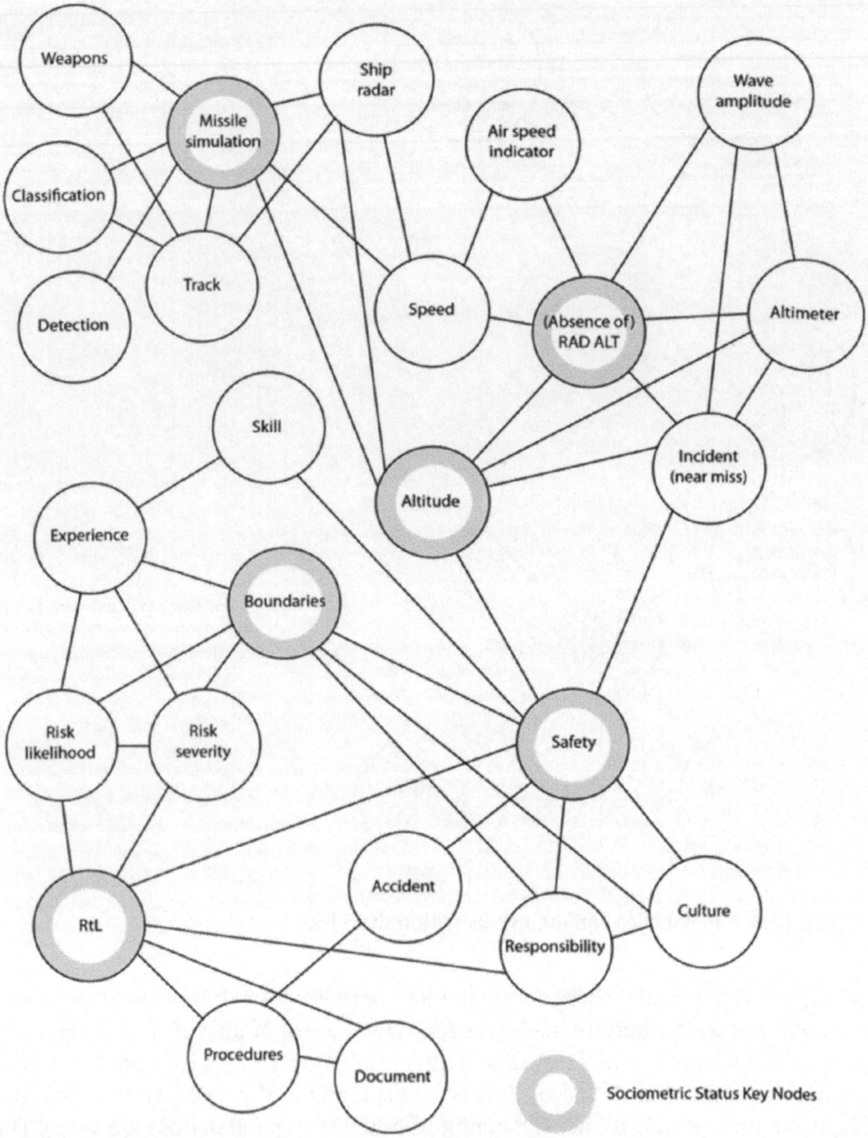

FIGURE 12.4 Information nodes and their links within the STS.

Six information nodes were identified as key nodes according to sociometric status: missile simulation, (absence of) Rad-Alt, altitude, boundaries, safety and RtL. Safety had the highest betweenness (315.8), standard closeness (.53) and B-L centrality scores (18.51), and this reflects the importance of this in the case study; the aim within the STS is to achieve a safe solution for missile simulation. Density and cohesion were relatively low, that is, compared with the social network, as the edges between nodes were single and non-directional.

BROKEN-LINKS ANALYSIS

Studies of networks have discussed the effects of removing one or more nodes from a network on the resilience of that network to systemic failures and the resulting destabilisation (Baber et al. 2013; Houghton et al. 2008; Stanton 2014). This has been used to explore the resulting influence on network structure, rather than as a method for predicting specific risks. Previously, the network diagrams in EAST have been used to provide a visual representation of a system to further the users' understanding of distributed cognition (Stanton 2014). In this study, however, the EAST network analysis was extended to identify and examine possible risks by 'breaking' the links between the various nodes, in a similar approach to the removal of nodes, to explore system effects. 'Broken links' represent failures in communication and information transfer between nodes in the networks and these failures can then be used to make predictions about the possible risks within the STS. Previously, broken links have only been investigated by EAST analysts when looking retrospectively at accidents to identify underlying causes. Griffin et al. (2010) demonstrated that the broken link between the Engine Vibration Indicator and the pilots in the cockpit was a causal factor in their failure to shut down the correct engine in the Kegworth accident. If this information had been communicated more effectively, it could have helped to prevent the crash. Similarly, the EAST method has been adapted to analyse incidents of fratricide (Rafferty et al. 2012), although this has been conducted as retrospective and concurrent, rather than predictive, analyses. The broken-links analysis was performed for the Hawk missile simulation case study on the social and task networks shown in Figures 12.2 and 12.3 respectively. The information network was not subject to the broken-links analysis because broken links between information nodes were not considered to represent risks as they are caused by a failure in either the social or task networks. In other words, information does not fail in isolation; it is the failure to use or communicate the information correctly, and in all cases, this can be attributed to social nodes, task nodes or both. For the social and task networks, each link was identified and documented in a table. The combined EAST networks diagram (see Figure 12.5) shows the information network tagged with the social networks nodes (to show who owns each information node in the network) and grouped by the task network nodes (to show which task each information node belongs to). Details on construction of the combined network have been reported by Stanton (2014). This combined network was used to identify what information (from the information network) should be communicated from the origin node to the destination node in the task and social networks and therefore what information would not be communicated if the links between the nodes in the task and social networks were removed.

Figure 12.5 also shows the combined information-task-social network as a single depiction of the entire STS. This shows the overlaps between the three networks, in other words, which information is being communicated by which agents in which tasks and how these nodes are interlinked.

In order to conduct the broken-links analysis, the social and task networks were compared to the combined information-task-social network in turn. For example, there is a reciprocal relationship between the duty holder and the pilot (in the social

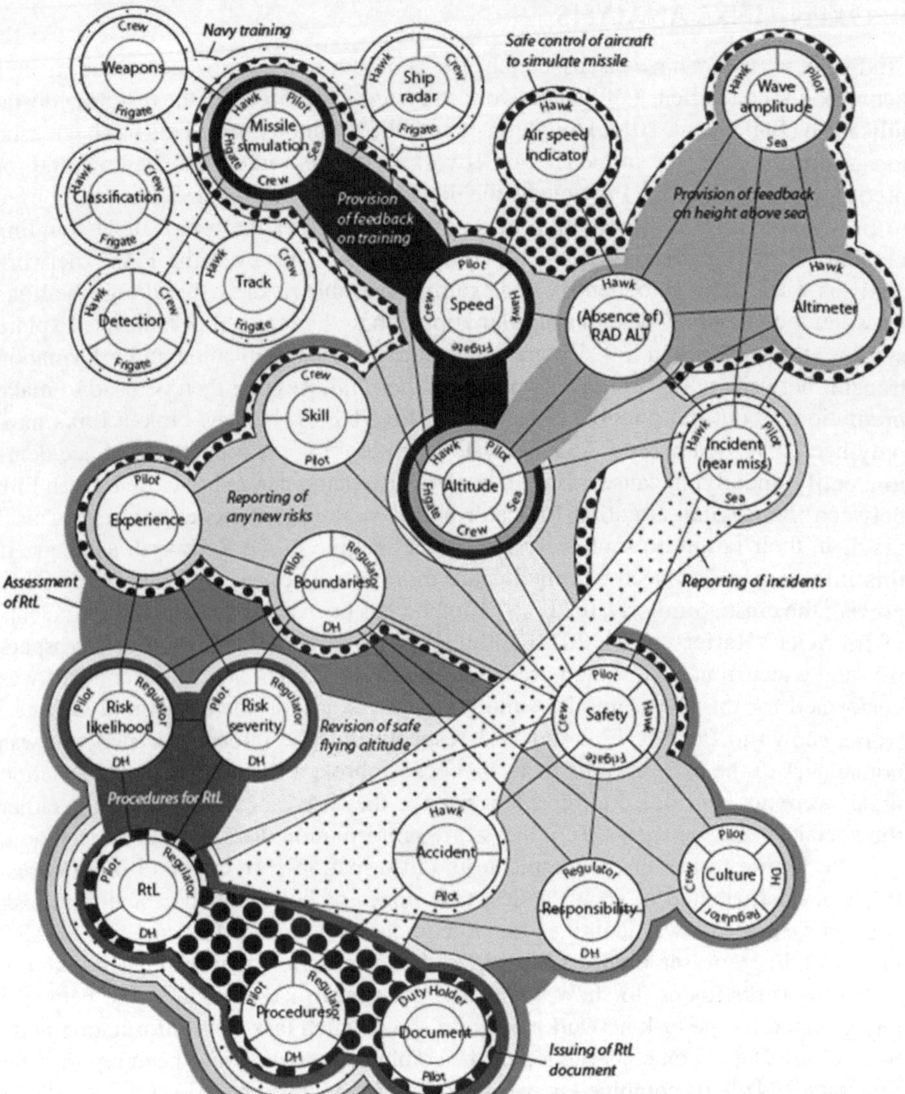

FIGURE 12.5 Combined information-task-social network for the Hawk case study (shading represents the different tasks being undertaken).

network shown in Figure 12.2), and the duty holder and the pilot share the nodes of boundaries, RtL, risk likelihood, risk severity, procedures, document, responsibility and safety (in the combined information-task-social network shown in Figure 12.5). The risk assessment procedure requires that the relationship between the duty holder and pilot be interrogated to see what would happen if each information element was not transmitted, as shown in Table 12.2. The pilot was identified as having the highest Sociometric Status in the analysis presented in Figure 12.2, so was chosen for the

TABLE 12.2
Extract from Broken-links Analysis for EAST Social Network

From (Agent)	To (Agent)	Information not Communicated	Resulting Risk	Mitigation Strategy
Duty Holder	Pilot	*Boundaries*	Pilots are not aware of the boundaries for flight operations and for the identification and reporting of risks within this.	Boundaries for risk reporting must be made clear to pilots as part of the RtL process.
Duty Holder	Pilot	*RtL*	Pilots are not made aware of the results and consequences of the RtL assessment process after it is conducted at DH level.	Results and consequences of the RtL assessment process must be effectively communicated to pilots.
Duty Holder	Pilot	*Risk likelihood*	Pilots are not made aware of risks assessed that their likelihood of occurrence.	Risks identified as having a high likelihood of occurrence must be reported to pilots.
Duty Holder	Pilot	*Risk severity*	Pilots are not made aware of risks assessed and their severity of impact.	Risks deemed as having a high severity of impact must be reported to pilots.
Duty Holder	Pilot	*Procedures*	Pilots are not aware of how the RtL process is conducted at DH level and of procedures for reporting incidents to the DH.	Pilots must be provided with clear procedures describing the assessment of RtL at DH level and the reporting of risks to DH.
Duty Holder	Pilot	*Document*	Pilots are not provided with documentation covering the RtL process and its results.	Pilots must be provided with documentation covering the RtL process and its results.
Duty Holder	Pilot	*Responsibility*	Pilots are not aware of the DH's or their own responsibilities for safety.	The responsibilities of both the pilot and DH for safety must be clearly defined and understood by pilots.
Duty Holder	Pilot	*Safety*	Pilots do not receive information about the safety of operations based on the RtL assessment process.	The safety of operations, as assessed during the RtL process, must be reported to the pilots.

(*Continued*)

TABLE 12.2 (CONTINUED)
Extract from Broken-links Analysis for EAST Social Network

From (Agent)	To (Agent)	Information not Communicated	Resulting Risk	Mitigation Strategy
Pilot	Duty Holder	*RtL*	The DH does not receive information about new risks identified by the pilots.	Pilots must clearly report all relevant risks to the DH.
Pilot	Duty Holder	*Risk likelihood*	The DH does not receive information about the likelihood of new risks identified by the pilots.	Pilots must report their estimate of the likelihood of occurrence of all relevant risks.
Pilot	Duty Holder	*Risk severity*	The DH does not receive information about the severity of new risks identified by the pilots.	Pilots must report their estimate of the severity of impact of all relevant risks.
Pilot	Duty Holder	*Incident (near miss)*	The DH does not receive information about incidents (or near misses) that occur during Hawk operations.	Pilots must clearly report all relevant incidents that occur during Hawk operations to the DH.
Pilot	Duty Holder	*Experience*	The DH cannot learn from the pilots' experience of Hawk operations and the risks encountered.	Pilots must clearly report their experience levels to the DH. Pilots must report their assessment of risks and any consequent assumptions, based on this experience.
Pilot	Duty Holder	*Skill*	The DH cannot learn from the pilots' skill in Hawk operations.	Pilots must clearly report their skill levels to the DH. Pilots must report their assessment of risks and any consequent assumptions, based on this skill level.
Pilot	Duty Holder	*Safety*	The DH does not receive information about the safety of Hawk operations.	Pilots must report their estimates of the safety impact of any risks identified in Hawk operations to the DH.
Pilot	Duty Holder	*Culture*	The DH is not aware of the culture of safety among the Hawk pilots.	Pilots must consider and report the estimated influence of safety culture on the risks to Hawk operations.

illustration of the broken-links analysis in Table 12.2. Although the pilot is linked to all other agents in the social network, for the purpose of illustration, just their reciprocal relationship with the duty holder is presented in Table 12.2.

Table 12.2 shows the risks resulting from the failure to pass relevant information between duty holder and pilot and vice versa. Anecdotal evidence from our SME suggests that there is variability in what individual pilots will chose to report back to the duty holder, as they have different interpretations of what they consider to be a risk and near miss. This shows that there is at least some face validity for the approach we have proposed.

In the similar manner to the social-information broken-links analysis shown in Table 12.2, there is a task-information broken-links analysis in Table 12.3. From the task network, there is a uni-directional relationship between the 'Issuing of RtL document' and the 'Revision of safe flying altitude' (in the task network shown in Figure 12.3), and they overlap in the combined information-task-social network (shown in Figure 12.5). The risk assessment procedure requires that the relationship between the 'Issuing of RtL document' and the 'Revision of safe flying altitude' be interrogated to see what would happen if each information element was not transmitted, as shown in Table 12.3. The 'Issuing of RtL document' was chosen as it has the highest Sociometric Status in the analysis presented in Figure 12.3. The 'Safe control of aircraft to simulate missile' was chosen for the same reason and is paired with 'Navy training' for the purposes of offering an illustrative example of the method in Figure 12.3.

Examination of the analysis in Tables 12.2 and 12.3 offers a systematic approach for examining a system of operation in a holistic manner. For example, increasing the safe flying attitude (see Figure 12.1) has led to the altitude profile of the Hawk not matching that of the low-flying missile (see Table 12.3). This has meant that reducing the risk for the Hawk pilots could have a negative effect on training of the crew on the frigate, ultimately increasing their risk. So, whilst the top of Table 12.3 is about improving the safety of the pilot – by increasing altitude, for example – the bottom of table three shows that this could reduce the safety of the Navy frigate crew as they do not receive realistic training. The benefit of systems approaches is that the emergent properties become more readily apparent.

DISCUSSION

This work aimed to explore the use of a modified version of EAST (network modelling and broken-links analysis, see Stanton 2014) in a case study of a Royal Navy training activity. First, the findings of this study are discussed in terms of criteria that were identified as essential for methods designed to analyse the human component of STS (Stanton et al. 2012; Harvey and Stanton 2014). This enabled comparisons to be made between the method and the current RtL procedure used in the Hawk missile simulation case study. Second, the modifications and extensions to EAST are discussed with reference to use of the method as an assessment of potential risks within a STS.

Aviation accidents, as with most accidents in STS, usually occur due to a conjunction of factors (Hodgson et al. 2013; Jenkins et al. 2010), and it is therefore essential that analysis methods are able to explore all of these factors by taking an integrated and holistic approach (Ramos et al. 2012; Salmon et al. 2011b). EAST specifically enables the exploration of the social, task and information components of the STS, allowing a high-level model of the STS to be created (visual diagrams) and analysed (social network metrics). This visual component is likely to help analysts and other stakeholders to understand the interactions within networks (Flach et al. 2015); this is an advantage over many other methods such as HAZOP and Fault Tree analysis, as well as the MAA's RtL/HRM (Hazard-Risk Matrix) approach. Baber et al. (2013) argued that it is sensible to speak of a 'useful' (rather than 'complete') network, as there will always be a possibility that some connections have been left out due to not being observed, reported and/or documented. This is certainly applicable to the networks generated by EAST, as it is impossible to know whether an analysis has been exhaustive, and it is therefore safest to assume that it has not. It is also particularly true in this case as the analysis was performed on an SME's reports of activities within the STS rather than communications between STS actors (as in Stanton 2014). A consequence of this approach is a lack of richness of information, although if the main contribution of EAST lies in its ability to visually represent an STS, then this may not be a significant issue. EAST includes the calculation of SNA metrics, which provide the analyst with quantitative values to represent various characteristics of the networks. In this way, the analysis encompasses all elements of a STS and provides the analyst with an understanding of the structure of a system as a whole and the relationships between individual system components. These metrics can provide potential insight into the resilience of the networks (Stanton et al. 2016).

The inclusion of particular agents in an accident model is dependent on the information put into the analysis and therefore on the analysts and SMEs involved. This also is true for more traditional HAZOP and error identification methods and the current RtL assessment process, as well as EAST. However, because HAZOP and RtL assessment essentially focus on a list of potential errors, there is no formal procedure for identifying the decision makers involved. In contrast, EAST enabled a visual representation of the decision-making agents (Flach et al. 2015) and their relationships with other nodes in a STS to be constructed, thereby encompassing the identification of decision makers into the analysis process. This can allow analysts to understand where responsibility for risks resides within the STS and so target mitigation strategies appropriately (Lundberg et al. 2010). This case study showed that EAST provided a useful visual representation of relationships between the various components of the STS. EAST examines the links between nodes and so is focused on communications, and therefore on the consequences of an action at a node, rather than its causes (Rafferty et al. 2012; Walker et al. 2010a).

In this case study, the analysts used a modified version of EAST, concentrating on the social, information and task networks (Stanton 2014). Guidance is provided on structuring a model of the STS under investigation, and there are numerous examples of previous EAST models (e.g. Griffin et al. 2010; Rafferty et al. 2012; Walker et al. 2010a; Stanton 2014) in the literature. The 'broken-link' process is very straightforward indeed and would be a useful addition to the current RtL assessment process (Haddon-Cave 2009). The current guidance from the MAA states that risks should be identified from a number of sources including HAZOP, error data and

TABLE 12.3
Extract from Broken-links Analysis for EAST Task Network

From (Task)	To (Task)	Information not Communicated	Resulting Risk	Mitigation Strategy
Issuing of RtL document	Revision of safe flying altitude	*Document*	The information contained in the RtL document does not trigger a revision of safe flying altitude.	The RtL document must be used by regulators to inform changes to regulations and safety guidance where appropriate.
Issuing of RtL document	Revision of safe flying altitude	*RtL*	The outcome of the RtL process outlined in the RtL document does not trigger a revision of safe flying altitude.	The outcomes of RtL assessment must be used by regulators to inform changes to regulations and safety guidelines where appropriate.
Issuing of RtL document	Revision of safe flying altitude	*Risk likelihood*	The outcome of the Risk likelihood assessment, conducted as part of the RtL process and outlined in the RtL document, does not trigger a revision of safe flying altitude.	The outcome of the Risk likelihood assessment, conducted as part of the RtL process and outlined in the RtL document, must be used to inform changes to regulations and safety guidelines where appropriate.
Issuing of RtL document	Revision of safe flying altitude	*Risk severity*	The outcome of the Risk severity assessment, conducted as part of the RtL process and outlined in the RtL document, does not trigger a revision of safe flying altitude.	The outcome of the Risk severity assessment, conducted as part of the RtL process and outlined in the RtL document, must be used to inform changes to regulations and safety guidelines where appropriate
Issuing of RtL document	Revision of safe flying altitude	*Safety*	The safety implications of the RtL process, outlined in the RtL document, do not trigger a revision of safe flying altitude.	The safety implications of RtL assessment must be used by regulators to inform changes to regulations and safety guidelines where appropriate.
Issuing of RtL document	Revision of safe flying altitude	*Responsibility*	Responsibility for the revision of safe flying altitude is not outlined in the RtL document.	Responsibility for changes to regulations and safety guidelines based on RtL assessment must be clearly assigned and accepted.

(*Continued*)

TABLE 12.3 (CONTINUED)
Extract from Broken-links Analysis for EAST Task Network

From (Task)	To (Task)	Information not Communicated	Resulting Risk	Mitigation Strategy
Safe control of aircraft to simulate missile	Navy training	*Missile simulation*	The overall control of the Hawk does not adequately simulate missile attack on the frigate to aid with training.	The operation of the Hawk must aid Navy training for missile attack situations.
Safe control of aircraft to simulate missile	Navy training	*Speed*	The speed profile of the Hawk does not adequately simulate missile attack on the frigate to aid with training.	The speed of the Hawk during missile simulation must be sufficiently realistic to aid Navy training for missile attack situations.
Safe control of aircraft to simulate missile	Navy training	*Altitude*	The altitude profile of the Hawk does not adequately simulate missile attack on the frigate to aid with training.	The altitude of the Hawk during missile simulation must be sufficiently realistic to aid Navy training for missile attack situations.
Safe control of aircraft to simulate missile	Navy training	*Track*	The track of the Hawk does not adequately simulate missile attack on the frigate to aid with training.	The track of the Hawk during missile simulation must be sufficiently realistic to aid Navy training for missile attack situations.

experience of previous events; however, there are no explicit instructions on how many of these methods to use and when to stop this analysis. This means that the RtL assessment may proceed without a comprehensive list of potential risks. It appears that EAST could be a useful model for ensuring that this does not happen; however, it is important to note that provision of guidance may not be sufficient for successful application of STS methods. The training requirements of these methods can often be high for practitioners, with many citing a lack of time and difficulty accessing new information as barriers to STS analysis (Underwood and Waterson 2013).

Stanton et al. (2012, 2014) previously suggested that EAST could be suitable for prospective analysis of STS risks; however, these studies only demonstrated the utility of methods for retrospective analysis (Salmon et al. 2011b; Waterson et al. 2015). In this study, EAST has been applied to a STS that is currently in operation in order to investigate the ability of methods to model the future state of a STS. The Hawk missile simulation STS has already experienced and been impacted by incidents (e.g. Hawk sea strike) and accidents (e.g. Nimrod), but this analysis focused on the prediction of a future state given the changes in the STS, such as the alteration in safe flying altitude for the Hawk and the effects of this on missile simulation for the frigate and crew. Having said this, the emphasis with EAST is not on predicting accidents per se; rather, it is about creating a comprehensive model of the links and information flows within the STS and by doing so making the analysts aware of potential breakdowns and failures that may occur in the future. This means that the success of EAST for prospective analysis is dependent on the participation in the assessment process of those who will be impacted by these failures and those that can apply the appropriate mitigation strategies.

In summary, this study used a modified version of EAST, following the examples in Stanton (2014). In this case, only the network analysis phases of EAST were applied (followed by the new paradigm of the broken-links analysis, which has not been previously reported) because the preliminary stages of EAST were negated by having already collected and represented the data via interviews with an SME. Furthermore, some of the EAST methods require communications data, which were lacking in this particular case study as the information came from a SME's account of the STS. Compared to Stanton's (2014) analysis of the operations within a submarine control room, the current study analysed activity at a macro level (Grote et al. 2014), using an SME's account of activities within the STS rather than a transcript of direct communications between STS actors. Recording and transcribing communications within a working system in real time is difficult, time-consuming and potentially disruptive to the STS under investigation. The approach presented in the current chapter would be easier for personnel within the STS itself to apply to support their own safety management and risk prediction activities, as it relies on a macro-level account of actions and relationships with a STS. This also allows these personnel to create a systems view of the STS of interest, which, as previously discussed, offers benefits over traditional taxonomic techniques. The absence of communications data obviously means that the analysis lacks detail and a richness of information that comes from speech data. This also meant that frequency of communications could not be represented in the same way as Stanton (2014). So, in effect, the network diagrams in this case study offer a basic visual representation of a STS that could aid understanding of the relationships between agents, tasks and information as well as their combination. Of greater importance is the extension to EAST presented in this chapter: the broken-links analysis. In order to identify potential risk in the STS, these links between nodes in the networks had to be examined in more detail; this was accomplished in the broken-links analysis. In this phase, each link between the task and social nodes was 'broken' to illustrate the effect of a communication breakdown between nodes. In this case study, 19 social links and 12 task links were broken and assessed against numerous information nodes, resulting in the identification of 137 risks in total. These breakdowns would result in a failure in information transfer, so each broken link was analysed against the information nodes to identify potential risks. This extension to the EAST method provides a structured method for identifying all of the risks within a STS. The broken links can be listed in table form, along with 'to' and 'from' information detailing the origin and destination nodes between which information is transferred. The broken-links analysis is concluded by developing mitigation strategies for each of the identified risks in a similar way to traditional human-error analysis methods.

CONCLUSIONS

This chapter presented a case study of the extension to the EAST method applied in the analysis of a STS, specifically Hawk missile simulation intended to aid with the training of Navy crew. The approach models the STS in two phases. In the first phase, the system is modelled as three connected networks (social, information and task), and the second phase, the social and task network links are systematically broken to reveal what risks are introduced by the failure to communicate information. The broken-links approach is a substantial and novel innovation over traditional human-error taxonomic approaches to assessing risks in systems. The approach is based on the premise that most, if not all, accidents and near misses are caused, at least in part, by the failure to communicate information between agents and tasks. By enabling the generation of a system model, EAST ensures that all of the components of interest within a STS have at been identified, and this should lead to a more comprehensive analysis of potential risks. The extension to EAST offers a holistic, structured and systematic approach to the identification of information communication failures in task and social networks. The EAST network broken-links approach is a new paradigm for risk assessment in systems. The approach can be applied to any STS in any domain where an EAST model has been constructed. Future work should explore the risks associated with multiple communication failures occurring simultaneously as well as considering the degree of resilience offered by different system network models.

ACKNOWLEDGEMENTS

The authors would like to thank Wing Commander Neil Bing (Bingo) of Air Cap SO1 Lightning, RAF High Wycombe, for his account of the Hawk RtL case study and his very valuable insights into the challenges faced by this complex sociotechnical system. This work was part-funded by the Defence Human Capability Science and Technology Centre (DHCSTC) grant reference TIN 2.002.

REFERENCES

Alexander, R. and Kelly, T. (2013). Supporting systems of systems hazard analysis using multi-agent simulation. *Safety Science*, 51, 302–318.

Baber, C., Stanton, N. A., Atkinson, J., McMaster, R. and Houghton, R. J. (2013). Using social network analysis and agent-based modelling to explore information flow using common operational pictures for maritime search and rescue operations. *Ergonomics*, 56, 889–905.

Benta, M. I. (2005). Studying communication networks with AGNA 2.1. *Cognition, Brain, Behaviour*, 9, 567–574.

Dekker, S. (2014). *The Field Guide to Understanding 'Human Error'*. Ashgate: Aldershot, UK.

Driskell, J. E. and Mullen, B. (2005). Social network analysis. In: N. A. Stanton, A. Hedge, K. Brookhuis, E. Salas and H. Hendrick (Eds.), *Handbook of Human Factors and Ergonomics Methods*. CRC Press: Boca Raton, FL.

Dul, J., Bruder, R., Buckle, P., Carayon, P., Falzon, P., Marras, W. S., Wilson, J. R. and Van Der Doelen, B. (2012). A strategy for human factors/ergonomics: Developing the discipline and profession. *Ergonomics*, 55, 377–395.

Embrey, D. E. (1986). SHERPA: A systematic human error reduction and prediction approach. Paper presented at the International Meeting on Advances in Nuclear Power Systems, Knoxville, TN.

Flach, J. M., Carroll, J. S., Dainoff, M. J. and Hamilton, W. I. (2015). Striving for safety: Communicating and deciding in sociotechnical systems. *Ergonomics*, 58(4), 615–634.

Griffin, T. G. C., Young, M. S. and Stanton, N. A. (2010). Investigating accident causation through information network modelling. *Ergonomics*, 53, 198–210.

Grote, G., Weyer, J. and Stanton, N. A. (2014). Beyond human-centred automation – Concepts for human–machine interaction in multi-layered networks. *Ergonomics*, 57(3), 289–294.

Haddon-Cave, C. (2009). *The Nimrod Review: An Independent Review into Broader Issues Surrounding the Loss of the RAF Nimrod MR2 Aircraft XV230 in Afghanistan in 2006*. HM Stationery Office: London, UK.

Harvey, C. and Stanton, N. A. (2014). Safety in System-of-Systems: ten key challenges. *Safety Science*, 70, 358–366.

H. M Government. (2010). *Securing Britain in an age of uncertainty: the strategic defence and security review*. London, UK: The Stationery Office.

Hodgson, A., Siemieniuch, C. E. and Hubbard, E.-M. (2013). Culture and the safety of complex automated sociotechnical systems. *IEEE Transactions on Human-Machine Systems*, 43, 608–619.

Hollnagel, E. (1998). *Cognitive Reliability and Error Analysis Method*. Elsevier Science: Oxford, UK.

Hollnagel, E., Woods, D. D. and Leveson, N. C. (Eds.) (2006). *Resilience Engineering: Concepts and Precepts*. Ashgate: Aldershot, UK.

Houghton, R. J., Baber, C., Cowton, M., Walker, G. H. and Stanton, N. A. (2008). WESTT (workload, error, situational awareness, time and teamwork): An analytical prototyping system for command and control. *Cognition, Technology and Work*, 10, 199–207.

Houghton, R. J., Baber, C., McMaster, R., Stanton, N. A., Salmon, P. M., Stewart, R. and Walker, G. H. (2006). Command and control in emergency services operations: A social network analysis. *Ergonomics*, 49, 1204–1225.

Jenkins, D. P., Stanton, N. A., Salmon, P. M. and Walker, G. H. (2010). A systemic approach to accident analysis: A case study of the Stockwell shooting. *Ergonomics*, 53, 1–17.

Leveson, N. G. (2012). *Engineering a Safer World: Systems Thinking Applied to Safety*. MIT Press: Cambridge, MA.

Lundberg, J., Rollenhagen, C. and Hollnagel, E. (2010). What you find is not always what you fix – How other aspects than causes of accidents decide recommendations for remedial actions. *Accident Analysis and Prevention*, 42, 2132–2139.

MAA (2010). RA 1210: Management of operating risk (risk to life). In: M.A. AUTHORITY (Ed.) *RA 1210*. Military Aviation Authority: London, UK.

MAA (2011). *Regulatory Instruction MAA RI/02/11 (DG) – Air Safety: Risk Management*. Military Aviation Authority: London, UK.

Ministry of Defence (MoD). (2007). *Defence standard* 00-56, issue 4, part 2. London, UK: MoD.

Moray, N., Groeger, J. and Stanton, N. A. (2017). Quantitative modelling in cognitive ergonomics: Predicting signals passed at danger. *Ergonomics*, 60(2), 206–220.

Noyes, J. and Stanton, N. A. (1997). Engineering psychology: Contribution to system safety. *Computing & Control Engineering Journal*, 8(3), 107–112.

Plant, K. L. and Stanton, N. A. (2012). Why did the pilots shut down the wrong engine? Explaining errors in context using Schema Theory and the Perceptual Cycle Model. *Safety Science*, 50(2), 300–315.

Rafferty, L., Stanton, N. A. and Walker, G. H. (2012). *The Human Factors of Fratricide*. Ashgate: Surrey, UK.

Ramos, A. L., Ferreira, J. V. and Barceló, J. (2012). Model-based systems engineering: An emerging approach for modern systems. *IEEE Transactions of Systems, Man and Cybernetics: Part C – Applications and Reviews*, 42, 101–111.

Rasmussen, J. (1997). Risk management in a dynamic society: A modelling problem. *Safety Science*, 27, 183–213.

Read, G. J. M., Salmon, P. M., Lenné, M. G. and Stanton, N. A. (2015). Designing sociotechnical systems with cognitive work analysis: Putting theory back into practice. *Ergonomics*, 58, 822–851.

Royal Navy. (2012). Fleet requirements air direction unit (FRADU) . Royal Navy. Available at: http://www.royalnavy.mod.uk/sitecore/content/home/the-fleet/air-stations/rnas-culdrose/fleet-requirements-air-direction-unit-fradu (Accessed 11 January 2013).

Salmon, P. M., Cornelissen, M. and Trotter, M. J. (2011a). Systems-based accident analysis methods: A comparison of AcciMap, HFACS, and STAMP. *Safety Science*, 50, 1158–1170.

Salmon, P. M., Lenné, M. G., Walker, G. H., Stanton, N. A. and Filtness, A. (2014). Using the event analysis of Systemic Teamwork (EAST) to explore conflicts between different road user groups when making right hand turns at urban intersections. *Ergonomics*, 57, 1628–1642.

Salmon, P. M., Stanton, N. A., Lenné, M., Jenkins, D. P., Rafferty, L. and Walker, G. H. (2011b). *Human Factors Methods and Accident Analysis*. Ashgate: Surrey, UK.

Shappell, S. and Wiegmann, D. (2001). Applying reason: The human factors analysis and classification system. *Human Factors and Aerospace Safety*, 1, 59–86.

Shorrock, S. T. and Kirwan, B. (2000). Development and application of a human error identification tool for air traffic control. *Applied Ergonomics*, 33, 319–336.

Stanton, N. A. (2006). Hierarchical task analysis: Developments, applications, and extensions. *Applied Ergonomics*, 37, 55–79.

Stanton, N. A. (2014). Representing distribution cognition in complex systems: How a submarine returns to periscope depth. *Ergonomics*, 57, 403–418.

Stanton, N. A., Baber, C. and Harris, D. (2008). *Modelling Command and Control: Event Analysis of Systematic Teamwork*. Ashgate: Surrey, UK.

Stanton, N. A., Harris, D. and Starr, A. (2016). The future flight deck: Modelling dual, single and distributed crewing options. *Applied Ergonomics*, 53, 331–342.

Stanton, N. A., Rafferty, L. A. and Blane, A. (2012). Human factors analysis of accidents in systems of systems. *Journal of Battlefield Technology*, 15, 23–30.

Stanton, N. A., Salmon, P., Harris, D., Marshall, A., Demagalski, J., Young, M. S., Waldmann, T. and Dekker, S. W. A. (2009). Predicting pilot error: Testing a new methodology and a multi-methods and analysts approach. *Applied Ergonomics*, 40(3), 464–471.

Stanton, N. A., Salmon, P. M., Rafferty, L. A., Walker, G. H., Baber, C. and Jenkins, D. P. (2013). *Human Factors Methods: A Practical Guide for Engineering and Design, Second Edition*. Ashgate : Aldershot, UK.

Stanton, N. A., Salmon, P. M., Walker, G. H., Baber, C. and Jenkins, D. P. (2005). *Human Factors Methods: A Practical Guide for Engineering and Design, First Edition*. Ashgate: Aldershot, UK.

Stanton, N. A. and Stevenage, S. (1998). Learning to predict human error: Issues of reliability, validity and acceptability. *Ergonomics*, 41(11), 1737–1756.

Stanton, N. A., Stewart, R., Harris, D., Houghton, R. J., Baber, C., McMaster, R., Salmon, P. et al. (2006). Distributed situation awareness in dynamic systems: Theoretical development and application of an ergonomics methodology. *Ergonomics*, 49, 1288–1311.

Stanton, N. A. and Walker, G. H. (2011). Exploring the psychological factors involved in the Ladbroke Grove rail accident. *Accident Analysis & Prevention*, 43(3), 1117–1127.

Stewart, R., Stanton, N. A., Harris, D., Baber, C., Salmon, P., Mock, M., Tatlock, K., Wells, L. and Kay, A. (2008). Distributed situation awareness in an airborne warning and control system: Application of novel ergonomics methodology. *Cognition, Technology and Work*, 10, 221–229.

Swain, A. D. and Guttmann, H. E. (1983). *A Handbook of Human Reliability Analysis with Emphasis on Nuclear Power Plant Applications, NUREG CR-1278*. USNRC: Washington, DC.

Underwood, P. and Waterson, P. (2013). System accident analysis: Examining the gap between research and practice. *Accident Analysis and Prevention*, 55, 154–164.

von Bertalanffy, L. (1950). An outline of general system theory. *British Journal for the Philosophy of Science*, 1, 134–165.

Walker, G. H., Gibson, H., Stanton, N. A., Baber, C., Salmon, P. and Green, D. (2006). Event analysis of systematic teamwork (EAST): A novel integration of ergonomics methods to analyse C4i activity. *Ergonomics*, 49, 1345–1369.

Walker, G. H., Stanton, N. A., Baber, C., Wells, L., Gibson, H., Salmon, P. M. and Jenkins, D. (2010a). From ethnography to the EAST method: A tractable approach for representing distributed cognition in air traffic control. *Ergonomics*, 53, 184–197.

Walker, G. H., Stanton, N. A., Sal mon, P.M. and Jenkins, D. P. (2008). A review of sociotechnical systems theory: A classic concept for command and control paradigms. *Theoretical Issues in Ergonomics Science*, 9, 479–499.

Walker, G. H., Stanton, N. A., Salmon, P. M. and Jenkins, D. P. (2009). *Command and Control: The Sociotechnical Perspective*. Ashgate: Aldershot, UK.

Walker, G. H., Stanton, N. A., Salmon, P. M., Jenkins, D. P. and Rafferty, L. A. (2010b). Translating concepts of complexity to the field of ergonomics. *Ergonomics*, 53, 1175–1186.

Waterson, P., Robertson, M. M., Cooke, N. J., Militello, L., Roth, E. and Stanton, N. A. (2015). Defining the methodological challenges and opportunities for an effective science of sociotechnical systems and safety. *Ergonomics*, 58(4), 565–599.

Wilson, J. R. (2012). Fundamentals of systems ergonomics. *Work*, 14, 3861–3868.

13 From CWA to SNA *Modelling Future Flight Decks*

With Don Harris and Alison Starr

INTRODUCTION

The trend in flight deck design over the past half century has been one of progressive 'de-crewing'. Fifty years ago, it was not uncommon for there to be five crew on the flight deck of a civil airliner (two Pilots, Flight Engineer, Navigator and Radio Operator). Today, just two pilots accomplish the same tasks once undertaken by five. Many functions are now wholly or partially automated. Consequently, the role of the pilot has changed from one of being a 'flyer' to one of being a systems/flight deck manager.

Airline personnel costs vary between about 11% of operating costs to nearly 25%, depending upon aircraft type, sector length and how much activity is outsourced (Ryanair 2009; easyJet plc 2013). Crew costs for smaller commercial aircraft can be between 15% and 35% of the aircraft direct operating costs (Alcock 2004). Annual accounts from a typical low-cost operator suggest that even for a larger airliner, the crew represent nearly 19% of operating costs (excluding fuel and propulsion – easyJet plc 2013). The scope to make significant cost savings with the current common configuration for aircraft (cylindrical fuselage with wings, rudder and tail plane) is now limited. This configuration is approaching the end of its development potential. Alternative configurations such as the blended wing body concept, which offer considerable structural and aerodynamic advantages, have met with limited enthusiasm from potential passengers. Problems of ensuring a safe and efficient means of passenger evacuation have also been identified (Galea et al. 2011). This configuration also needs a great deal of development in other areas (such as flight control systems and structural testing) before it will be suitable for service entry. Reducing the number of crew on the flight deck to just a single pilot will produce significant cost savings, especially in smaller commercial aircraft operated on shorter and 'thinner' (lower-demand) routes.

Some manufacturers (e.g. Embraer) are already developing the technology for a single-crew aircraft, as are avionics suppliers (e.g. Honeywell – see Keinrath et al. 2010). The approaches being adopted in these instances centre upon the development of sophisticated airborne technology to assist the pilot (e.g. Intelligent Knowledge-Based Systems and adaptive automation). This approach is also being adopted in other research programmes looking at flight deck automation and crewing, for example, the development of an Electronic Standby Pilot (ESP) as part of the Advanced Cockpit for the Reduction of Stress and Workload (ACROSS) project (see http://www.across-fp7.eu/). The medium-term objectives of the ACROSS project are concerned with reducing the number of flight deck crew in the cruise phase in long-haul flights to permit crew to rest and help prevent fatigue. It is anticipated that the same technology will aid in the case of partial (or even full) flight crew incapacitation. The longer-term objectives of the project are to form the basis for potential single-crew operations.

A similar approach has been adopted by other researchers in the past, particularly in the military domain, but with only mixed success (e.g. the COGnitive cockPIT – COGPIT programme – Bonner et al. 2000; Taylor et al. 2000; and the Cockpit Assistant Military Aircraft – CAMA programme – Schulte and Stütz 2001; Stütz and Schulte 2001). CASSY (the Cockpit ASsistant System) was a civil aircraft version of the latter developed by the same team (see Onken 1994; Onken 1997). The Cognitive Adaptive Man–Machine Interface (CAMMI) project (Keinrath et al. 2010) also makes use of extensive Artificial Intelligence (AI) software in its approach to adaptive automation. A later requirements analysis for developing concepts for single-pilot operations was also predicated upon the notion of incorporating extensive pilot automated assistance on the flight deck, particularly synthetic vision systems, data linking and direct voice input/output systems (Deutch and Pew 2005). The main arguments for the use of two members of flight deck crew centre around issues concerned with pilot workload (specifically instances of workload peaks), the reduction of flight crew error and pilot incapacitation. Many of the assumptions are either questionable or are becoming out-dated as discussed in the following three paragraphs.

From the perspective of the person in command of any aircraft, there is a workload 'cost' associated with the management of crew on the flight deck. The requirement to coordinate crew, cooperate and communicate on the flight deck itself has workload associated with it. Doubling the number of crew does not half the workload. Furthermore, modern flight decks are already certificated so that they can be operated by a single member of flight deck crew (see FAR/CS 25.1523). Automated flight deck systems have already considerably reduced pilot workload (Weiner and Curry 1980; Harris 2003).

While the second crew member may distribute the workload around the flight deck somewhat, it can also be argued that they actually introduce an error mode. Poor CRM (Crew Resource Management) has been implicated as a contributory factor in nearly 23% of all fatal commercial jet aircraft accidents (CAA] 2008). The effectiveness of the second pilot as an 'error checker' is also questionable. Omission of action or inappropriate action was implicated in 39% of accidents, and an incorrect application of procedures or a deliberate non-adherence to procedures was implicated in a further 13% (CAA 2008). Becoming 'low and slow' (a failure to cross-monitor the flying pilot) was a factor in 12% of accidents. As a cross-check on the position of the aircraft, the PM's (Pilot Monitoring) effectiveness would also seem to be questionable, as a lack of positional awareness was identified as a causal factor in 27% of cases (CAA 2008). This is quite a crude analysis, however. It is acknowledged that what these data do not show is in how many cases the second pilot trapped an error made by the other pilot and avoided an accident; this is unknown and unknowable. However, observational data obtained from routine flights reported that 47.2% of errors committed by Captains involved intentional non-compliance with Standard Operating Procedures (SOPs) or regulations; 38.5% were unintentional procedural non-compliance (Thomas 2003). Thomas also reports that in observations of line operations, crews did not demonstrate effective error detection, with more than half of all errors remaining undetected by one or both of the flight crew. As a result, it can be argued that removing one of the pilots actually reduces the scope for accidents occurring as a result of miscommunication or misunderstanding between the pilots, and that removing the PM does not double the workload on the flight deck.

Perhaps the greatest concern for the development of a single-crew aircraft is that associated with pilot death, incapacitation or impairment. However, such instances are very rare. A study of in-flight medical incapacitations in US airline pilots between 1993 and 1998 found only 39 instances of incapacitation and 11 instances of impairment (DeJohn et al. 2004). The rate of in-flight medical events (encompassing both types) was 0.058 per 100,000 flight hours. The probability that one of these events would *subsequently* result in an accident was calculated to be 0.04. DeJohn et al. (2004) observed that the safety of the flight was seriously impacted in only seven cases and resulted in two non-fatal accidents. A later study of UK commercial pilots by Evans and Radcliffe (2012) suggest that the annual in-flight incapacitation rate was 0.25%; however, this study is seriously flawed in than it was not weighted by flight hour and the rate is expressed as a percentage of all UK registered pilots (irrespective of flight hours accumulated by each, per year).

It is argued that with the judicious use of existing equipment, there are no major reasons why a single-pilot-operated commercial aircraft is not feasible in the very near future using existing technology. Military aviation has flown complex, high-performance single-crew aircraft for many years, and Unmanned Air Vehicles (UAVs) are now commonplace. UAV technology has matured, and such aircraft are now regularly being used for national border and port security, homeland surveillance, scientific data collection and telecommunications services (Harris 2007). Airworthiness standards for their design and operation in civil airspace are being developed on both sides of the Atlantic (e.g. UK Civil Aviation Authority 2010 – *Unmanned Aircraft System Operations in UK Airspace – Guidance [CAP 722]*). Several UAVs are now the size of a small aircraft, with performance similar (or exceeding) that of a conventional aeroplane. It is worth noting that UAVs still have a designated 'Pilot in Command' International Civil Aviation Organisation Annex 2, 'Rules of the Air', states that the 'Responsibility of the pilot-in-command' of an aircraft shall, whether manipulating the controls or not, be responsible for the operation of the aircraft in accordance with the rules of the air, except that the pilot-in-command may depart from these rules in circumstances that render such departure absolutely necessary in the interests of safety'.

The greatest obstacle to the operation of civilian, single-pilot aircraft is not the technology per se. Rather, the barriers are combining the ground and airborne technologies, designing the user interfaces and developing new concepts of operations to make such an aeroplane safe and useable in a wide range of normal and non-normal operating situations (when flown by a typical commercial pilot). That is to say that the Human Factors requirements are the prime driver in this case, not the technology. The concept evaluated in this chapter is based upon an alternative design approach to that of utilising a large amount of on-board, complex computing (e.g. that using agent-based software) first described by Harris (2007). The concept uses a sociotechnical systems-based design philosophy utilising a great deal of currently existing technology. In this case, the control and crewing of the aircraft is distributed in real time across both the aircraft's flight deck and ground stations (see also Stanton et al. 2014). The second pilot is not *replaced* by on-board AI or Intelligent Knowledge-Based Systems, which would be both difficult to develop and challenging to certificate; they are merely *displaced.*

DESIGN APPROACH

The proposed approach regards a future single-crew aircraft as just one part of a wider operating system, a radical change from the operation of current generation airliners. The initial high-level design architecture proposed for operating the single-crew aircraft consists of several discrete elements (Stanton et al. 2014):

- The aircraft itself (including pilot)
- Ground-based components including
  - 'Second pilot' support station/office
  - Real-time engineering support
  - Navigation/flight planning support
- System 'Mirror'.

This is the initial envisaged instantiation of the system upon which to base a cognitive work analyses in order to evaluate its efficacy and to identify required technology development paths. The design of this single-crew aircraft operations system is intended to form the basis of incremental developments, incorporating subsequent airborne pilot-support technology as it is developed. This will allow the further rapid development of the concept.

AIRCRAFT COMPONENT

In many ways, the aircraft component will be little different to current types. The requirements are that aircraft should be able to function in all types of airspace without any special Air Traffic Control/Air Traffic Management (ATC/ATM) procedures and should be able to be flown by regular Airline Transport Pilots Licence (ATPL) qualified professional pilots without extraordinary training. It should exhibit an equivalent level of safety to fourth-generation modern airliners. Initially, such an aircraft will be optimised for shorter-range, low-cost operations (including cargo operations) and for 'thinner' routes where cost of operation is a critical factor. Emphasis will be placed upon reduction of workload and error by simplification of operation.

Ground-Based Component

The ground-based component's primary functions are to support the pilot (e.g. in navigation, system management, ATC/ATM support or fault diagnosis), not necessarily to duplicate their skills and functions. Suitably skilled personnel on the ground may simultaneously support several aircraft (resulting in economy of scale and reduction in duplication of effort). Furthermore, these personnel need not necessarily have the same skill set as a pilot; rather, they could be specialists in navigation, communications or avionics systems. In the case of the single-crew aircraft, *control* from a ground station would only be required in the advent of pilot incapacitation, although *assistance* from the ground may be required in other circumstances, such as high workload (e.g. take-off/landing) or abnormal (e.g. re-routing/bad weather/emergency) situations. This approach would allow ground-based operators to undertake many of the key roles of the second pilot but to do so for several aircraft simultaneously. This emphasises that the design of a single-crew aircraft is not simply concerned with the re-design of the flight deck. It requires a change in the overall operating philosophy.

'Second Pilot' Support Station/Office The primary functions of the ground-based 'second-pilot' station/office is to provide real-time support for the pilot on the aircraft as required and to assist in flight planning and pre-flight preparation. If the hierarchy of tasks when flying an aircraft can be conceptualised as 'Aviate', 'Navigate', 'Communicate' and 'Manage', whenever possible the 'Aviate' component will remain on the flight deck itself. However, if necessary, it will be possible to operate the aircraft remotely from this position. It is envisaged that this support station will be able to support a number of aircraft.

The function of this aspect of the ground-support element is four-fold:

- Prior to flight, the ground-based support function provides a cross-check of flight planning data.
- When pilot workload is high, if required, the operator in the ground-based workstation may assume many of the duties of the second pilot normally on board the aircraft.
- During non-normal, abnormal or emergency situations, the operator in the ground-based workstation will also assume the duties normally undertaken by the second pilot; however, this will also be in conjunction with other support (provided as required) by engineering support and navigation/flight planning support elements.
- In the event of pilot incapacitation, the aircraft can be 'flown' from the ground station (in the same manner as a UAV). It is envisaged that any direct operation from this position will be undertaken via the aircraft's automation; direct operation via the primary flight controls will be a last resort.

The main focus of ground-based support from this operator station will be on aircraft configuration management and short-term navigation and communication support to the pilot. Emphasis will be placed upon decision-aiding aspects of pilot support.

This station will be crewed by pilots qualified on the single-crew aircraft, who 'rotate' through the post in between actually flying the aircraft. This should enhance the shared Situation Awareness between ground and air components.

Real-Time Engineering Support: The single-crew aircraft will routinely pass system operation information to the ground for automated health monitoring. This is already done routinely for engine data but may be expanded to encompass all critical aircraft systems in real time. System information will be monitored for significant deviations from normal on the ground. In the event of a non-normal situation being detected, ground-based automated systems and engineering staff will evaluate the system failure and either provide advice to the pilot or re-configure the system remotely. Emphasis will be upon stabilising the situation and evaluating consequences rather than attempting to rectify the problem in flight.

In the event of a serious system failure, the implications for the continuation of the flight will be evaluated. If it is necessary to perform an immediate descent or diversion, the appropriate navigation/flight planning facility will be notified.

Engineering support will be provided by engineers qualified on the aircraft and trained to provide remote real-time support in the advent of an airborne system malfunction.

Navigation/Flight Planning Support: This is an extension of the existing airline flight planning functions to encompass real-time, in-flight planning/re-planning facilities. This facility will be expanded to encompass an 'aircraft-centric' point of view and will be able to access the on-board flight management computers to up-link new routing information. Support (long and short term) will be provided by flight planning specialists.

System 'Mirror'

The system mirror will be an independent, ground-based (software) representation of the aircraft system states (in particular the Flight Management System [FMS], autopilot system and autothrust systems and the general configuration of the aircraft – flaps, slats, etc.). Ground-based system elements (pilot support station, flight/navigation panning and engineering support) will be able to interact with the ground-based system mirror without directly affecting aircraft systems (if required). In the advent of a datalink failure, the system mirror will contain the last known configuration of the aircraft systems and will be able to update aircraft systems, if required. During normal operations, the automation mirror will normally be 'transparent' to all operators in the system.

MODELLING AND ANALYSIS OF SYSTEM CONFIGURATIONS

Complex sociotechnical systems (such as flight operations) are made up of numerous interacting parts, both human and non-human, operating in a dynamic, ambiguous and safety–critical domain (Harris and Stanton 2010). The complexity embodied in these systems presents significant challenges for modelling and analysis and requires Systems Ergonomics methods for the effective design of future work systems (Wilson and Carayon 2014). Cognitive Work Analysis (CWA) is a structured framework specifically developed for considering the development and analysis of such complex sociotechnical systems (Rasmussen et al. 1994; Vicente 1999; Jenkins et al. 2009). The framework leads the analyst to consider the environment within which the task takes place and the effect of the imposed constraints on the way work can be conducted. The framework guides the analyst through the process of answering the question of why the system exists and what activities can be conducted within the domain, as well as how these activities can be achieved and who can perform them, also identifying the competencies required. Recent case studies have been conducted in the assessment of risk associated with nuclear decommissioning (Walker et al. 2014) and command team activities in control of a submarine (Stanton and Bessell 2014).

The analyses in this chapter commence with a generic CWA concerning the operation of various options for system configurations in a number of operating scenarios. These analyses are supplemented by a Social Network Analysis (SNA) which provides an indication of the resilience of the operational networks in each case (Diskell and Mullen 2004; Baber et al. 2013). This approach has previously been used to examine the characteristics of terror cells (Kenney et al. 2013) and command networks (Walker et al. 2009). These analyses can be used to examine options for the proposed distributed system architecture for the operation of a single-crew aircraft. They also help to define more specific aspects of new on-board automation requirements for the single-crew aircraft and the role of the Ground Support Station/Office.

Analyses commence with a 'conventional' baseline aircraft (Airbus A320 and variants) being flown by two crew, followed by an analysis of the same aircraft being operated by a single pilot. Note that it is a certification requirement that all aircraft must be capable of being safely operated by a single crew member (FAR/CS 25.1523 – Minimum Flight Crew). These initial analyses were followed by a comparative analysis of four different configuration options for a single-crew-operated aircraft being operated as part of a distributed system. Four potential versions of single-pilot flight operations are presented for comparison (as shown in Table 13.1), that is, a single-pilot aircraft (A); a single-pilot aircraft with an additional pilot at a ground station who can be called upon at times of need (C); a single-pilot aircraft with an automation mirror on the ground that cross-checks the inputs and outputs independently of the aircraft automation (B); and a single-pilot aircraft with an additional pilot at a ground station and also with an independent automation mirror (D).

TABLE 13.1
Comparison of Four Potential Versions of Single-Pilot Flight Operations

Four Versions of the Future		Pilot in Ground Station	
Automation		NO	YES
Mirror	NO	Single-pilot aircraft (Option A)	Single-pilot aircraft with additional pilot on the ground (Option C)
	YES	Single-pilot aircraft with additional automation mirror on the ground (Option B)	Single-pilot aircraft with additional automation mirror on the ground and additional pilot on the ground (Option D)

Material for undertaking the CWA and SNA was drawn from a number of sources, including operations manuals and SOPs (Airbus Industrie n.d.). These were complemented by a structured de-brief of an experienced qualified Test Pilot who was also type rated on a number of Airbus types (and is a training Captain for a major airline), who also helped to devise the various operating scenarios. The CWA material will be presented first as the formative design of the future system alternatives. Much has been made of CWA as a formative approach (Vicente 1999; Jenkins et al. 2009; Naikar 2013) but little is seen in practice. CWA served as the basis for identifying the functional loading on agents in the system as well as providing the data for the SNA.

Cognitive Work Analysis (CWA)

In the first phase of CWA, the Abstraction Hierarchy (AH) is used to model the work domain as follows.

Work Domain Analysis (WDA)

WDA is the most commonly used component within CWA (McIlroy and Stanton 2011), and it identifies the constraints imposed on workers' behaviour by the purposive and physical context, or problem space, in which workers operate (Naikar 2006a). WDA is conducted at the functional rather than behavioural level; it is used to define the environment within which the activity is conducted. WDA identifies a fundamental set of constraints on the actions of any system component, thus providing a solid foundation for subsequent phases of the development of the aircraft (McIlroy and Stanton 2011). The abstraction component of the diagram models the same system at a number of levels; at the highest level, the overall functional purpose of the system is considered, while at the lowest level, the individual components within the system are described. Generally, five levels of abstraction are used:

- Functional Purposes (the purposes of the work system and the external constraints on its operation)
- Values and Priority Measures (the criteria that the work system uses for measuring its progress towards the functional purposes)
- Purpose-Related Functions (the general functions of the work system that are necessary for achieving the functional purposes)
- Object-Related Processes (the functional capabilities and limitations of physical objects in the work system that enable the purpose-related functions)
- Physical Objects (the physical objects in the work system that afford the object-related processes).

Functional Purposes

The AH was created in both a top-down and a bottom-up fashion. First, the overall functional purpose of flight operations was specified. This is the reason why the system exists. These purposes are independent of time; they exist for as long as the system exists. In this case, it is to 'transport people and cargo safely from A to B' and 'increase

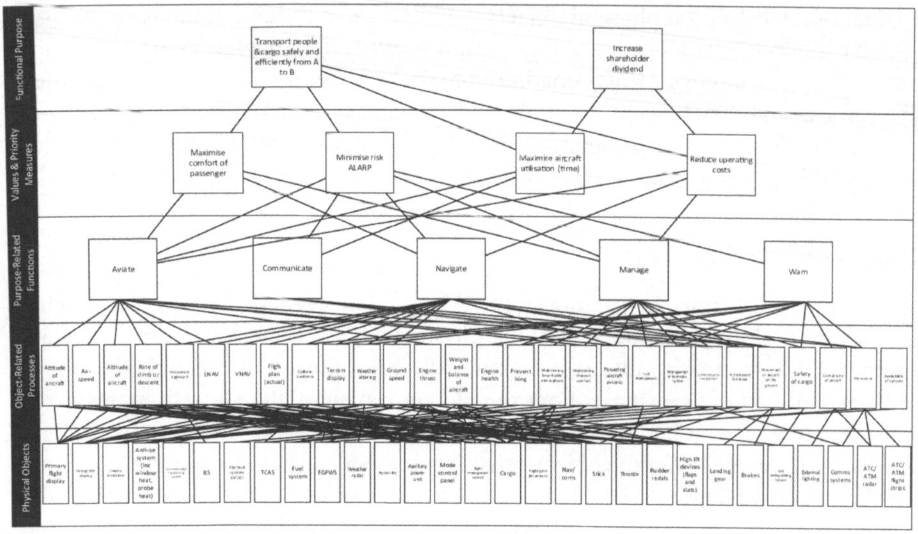

FIGURE 13.1 Work domain analysis showing the five levels and the means-ends-links.

shareholder dividend' (i.e. ensure that the aircraft was economic to operate). These purposes were represented as two nodes at the top of the WDA in Figure 13.1.

Values and Priority Measures

The next stage of the construction was to consider the level below the functional purposes, the values and priority measures, as shown in Figure 13.1. Here, further constraints on the system were more explicitly listed; these were measures for determining how well the intended system achieved its functional purposes. In this case, they were identified as 'maximise comfort of passengers' (in order to enhance the customer experience), 'minimise risk' (to keep the system as safe as possible), 'maximise aircraft utilisation' (to avoid storage costs and enhance revenue) and 'reduce operating costs' (to avoid wastage). The way that the future aircraft system is configured to meet these needs is likely to be heavily contextually dependent.

Physical Objects

As previously stated, a mixture of a top-down and a bottom-up approach has been employed in this analysis. Moving down to the lowest level of the hierarchy, the physical objects within the system were listed. The boundaries of this analysis limited this list to the systems of direct relevance to the operation of a future single-crew aircraft rather than every single object. The boundaries of the analysis indicate the levels of fidelity applied here. In an attempt to keep the analysis manageable, the boundary has omitted individual system elements. Whilst it would have been possible to decompose many of the listed objects into their component parts and describe their affordances more concisely, this was outside the remit of this initial evaluation. A list of the physical objects analysed can be found at the base of Figure 13.1. This list comprises primary flight display, navigation display, engine indication, anti-ice system, environmental conditioning system, electrical systems, fuel system, hydraulics, auxiliary power unit, flight management system, nav/comms (navigation/communication), throttle, rudder pedals, high-lift devices, landing gear, brakes, fire-extinguishing system and external lighting.

Object-Related Processes

The second level from the bottom of the AH, the object-related processes, captures the processes that are conducted by the physical objects to perform purpose-related functions. Most importantly, they capture the affordances of the physical objects independently of their purpose. For example, movement of the aircraft on the ground is afforded by the combination of the throttle, landing gear (nose wheel steering) and brakes. A complete list of the object-related processes can be found at the level above the base level of Figure 13.1 along with the links indicating to which object they relate. To aid readability, this list comprises attitude of aircraft, airspeed, altitude of aircraft, rate of climb or descent; Instrument Landing System (ILS) indications, current heading of aircraft, flight plan, collision avoidance, terrain display, weather alerting, ground speed, engine thrust; weight and balance of aircraft; engine health; preventing icing; maintaining breathable atmosphere, maintaining thermal comfort, powering aircraft avionics, fuel management, management of hydraulics, communication management, performance/cost index, movement of aircraft on the ground, safety of cargo and conspicuity of aircraft.

Purpose-Related Functions

In the middle of the AH, the purpose-related functions are listed. These functions have the ability to influence one or more of the values and priority measures. They link the purpose-independent processes with the object-independent functions. They are listed as 'aviate' (i.e. keeping the aircraft airborne); 'communicate' (i.e. staying in touch with ATC/ATM and company operations as well as communicating with crew and passengers); 'navigate' (i.e. plan, change and check the route for the aircraft as required); 'manage' (i.e. manage the avionics system; and 'warn' (i.e. indicate to aircrew and company operations when systems are outside their tolerances).

The use of the means-ends-links and the utility of the AH can be described with the example shown in Figure 13.1. Figure 13.1 shows the node 'aviate' in the purpose-related functions level. Following the links out of the top of this node answers the question 'why is this needed?' – in this case, to 'maximise comfort of passengers' and 'maximise aircraft utilisation'. Following the links down from the 'aviate', it is possible to answer the question 'how can this be achieved?' – in this case, by 'attitude of aircraft', 'airspeed', 'altitude of aircraft', 'rate of climb', ILS indications' and 'current heading/track of aircraft'.

The diagram does not prescribe a particular arrangement for providing this functionality; rather, it lists all of the components that can affect it. In this case, there is redundancy in the system.

Control Task Analysis (ConTA)

The second phase of the CWA framework, Control Task Analysis (ConTA), allows the requirements associated with known, recurring classes of situations to be identified. Naikar et al. (2006a, 2006b) have developed the Contextual Activity Template (CAT) (see Figure 13.2) for use in this phase of the CWA. This template is one way of representing activity in work systems that are characterised by both work situations and work functions. Work situations can be decomposed based on recurring schedules or specific locations. Rasmussen et al. (1994) describe work functions as being activity characterised by its content, independent of its temporal or spatial characteristics. These functions can often be informed by the abstraction hierarchy. Rasmussen et al. (1994) recommend that the analyst decompose on either work functions or work situations; however, Naikar et al. (2006a, 2006b) plot these on two axes so that their relationship can be investigated, allowing the representation of activity in work systems that are characterised by both work situations and work functions. Typically, the work situations (in this case the phases of flight as defined by the International Civil Aviation Organization [ICAO]) are shown

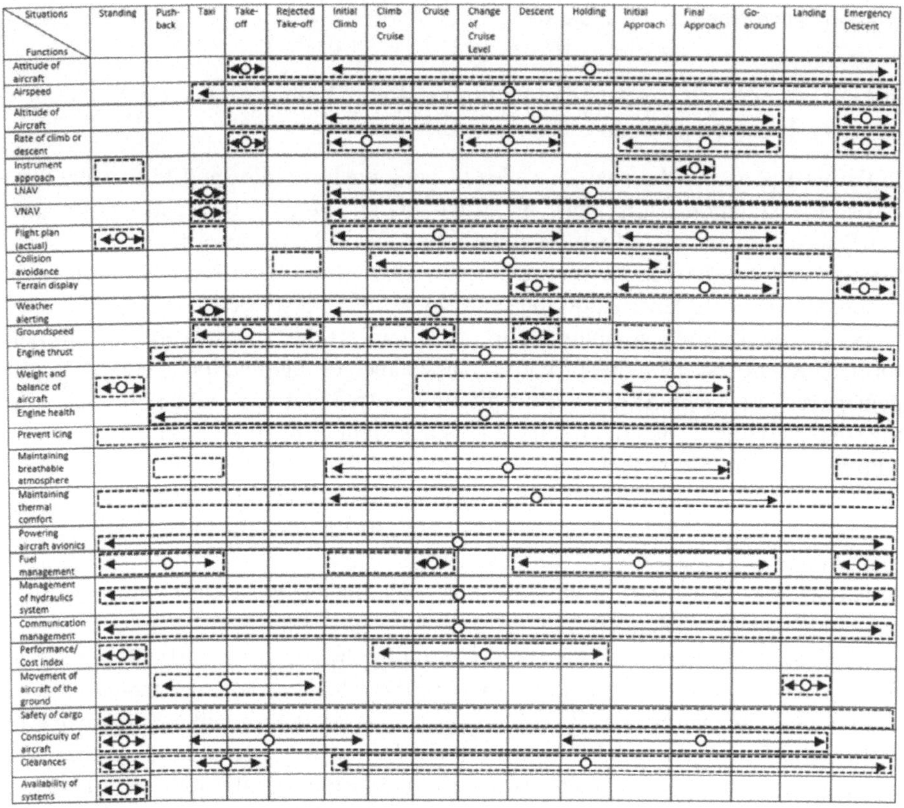

FIGURE 13.2 CAT for object-related processes.

along the horizontal axis, and the work functions (associated with controlling and managing the aircraft) are shown along the vertical axis of the CAT. The circles indicate the work functions with the bars showing the extent of the table in which the activity typically occurs. The dotted boxes around each circle indicate all of the work situations in which a work function *can* occur (as opposed to *must* occur), thus capturing the constraints of the system. Figure 13.2 shows the CAT for the purpose of related functions derived from the abstraction hierarchy (see Figure 13.1). The situations follow the main phases of flight operations, from the aircraft standing at a gate to landing on a runway (assuming that taxiing and standing will be similar for both take-off and landing), along with the addition of an emergency descent.

Looking at Figure 13.2, perhaps the most salient feature is that in an aircraft during flight operations, some function constraints are notably contextually influenced (e.g. rate of climb or descent, terrain display, ground speed, and movement of aircraft on the ground) whereas others are not (e.g. engine thrust, engine health, powering aircraft avionics, management of hydraulics and communication systems). To read Figure 13.2, note that an empty cell means that the function (vertical axis) cannot be performed in that situation (horizontal axis); a dashed box means that the function *can* be performed in the situation, and a ball-and-whisker means that the function is *typically* performed in the situation.

Social Organisational Cooperation Analysis – Contextual Activity Template (SOCA-CAT)

In the next phase of the analysis, roles of personnel are allocated to the functions across the situation. This phase is called Social Organisation and Cooperation Analysis – Contextual Activity Template (SOCA-CAT). A list of the key roles and their related coding can be seen in Figure 13.3. One of the roles is non-human, denoted as Aircraft Automation in Figure 13.3 (the light speckled coding).

As Figure 13.3 shows, many of the functions in the baseline aircraft are performed either exclusively by, or with the assistance of, automation. There is a clear demarcation between the roles of the pilot flying (red coding) and the pilot monitoring (green coding). The pilot monitoring is oversight of the pilot flying (red coding), automation (blue coding) and ATC/ATM (pink coding). It is striking that each function tends to be assigned to the same set of roles throughout all situations (where that function applies). Of the 25 functions identified, 10 are typically the responsibility of just a single agent, whereas 15 are joint responsibility (most being the responsibility of two agents). Some 240 of the 400 cells (functions against situations) have been assigned to agent roles in SOCA-CAT.

SOCA-CAT has provided the data for building the social networks. This method has previously been proven by Baber et al. (2013) when developing operational concepts for search and rescue missions. In essence, every within-cell and between-cell relationship becomes a link between agents in the network (except for the emergency descent). In this way, it is possible to step from a formative allocation of function to a formative SNA. This is a useful extension of the CWA framework, as it enables quantitative as well as qualitative exploration of future operational concepts.

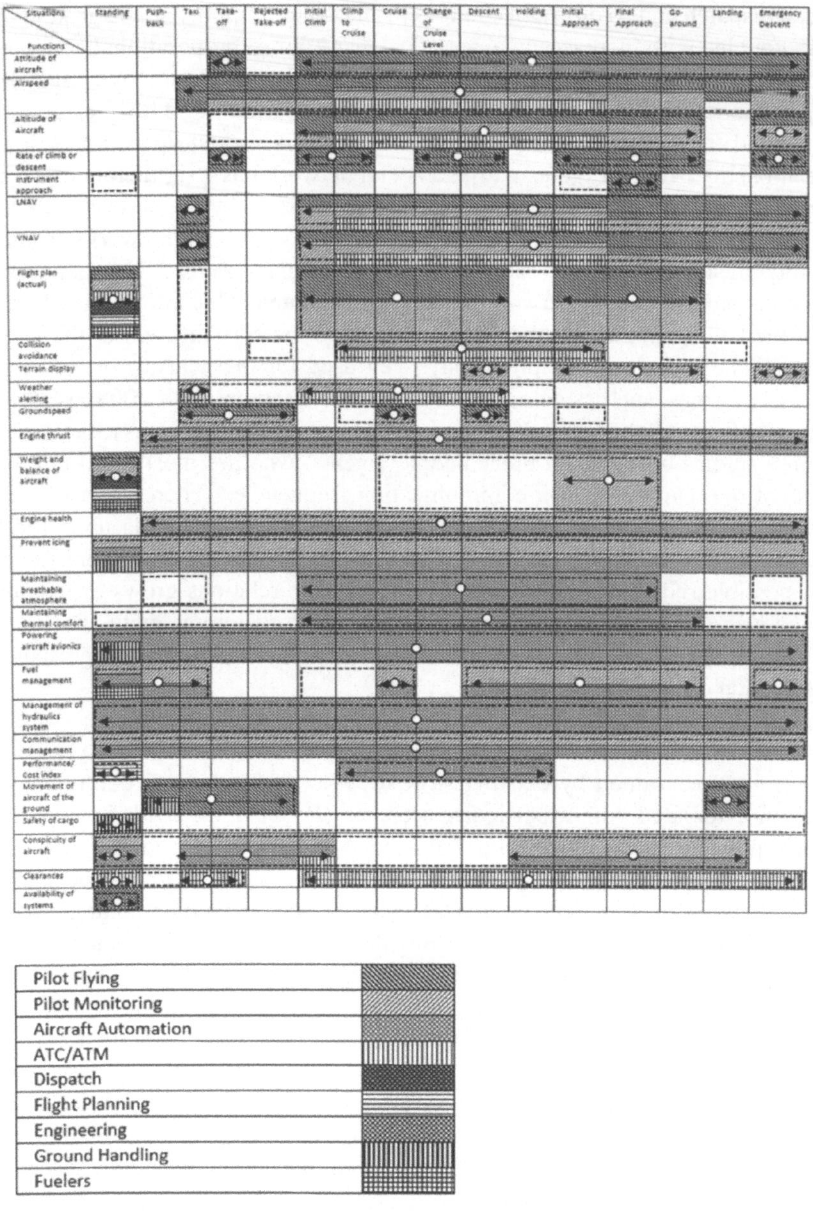

FIGURE 13.3 Social organisation and SOCA-CAT for current flight operations.

Social Network Analysis

SNA is used to analyse and represent the relationships between agents and artefacts within a network. A *social network* is defined as a set or team of agents that possess relationships with one another (Driskell and Mullen 2004). In this case, the social network extends beyond the aircraft to other agencies such as air traffic control, fuellers, dispatch and so on. With changes in automated support on the ground and in the air, coupled with increases in air-traffic density, aircraft operational concepts are constantly changing. Direct routing and self-assured separation ('free flight') have vastly expanded the Joint Cognitive System boundaries around airliner operations well beyond the flight deck (Hollnagel 2007). SNA is based upon the notion that the relationship between agents within a social network has a significant effect upon the actions performed and also the performance achieved by the network. SNA uses both graphical and mathematical procedures to represent social networks. Typically, centrality measures are calculated for each agent (e.g. degree, betweenness and closeness), and the overall network density is calculated. This allows the identification of the key agents within the network and also the classification of the network structure. The technique has previously been used for the analysis of networks in a number of areas, such as command and control (Stanton et al. 2008), search and rescue operations (Baber et al. 2013) and submarine command teams (Stanton 2014). All three studies have shown those people (agents) who are most important to the success of the team, based on communication frequencies. There are a number of metrics associated with the analysis of whole networks, depending upon the type of evaluation that is being performed. The size of the network determines the number of possible relations, and the number of possible relations grows exponentially with the size of the network. This defines the network's complexity. The latter is expressed in the form of social relations that are actually observed, represented as some fraction of the total possible.

A major advantage of networks is that they do not differentiate between different types of node (e.g. artefacts and/or people), so that from a modelling perspective, they are not constrained by existing structures but rather help to define the tasks allocation associated with a particular scenario (Baber et al. 2013; Stanton 2014). It is also possible to model the temporal aspects of networks by identifying critical moments in the sequence of activity. To do this, the scenario is divided into task phases, allowing active and non-active elements to be specified and represented.

The first step in a SNA involves defining the network that is to be analysed. Once the overall aircraft operation network is specified, the people and/or artefacts need to be defined. In the case of the development of the single-crew aircraft, a number of different networks were identified for analysis over a number of different flight scenarios. In this case, the agents identified were the pilot flying, the pilot monitoring, ATC/ATM, dispatch, flight planning, engineering, ground handling and the fuellers. For comparison, a description of current operations of two crew flying the baseline aircraft is provided (see Table 13.2), with which the envisaged single crew commercial aircraft options are contrasted. The initial analyses pertain to aircraft operations as a whole, not to any particular normal, non-normal or emergency situation. The matrix represents the frequency of associations between each agent in the network. This matrix shows whether or not an agent within the current aircraft system of operation can be associated with any other agent in the proposed architecture for the single-crew aircraft, specifically through frequency of communications. As an example, the association matrix for the generic aircraft operational scenario is presented in Table 13.2.

An example social network diagram for the takeoff task is presented in Figure 13.4. Comparison with the network archetypes shows that the network is mesh-like,

TABLE 13.2
Association Matrix for Baseline Aircraft Generic Operational Scenario

Network Agents: From/to	PF	PM	AA	ATC/ ATM	D	FP	GH	F
Pilot Flying (PF)	–	62	12	33	2	2	2	2
Pilot Monitoring (PM)	62	–	82	48	2	2	2	4
Aircraft Automation (AA)	12	82	–	2	0	0	4	2
ATC/ATM	33	48	2	–	1	2	0	0
Dispatch (D)	2	2	0	1	–	2	0	0
Flight Planning (FP)	2	2	0	2	2	–	0	0
Ground Handling (GH)	2	2	4	0	0	0	–	0
Fuelers (F)	2	4	2	0	0	0	0	–

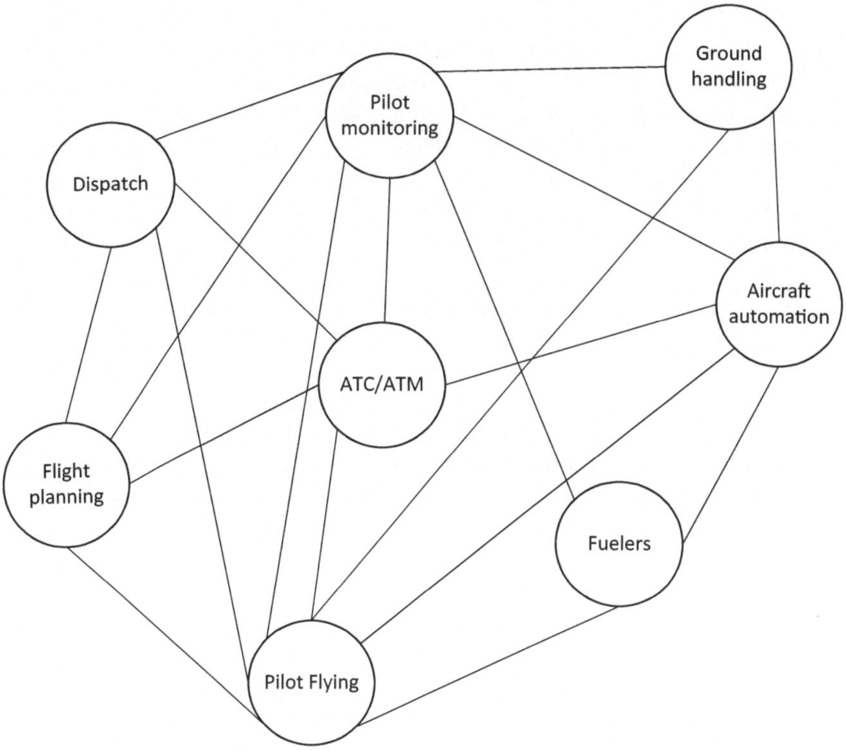

FIGURE 13.4 Network diagram for baseline aircraft generic operational scenario.

with dispatch and flight planning on the left-hand side; pilot monitoring, ATC/ATM and pilot flying in the middle; and ground handling, aircraft automation and fuellers on the right-hand side. Mesh structures are considered to be generally robust (Stanton et al. 2008).

TABLE 13.3
Social Network Analysis Statistics

Nodal statistics	PF	PM*	AA	ATC ATM	D	FP	GH	F
Emission/Reception	115	**202**	102	86	7	8	8	8
Eccentricity	1	**1**	2	2	2	2	2	2
Sociometric Status	25.55	**44.88**	22.66	19.11	1.55	1.77	1.77	1.77
Centrality (B-L)	5.28	**5.28**	4.11	4.11	3.70	3.70	3.36	3.36
Closeness	1.28	**1.28**	1.00	1.00	0.90	0.90	0.82	0.82
Farness	7	**7**	9	9	10	10	11	11
Betweenness	7.33	**7.33**	2.00	1.33	0	0	0	0

* Bold highlighting of possible role to be displaced to a ground station

Counting the number of collaborative (i.e. within-cell) and cooperative (i.e. between-cells) links gives rise to the data presented in Table 13.2 (which assumes a symmetrical network, with every link reciprocated, such as with a read-back or acknowledgement of each and every communication). This network forms the basis for the numerical analysis using SNA statistics (see Table 13.3). These statistics can either look at the network as a whole (e.g. the density of a network is defined by the number of social relations that are actually observed and can be represented as some fraction of the total possible) or analyse the individual nodes, e.g.:

- *Emission* and *reception* degree are the number of ties emanating from, and going to, each agent in the network.
- *Eccentricity* is defined by the largest number of hops an agent has to make to get from one side of the network to another.
- The *sociometric status* of each agent refers to the number of communications received and emitted, relative to the number of nodes in the network.
- Agent *centrality* is calculated in order to determine the central or key agent(s) within the network. There are a number of different centrality calculations that can be made; for example, agent centrality can be calculated using Bavelas-Leavitt's index.
- *Closeness* is the inverse of the sum of the shortest distances between each individual and every other person in the network. It reflects the ability to access information through the 'grapevine' of network members.
- *Farness* is the index of centrality for each node in the network, computed as the sum of each node to all other nodes in the network by the shortest path.
- *Betweenness* is defined by the presence of an agent between two other agents, which may be able to exert power through its role as an information broker.

The same principles were used to develop the four networks associated with each of the single-pilot options, each representing one of the four visions for future flight operations. Again, the networks appeared mesh-like, with fewer connections in the network in Option A (single-pilot aircraft) and more connections in Option D (a single-pilot aircraft with an additional pilot at a ground station and also with an independent automation mirror).

More robust networks generally have more connections, but that does not necessarily mean that everything should be connected to everything else. Unnecessary connections in a network can lead to greater coordination problems and could also mean that irrelevant information is being communicated (Rafferty et al. 2012; Sorensen and Stanton 2013). Analysis of the SOCA-CATs shows some differences in the functional loading of the single-pilot configurations compared to the current version of flight operations, as revealed in Table 13.4.

As Table 13.4 shows, removing the pilot monitoring from the network dramatically increases the functional loading on the remaining pilot on the flight deck as well as reducing the density of the network (where 0 means no connections, and 1 means everything is connected to everything else). The single pilot with a pilot on the ground (Options C and D) has the closest resemblance to current operations (although the pilot on the ground would have responsibility for monitoring several aircraft in different phases of flight). This comparison is for both the loading of functions and the network density. The most robust network appears to be option D, with the single pilot with the pilot on the ground and the automation mirror (which replicates the aircraft automation, using the same inputs and comparing with the results of the aircraft automation – only reporting when a discrepancy is found), as this network has the highest density (Walker et al. 2009)

The SOCA-CAT cooperation and collaboration interactions were compiled for comparative analysis. As before, the analysis assumes that the networks were

TABLE 13.4
Functional Loading for the Generic Scenario

Function Loading/Options	Current System	Option A	Option B	Option C	Option D
Pilot Flying (PF)	90	217	217	90	90
Pilot Monitoring (PM) or Pilot on the Ground (PG)	160	–	–	160	160
Aircraft Automation (AA)	143	143	143	143	143
System Mirror (SM)	–	–	143	–	143
Air Traffic Control (ATC) and Air Traffic Management (ATM)	51	51	51	51	51
Dispatch (D)	2	2	2	2	2
Flight Planning (FP)	3	3	3	3	3
Ground Handling (GH)	4	4	4	4	4
Fuelers (F)	3	3	3	3	3
Network Density*	0.422	0.266	0.377	0.422	0.555

* Network density is calculated as a fraction of total possible connections in the network.

symmetrical – that each input is met with a corresponding output. For human–human interaction, this would mean that a request is met with a response or that an instruction is met with confirmation (this could be verbal or non-verbal, such as moving a lever that is seen by the person issuing the instruction). In human–machine interaction, this would mean that depressing a switch or typing an instruction into the automation is met with some form of output, such as a change in the display status or direct voice output (Table 13.5).

The analysis of the networks across the various metrics confirm that the two networks on the right-hand side of the figures (i.e. the single pilot with additional pilot on the ground – top right – Option C) and single pilot with additional pilot on the ground and automation mirror on the ground (bottom right – Option D) are the most robust versions of the network as there is less dependency on the single role at the centre of the network (i.e. the pilot flying). This is revealed in several ways. First, the eccentricity of the nodes in the network (i.e. the largest number of hops an agent has to make to get from one side of the network to the other) where the pilot flying and the pilot on the ground have only one hop in the version on the right-hand side of the figure, making them all connected in the terms of the network archetypes (as shown by the number of ties). The pilot on the ground has the highest nodal degree and sociometric status on the right-hand side of the figure, which is due in part to the number of ties and in part to the greatest number of emission and reception connections with other nodes in the network. The pilot flying and the pilot on the ground are highest on the centrality and closeness metrics because they are both at the centre of the network and have the shortest distance to every other node. This suggests that they have high importance in the network, which is not very surprising (Baber et al. 2013; Harris et al. 2015; Kenney et al. 2013). They are also lowest on farness for the same reasons. On the betweenness metric, the pilot flying exerts less influence over the network because this influence is shared with the pilot on the ground. The combination of more links and less dependency on any one agent makes for a more robust network (Stanton 2014). In summary, SNA can be used to determine the importance of different agents within a social network and also to classify the network type. SNA offers a comprehensive analysis of the network in question. The key agents within the network are specified, as are the frequency and direction of communications within the network. As such, it proves useful for comparing the different networks that result from the new single-pilot concepts.

As a development of method, the relationships between CWA and SNA are not necessarily obvious but have allowed novel exploration of the differences between dual, single and distributed crewing options in this chapter. To that extent, we have further developed the approach reported by Houghton et al. (2006) and Baber et al. (2013). The SOCA-CAT phase of CWA offers a formative allocation of functions against situations (phases of flight in the present case). As such, it has enabled us to consider the possible communications between the agents in the system for all four of our proposed options against the current dual crewing operations. As safety is paramount in aviation, we require any future option to be as least as safe as contemporary crewing. SNA offers the opportunity to consider the likely effects on the communication networks from a quantitative perspective.

TABLE 13.5
Social Network Analysis of the Four Future Concepts

OPTION A	PF	AA	ATC ATM	D	FP	GH	F
Emission/ Reception	151	102	52	5	6	6	6
Eccentricity	1	2	2	2	2	2	2
Sociometric Status	33.55	22.66	11.55	1.11	1.33	1.33	1.33
Centrality (B-L)	5.00	3.75	3.75	3.33	3.33	3.00	3.00
Closeness	1.5	1.15	1.12	1.00	1.00	0.90	0.90
Farness	6	8	8	9	9	10	10
Betweenness	13	3	2	0	0	0	0

OPTION B	PF	AA	SM	ATC ATM	D	FP	GH	F
Emission/ Reception	245	204	204	54	5	6	10	8
Eccentricity	1	2	2	2	2	2	2	2
Sociometric Status	54.44	45.33	45.33	12.00	1.11	1.33	2.22	1.77
Centrality (B-L)	5.57	4.33	4.33	4.33	3.54	3.54	3.54	3.54
Closeness	1.28	1.00	1.00	1.00	0.81	0.81	0.81	0.81
Farness	7	9	9	9	11	11	11	11
Betweenness	14	2	2	4	0	0	0	0

OPTION C	PF	PG	AA	ATC ATM	D	FP	GH	F
Emission/ Reception	115	202	102	86	7	8	8	8
Eccentricity	1	1	2	2	2	2	2	2
Sociometric Status	25.55	44.88	22.66	19.11	1.55	1.77	1.77	1.77
Centrality (B-L)	5.28	5.28	4.11	4.11	3.70	3.70	3.36	3.36
Closeness	1.28	1.28	1.00	1.00	0.90	0.90	0.82	0.82
Farness	7	7	9	9	10	10	11	11
Betweenness	7.33	7.33	2.00	1.33	0	0	0	0

OPTION D	PF	PM	AA	SM	ATC ATM	D	FP	GH	F
Emission/ Reception	127	284	204	204	88	7	8	12	10
Eccentricity	1	1	2	2	2	2	2	2	2
Sociometric Status	28.22	63.11	45.33	45.33	19.55	1.55	1.77	2.66	2.22
Centrality (B-L)	5.87	5.87	4.7	4.7	4.7	3.9	3.9	3.9	3.9
Closeness	1.12	1.12	0.9	0.9	0.9	0.75	0.75	0.75	0.75
Farness	8	8	10	10	10	12	12	12	12
Betweenness	8.16	8.16	1.5	1.5	2.66	0	0	0	0

KEY: single pilot (top left – Option A), single pilot with system mirror on the ground (bottom left – Option B), single pilot with additional pilot on the ground (top right – Option C) and single pilot with additional pilot on the ground and system mirror on the ground (bottom right – Option D).

CONCLUSIONS

Previous approaches to developing a single-crew aircraft have focused on the development of sophisticated airborne technology to assist the pilot, but they have achieved only mixed success (Harris 2007). However, there are no major reasons why a single-pilot-operated commercial aircraft is not feasible in the very near future using existing technology. Military aviation has flown complex, high-performance, single-crew aircraft for many years, and Uninhabited Air Vehicles (UAVs) are now commonplace. It is time for these technologies to be spun out further into the commercial domain, where they may be applied to financial advantage. The greatest obstacle to the operation of civil, single-pilot aircraft is not the technology per se. It is combining the technology, designing the user interfaces and developing a new concept of operations to make such an aircraft safe and useable in a wide range of normal and non-normal operating situations when flown by a typical commercial pilot. The Human Factors requirements are the prime driver in this case, not the technology (Harris 2007).

Instead of developing a concept of single-pilot operations based upon equipping an aircraft with complex automation to aid the pilot, the approach described in this chapter begins to demonstrate how by using a distributed, air/ground, sociotechnical system, such an aircraft could be ready for service entry within a decade. Instead of replacing many of the functions of the second pilot, they are simply displaced to assistance on the ground (q.v. the operation of UAVs). Furthermore, in many circumstances, the assistance provided from the ground may be of higher quality and better targeted than that normally available from the second seat on the flight deck as a result of being able to draw upon a wider range of engineering and flight planning resources.

CWA, especially the SOCA-CAT phases, has been useful as a formative system design approach to aid understanding of the function allocation distribution between the agents and the functional loading on them. Extending this analysis to understand communication interactions in SNA has revealed Option D (single pilot with additional pilot on the ground and system mirror on the ground) to be the most resilient, even more than the current dual-pilot cockpit. It is debatable whether or not this option could have prevented the recent tragedy with Germanwings (BBC 2015), which has revealed a further weakness with current operations. Nevertheless, it is most likely that reduced crewing options are most likely to be implemented in short-haul cargo operations before they ever become a possibility for passenger-carrying aircraft.

The baseline descriptions for general operations of a conventional two-crew airliner (using CWA and SNA) and the same airliners being operated by just a single crew member give an overall idea of the level of complexity and resilience in the current system that need to be achieved in future single-crew operations. There is a natural split between the PF and PM functions that have evolved over decades. The main question being addressed here is whether these roles have to be co-located. In particular, the CWA and SNA describe the functions required irrespective of their location. The social network statistics produced for the various configurations of single-crew operations indicate that equivalent (or even enhanced) levels of system resilience may be achieved.

Future research will extend the analysis into contemporary issues such as trajectory, conflict and system-wide information management. Another important goal of this research is to validate the SOCA-CAT and SNA models in flight simulators. This would require developing the scenarios and structures as presented in the four options and testing against the baseline with pilot crews. This analysis would also enable experiential feedback from pilots to be gathered to supplement the performance data, offering an opportunity to understand how alternative crewing options might affect the duty of care the pilots have for the passengers that they are carrying. The immediate future of the flight deck is unlikely to change dramatically, but we can foresee a time within the next 20 years when the distributed crewing concept could enter short-haul cargo operations. When this is proven to be safe and effective, it might even enter into service for short-haul passenger aircraft.

REFERENCES

Airbus Industrie (n.d.). *A319/A329/A321 Flight Crew Operating Manual (Volume 3: Flight Operations) – Revision 36.* Author: Blagnac, France.

Alcock, C. (2004). New turboprop push tied to rising fuel costs. *Aviation International News*, 29 September 2004. Available at HYPERLINK "http://www.ainonline.com/Publications/era/era_04/era_newturbop18.html" http://www.ainonline.com/Publications/era/era_04/era_newturbop18.html (Accessed 30 January 2007).

Baber, C., Stanton, N. A., Atkinson, J., McMaster, R. and Houghton, R. J. (2013). Using social network analysis and agent-based modelling to explore information flow using common operational pictures for maritime search and rescue operations. *Ergonomics*, 56, 889–905.

BBC (2015). Germanwings crash latest updates. Available at http://www.bbc.com/news/live/world-europe-32070570 (Accessed 4 June 2015).

Bonner, M., Taylor, R., Fletcher, K. and Miller, C. (2000). Adaptive automation and decision aiding in the military fast jet domain. In: D. B. Kaber and M. R. Endsley (Eds.), *Human Performance, Situation Awareness and Automation: User-Centred Design for the New Millennium* (pp. 154–159). Omnipress: Madison, WI.

Civil Aviation Authority (2008). *Global Fatal Accident Review 1997–2006 (CAP 776).* Civil Aviation Authority: London, UK.

DeJohn, C. A., Wolbrink, A. M. and Larcher, J. G. (2004). *In-Flight Medical Incapacitation and Impairment of U.S. Airline Pilots: 1993 to 1998 (DOT/FAA/AM-04/16).* US Department of Transportation, Federal Aviation Administration, Office of Aerospace Medicine: Washington, DC.

Deutch, S. and Pew, R. W. (2005). *Single Pilot Commercial Aircraft Operation (BBN Report No. 8436).* BBN Technologies: Cambridge, MA.

Driskell, J. E. and Mullen, B. (2004). Social network analysis. In N. A. Stanton et al. (Eds.), *Handbook of Human Factors and Ergonomics Methods* (pp. 58.1–58.6). CRC Press: London, UK.

easyJet plc (2013). Annual report and accounts 2013. Available at: http://corporate.easyjet.com/~/media/Files/E/Easyjet-Plc-V2/pdf/investors/result-center-investor/annual-report-2013.pdf (Accessed 27 June 2014).

European Aviation Safety Agency (2014). *Certification Specifications for Large Aeroplanes (CS-25): Amendment 15.* EASA: Cologne. Available at: http://easa.europa.eu/system/files/dfu/00%20Annex%20to%20ED%20Decision%202014-026-R.pdf.

Evans, S. and Radcliffe, S. A. (2012). The annual incapacitation rate of commercial pilots. *Aviation, Space and Environmental Medicine*, 83(1), 42–49.

Galea, E. R., Filippidis, L., Wang, Z., Lawrence, P. J. and Ewer, J. (2011). Evacuation analysis of 1000+ seat Blended Wing Body aircraft configurations: Computer simulations and full-scale evacuation experiment. In: R. D. Peacock, E. D. Kuligowski and J. D. Averill (Eds), *Proceedings Pedestrian and Evacuation Dynamics 2010: 5th International Conference* (pp. 151–162). Springer: New York, NY.

Harris, D. (2003). The human factors of fully automatic flight. *Measurement and Control*, 36, 184–187.

Harris, D. (2007). A human-centred design agenda for the development of a single crew operated commercial aircraft. *Aircraft Engineering and Aerospace Technology*, 79, 518–526.

Harris, D. and Stanton, N. A. (2010). Aviation as a system of systems. *Ergonomics*, 53(2), 145–148.

Harris, D., Stanton, N. A. and Starr, A. (2015). Spot the difference: Operational event sequence diagrams as a formal method for work allocation in the development of single-pilot operations for commercial aircraft. *Ergonomics*, 58 (11), 1773–1791.

Hollnagel, E. (2007). Flight decks and free flight: Where are the system boundaries? *Applied Ergonomics*, *38*, 409–416.

Houghton, R. J, Baber, C., McMaster, R., Stanton, N. A., Salmon, P. M., Stewart, R. and Walker, G. H. (2006). Command and control in emergency services operations: A social network analysis. *Ergonomics*, 49(12–13), 1204–1225.

Jenkins, D. P., Stanton, N. A., Walker, G. H. and Salmon, P. M. (2009). *Cognitive Work Analysis: Coping with Complexity.* Ashgate: Aldershot, UK.

Keinrath, C., Vašek, J. and Dorneich, M. (2010). A cognitive adaptive man–machine Interface for future flight decks. In: A. Droog and M. Heese (Eds), *Performance, Safety and Well-being in Aviation Proceedings of the 29th Conference of the European Association for Aviation Psychology* (20–24 September 2010, Budapest, Hungary). European Association of Aviation Psychology.

Kenney, M., Horgan, J., Horne, C., Vining, P., Carley, K. M., Bigrigg, M. W., Bloom, M. and Braddock, K. (2013). Organisational adaptation in an activist network: Social networks, leadership, and change in al-Muhajiroun. *Applied Ergonomics*, 44(5), 739–747.

McIlroy, R. C. and Stanton, N. A. (2011). Getting past first base: Going all the way with cognitive work analysis. *Applied Ergonomics*, 42(2), 358–370.

Naikar, N. (2006a). Beyond interface design: Further applications of cognitive work analysis. *International Journal of Industrial Ergonomics*, 36, 423–438.

Naikar, N. (2006b). An examination of the key concepts of the five phases of cognitive work analysis with examples from a familiar system. *Proceedings of the Human Factors and Ergonomics Society 50th Annual Meeting*, 447–451.

Naikar, N. (2013). *Work Domain Analysis: Concepts, Guidelines, and Cases.* CRC Press: London, UK.

Onken, R. (1994). Human-centred cockpit design through the knowledge-based cockpit assistant system CASSY. In: *NATO DRG Panel 8 Workshop on Improving Function Allocation for Integrated System Design*. TNO Institute for Perception: Soesterberg, NL.

Onken, R. (1997). Functional development and field test of CASSY – A knowledge based cockpit assistant system. *Knowledge-Based Functions in Aerospace Systems, AGARD Lecture Series 200*, AGARD, Neuilly-sur-Seine, France.

Rafferty, L. A., Stanton, N. A. and Walker, G. H. (2012). *Human Factors of Fratricide.* Ashgate: Aldershot, UK.

Rasmussen, J., Pejtersen, A. and Goodstein, L. P. (1994). *Cognitive Systems Engineering.* Wiley: New York, NY.

Ryanair (2009). Full Year Results, 2009. Available at: http://www.ryanair.com/doc/investor/2009/q4_2009_doc.pdf (Accessed 17 August 2010).

Schulte, A. and Stütz, P. (2001). Cognitive concepts in mission management for air-to-ground attack aircraft. In D. Harris (Ed.), *Engineering Psychology and Cognitive Ergonomics* (pp. 23–30), Volume 5. Ashgate: Aldershot, UK.

Sorensen, L. J. and Stanton, N. A. (2013). Y is best: How distributed situational awareness is mediated by organisational structure and correlated with task success. *Safety Science*, 56, 72–79.

Stanton, N. A., Baber, C. and Harris, D. (2008). *Modelling Command and Control: Event Analysis of Systemic Teamwork*. Ashgate: Aldershot.

Stanton, N. A. (2014). Representing distributed cognition in complex systems: How a submarine returns to periscope depth. *Ergonomics*, 57(3), 403–418.

Stanton, N. A. and Bessell, K. (2014). How a submarine returns to periscope depth: Analysing complex socio-technical systems using cognitive work analysis. *Applied Ergonomics*, 45(1), 110–125.

Stanton, N. A., Harris, D. and Starr, A. (2014). Modelling and analysis of single pilot operations in commercial aviation. *Proceedings of HCI Aero 2014*. 30 July–1 August 2014, Santa Clara, CA.

Stütz, P. and Schulte, A. (2001). Evaluation of the Cockpit Assistant Military Aircraft (CAMA) in flight trials. In D. Harris (Ed.), *Engineering Psychology and Cognitive Ergonomics* (pp. 15–22), Volume 5. Ashgate: Aldershot, UK.

Taylor, R. M., Howells, H. and Watson, D. (2000). The cognitive cockpit: Operational requirement and technical challenge. In: P. T. McCabe, M. A. Hanson and S. A. Robertson (Eds.), *Contemporary Ergonomics 2000* (pp. 55–59). Taylor and Francis: London, UK.

Thomas, M. J. W. (2003). Improving organisational safety through the integrated evaluation of operational and training performance: An adaptation of the Line Operations Safety Audit (LOSA) methodology. *Human Factors and Aerospace Safety*, 3, 25–45.

Vicente, K. J. (1999). *Cognitive Work Analysis: Toward Safe, Productive, and Healthy Computer-based Work*. Lawrence Erlbaum Associates: Mahwah, NJ.

Walker, G. H., Cooper, M., Thompson, P. and Jenkins, D. (2014). Practitioner versus analyst methods: A nuclear decommissioning case study. *Applied Ergonomics*, 45(6), 1622–1633.

Walker, G. H., Stanton, N. A., Salmon, P. M., Jenkins, D., Stewart, R. and Wells, L. (2009). Using an integrated methods approach to analyse the emergent properties of military command and control. *Applied Ergonomics*, 40(4), 636–647.

Wiener, E. L. and Curry, R. E. (1980). Flight deck automation: Promises and problems. *Ergonomics*, 23, 995–1011.

Wilson, J. R. and Carayon, P. (2014). Systems ergonomics: Looking into the future – Editorial for special issue on systems ergonomics/human factors. *Applied Ergonomics*, 45(1), 3–4.

14 Future Directions for the Event Analysis of Systemic Teamwork

INTRODUCTION

This book has presented a series of studies in which the Event Analysis of Systemic Teamwork (EAST) framework has been used to describe, understand, and even forecast the behaviour of complex systems. EAST has shown itself capable of dealing with highly complex, multi-agent, distributed systems, and in most cases the outputs were used to identify interventions to enhance performance or safety. Whilst this is encouraging, however, the systems in which human factors and ergonomics practitioners work are arguably becoming more complex and technology driven (Grote et al. 2013; Dekker et al. 2013; Salmon et al., 2017; Walker et al. 2017; Woods and Dekker 2000). This, coupled with a dramatic shift towards the systems thinking paradigm, is beginning to stretch the capabilities of many human factors and ergonomics methods.

We are certainly not alone in expressing these concerns. Dekker (2014), Leveson (2011), Salmon et al. (2011; 2017) and Walker et al. (2017) discuss the increasing complexities of modern-day systems and the extent to which they are rapidly outpacing the capabilities of our methodological toolkit. Salmon et al. (2017) in particular outlines five key challenges that human factors and ergonomics methods face. It is our contention that the EAST framework can respond to these challenges.

The aim of this chapter is to demonstrate this by examining the contemporary ergonomics problem space and the extent to which EAST can meet the challenges posed. Specifically, we outline Salmon et al.'s five key challenges, and discuss the capacity of EAST to respond to them. The five challenges include (Salmon et al., 2017):

- normal performance as a cause of accidents (Dekker 2011; Leveson 2004; Rasmussen 1997);
- accident prediction (Salmon et al. 2014a);
- system migration (Rasmussen 1997);
- systems concepts (Hutchins 1995; Stanton et al. 2006); and
- human factors and ergonomics in design.

NORMAL PERFORMANCE AS A CAUSE OF ACCIDENTS

Significant progress has been made in understanding accident causation in safety critical systems. Systems models, in particular, are now widely accepted (Leveson 2004;

Rasmussen 1997), and there are a range of methods that enable accidents to be analysed from this perspective (e.g. Hollnagel 2012; Svedung and Rasmussen 2002; Leveson 2004). Accidents are now widely acknowledged to be systems phenomena, just as safety is (Hollnagel 2004; Dekker 2011). Both safety and accidents, therefore, are emergent properties arising from non-linear interactions between multiple components distributed across a complex web of human and machine agents and interventions (e.g. Leveson 2004).

This form of thinking, and the evolution of it, brings the methods we currently use into question. Salmon et al. (2017) point out that, despite the great progress in safety performance that has been made in most safety-critical sectors since the Second World War, significant trauma still occurs, and in some areas, progress may be slowing. Worse still, in some domains, incidents appear to be rising again. Leveson (2011) agrees, and suggests that one reason is that our methods do not fully uncover the underlying causes of accidents. Salmon et al. (2017) argue that a major the issue may be that the evolution in accident causation models is not reflected in current accident analysis methods.

One of the fundamental advances provided by state-of-the-art accident causation models relates to the idea that the behaviours underpinning accidents do not necessarily have to be errors, failures or violations (e.g. Dekker 2011; Leveson 2004; Rasmussen 1997). As Dekker (2011) points out, systems thinking is about how accidents can happen when no parts are broken. 'Normal performance' plays a role too (Perrow 1984). Salmon et al. (2017) outlines two tenets that have important implications for the methods that we use. First, normal performance plays a role in accident causation; and second, accidents arise from the very same

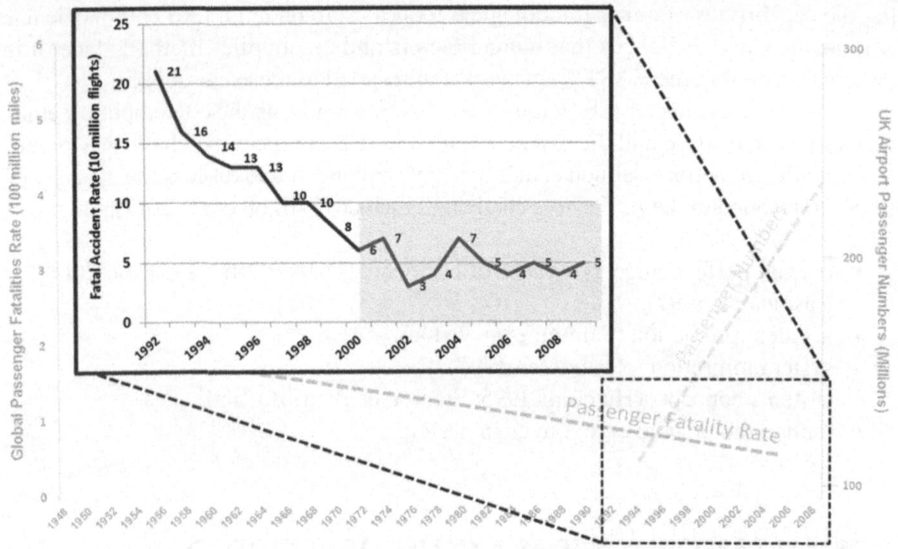

FIGURE 14.1 The pattern of global passenger fatalities per 10 million passenger miles on scheduled commercial air transport since 1993. The graph shows that the precipitous drop in fatality rates since 1945 has, since 2003, levelled off.

behaviours and processes that create safety. Dekker (2011), for example, argues that the seeds for failure can be found in 'normal, day-to-day processes' (p. 99) and are often driven by goal conflicts and production pressures. These normal behaviours include workarounds, improvisations, and adaptations (Dekker 2011) but may also just be normal work behaviours routinely undertaken to get the job done. It is only with hindsight and a limited investigation method that these normal behaviours are treated as failures.

The tenets described above provide an interesting shift in the requirements for accident analysis methods. Dekker (2014) argues that practitioners should not look for the known problems that appear in incident reporting data or safety management systems. Instead, he argues that the focus should be in the places where there are no problems, in other words, normal work. In addition, the burgeoning concept of Safety II (Hollnagel et al. 2013) argues that safety management needs to move away from attempting to ensure that as little as possible goes wrong and towards ensuring that as much as possible goes right. A key part of this involves understanding performance when it went right as well as when it went wrong.

Salmon et al. (2017) discuss how, out of three popular contemporary accident analysis methods (Accimap, STAMP, HFACS), only the Accimap method (Rasmussen, 1997) is capable of describing normal performance and of considering the role of normal performance in accident causation. AcciMap is used to describe accidents in terms of contributory factors and the relationships between them. This enables a comprehensive representation of the network of contributory factors involved. It does this by decomposing systems into six levels, across which analysts place the decisions and actions that enabled the accident in question to occur (although the method is flexible in that the number of levels can be adjusted based on the system in question). Interactions between the decisions and actions are subsequently mapped onto the diagram to show the relationships between contributory factors within and across the six levels. A notable feature of AcciMap is that it does not provide analysts with taxonomies of failure modes; rather, analysts have the freedom to incorporate any factor deemed to have played a role in the accident in question.

Both HFACS and STAMP provide taxonomies of error and failure modes that are used to classify the behaviours involved in accident scenarios, which in turn means that there is little scope for analysts to include behaviours other than those deemed to have been failures of some sort. There is no opportunity for analysts to incorporate normal behaviours in their descriptions of accidents – they have to force-fit events into one of the error or failure modes provided. A consequence is that the normal behaviours that contribute to accidents are not picked up during accident analysis efforts. This may impact accident prevention activities by providing a false sense of security that nothing else is involved and thus nothing needs modification (apart from error-producing human operators).

AcciMap, on the other hand, does not use a taxonomy of failure or error modes and so enables analysts to incorporate normal performance and to show its relationship with other behaviours (Salmon et al., 2017). Despite this, AcciMap descriptions still tend to incorporate contributory factors that relate only to failures, and few Accimap analyses actually incorporate normal performance.

The EAST framework provides an approach that can be used to integrate analyses of normal performance and accident causation. As demonstrated in this book, the EAST framework is used primarily to describe 'normal performance' by describing the task, social and information networks underpinning behaviour. In addition, EAST has also been used for accident analysis purposes (Griffin et al., 2008; Rafferty et al., 2012; Salmon et al., 2011; 2016). Since EAST was originally developed to describe normal performance, and has since been applied for accident analysis purposes, it seems suited to incorporating both normal performance and failures in descriptions of accidents. It is therefore recommended that further applications of EAST for accident analysis purposes are undertaken. In particular, within the three networks a focus should be placed on how normal performance either creates or interacts with failures. This will allow rich, in-depth descriptions of the role of both normal performance and failures in accidents.

The conclusion, then, is that there is room for improvement in our accident analysis methods. Not all state-of-the-art methods are consistent with our current understanding of accident causation. Further, even for the methods that are, such as AcciMap, it is questionable whether they are being used in a manner that is consistent with contemporary models of accident causation. This represents a key issue for human factors and ergonomics researchers and practitioners and for safety science generally. On the one hand, there is now a widespread understanding that the role of normal performance in accidents is apparent and needs to be understood (Dekker 2011, 2014; Leveson 2011; Rasmussen 1997). On the other hand, accident analysis efforts, regardless of domain, do not seem to be dealing particularly well with this feature. This means our understanding of accidents may be incomplete. The EAST framework described in this book provides can be used to respond to this challenge.

ACCIDENT PREDICTION

Forecasting accidents before they occur has been labelled the final frontier for human factors and ergonomics (Moray 2008; Salmon et al. 2011; Stanton and Stammers 2008). Although there have been various attempts at developing accident prediction models (e.g. Deublein et al. 2013), most are statistical models that are unable to identify and describe how behaviours across overall sociotechnical systems might combine to create failure scenarios. Other predictive methods are available, such as those that can be used to predict the 'human errors' that lead to accidents (see Stanton et al. 2013). Indeed, some of these methods have been shown to achieve acceptable levels of reliability and validity (e.g. Stanton et al. 2009). The problem is that they predict what is likely the last behaviour in a long and complex network of interacting and emergent behaviours occurring across various parts of the system. They predict consequences, not causes, and they do not identify the network of contributory factors that might co-occur to create accidents. Accident prevention efforts are better served by looking at the interactions that occur before the human operator makes the error. In short, it is the entire accident scenario, including interacting factors and emergent behaviours, that are important for understanding how to prevent accidents.

A Systems Approach to Prediction

A key requirement, then, is a prediction method that is underpinned by systems thinking. Error prediction methodologies can be thought of as reductionist (although they are not entirely reductionist as they do focus on human–machine interactions). Reductionist approaches, those that rely on taking the system apart in order to understand the components and then reassembling the components back into the complete system (on the tacit assumption that the whole cannot be greater or less than the sum of its parts), do not allow us to detect the emergent properties associated with the types of risk issues upon which we wish to make progress (see Walker et al. 2009). Systems approaches such as EAST do. One means by which they can enable forecasting and prediction is to consider the causal texture of the systems environment and the system's movement through that environment.

As discussed, the systems approach has become popular in part because of various systems analysis methodologies that can, to some extent at least, do this (e.g. Rasmussen 1997). These methods, for example AcciMap (Rasmussen 1997), are becoming increasingly popular for accident investigation purposes. A major limitation of these methods is that, so far, they have not been used predictively; organisations are effectively waiting for loss events to occur before they can work on prevention strategies. The lack of data resulting from improved safety trends combined with greater operational intensity and risk exposure means that, if anything, loss events are more likely to be large scale and unexpected, meaning that 'learning from disasters' is becoming increasingly dubious from an ethical perspective. The need for systems-based prediction approaches is discussed extensively in the literature (e.g. Moray 2008; Salmon et al. 2011; 2017; Stanton and Stammers 2008) but as yet, a credible approach has yet to emerge.

EAST and Accident Prediction

Encouragingly, the EAST framework is beginning to demonstrate its utility for predicting system behaviour (Stanton & Harvey, 2017).

Stanton & Harvey (2017 See chapter 10) first explored the predictive capacity of EAST by using it to assess the risks associated with Royal Navy low flying hawk training scenarios designed to simulate a sea skimming missile attack. This involved first using EAST to describe the task, social and information networks involved, and then using a breaking links process to identify potential risks. This involved adding and breaking links and nodes within the networks to explore different system states. The effect of this make/break-link process is to create 'short circuits', 'long circuits' or 'no circuits', all of which put systems into new configurations (see Figure 14.2). For example, adding and breaking links within the networks reveals instances where a human operator may or may not be aware of a particular piece of information, where a task will or will not be fulfilled by a human operator or piece of technology or where a required communication may or may not be made. By these means, it is possible to model the majority of possible accident pathways in a given system model under a wide range of different permutations. On the other hand, adding and breaking links may reveal aspects of normal performance that move towards the

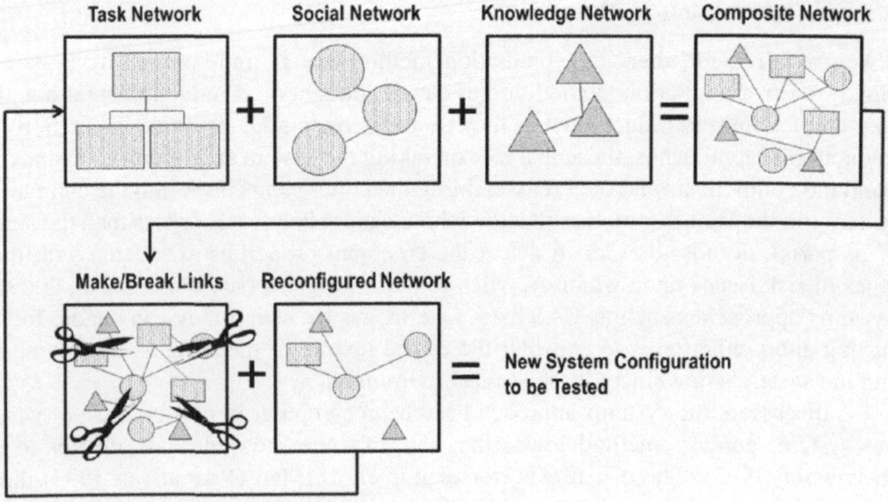

FIGURE 14.2 The EAST models sociotechnical systems by combining task, social and knowledge networks into composite networks that can be systematically degraded into all possible configurations.

boundary of safe operations (Rasmussen 1997). This systems level focus encourages a different approach to forecasting resilience: to examine persistent and emergent patterns that arise even though the boundaries of the system cannot always be fully known. Further testing of EAST in this capacity is currently being undertaken by the authors, and it is recommended that further predictive applications of EAST are undertaken.

MIGRATION TOWARDS SAFETY BOUNDARIES

Rasmussen's risk management framework (Rasmussen 1997; see Figure 14.3) is becoming one of the most popular safety and risk management models of our time (e.g. Salmon et al. 2014c; Underwood and Waterson 2014; Goode et al. 2014). In his seminal article, Rasmussen outlined the concept of migration based on the 'Brownian movements' of gas molecules, describing how organisations shift towards and away from safety and performance boundaries due to various constraints including financial, production and performance pressures. According to Rasmussen, there is a boundary of economic failure: the financial constraints on a system that influence behaviour towards greater cost efficiencies. There is also a boundary of unacceptable workload: the pressures experienced by people and equipment in the system as they try to meet economic and financial objectives. The boundary of economic failure creates a pressure towards greater efficiency, which works in opposition to a similar pressure against excessive workload. As systems involve human as well as technical elements, and because humans are able to adapt situations to suit their own needs and preferences, these pressures inevitably introduce variations in behaviour that are not explicitly designed and can lead to increasingly emergent system behaviours, both good and bad (Qureshi 2007; Clegg 2000). Over time, this adaptive

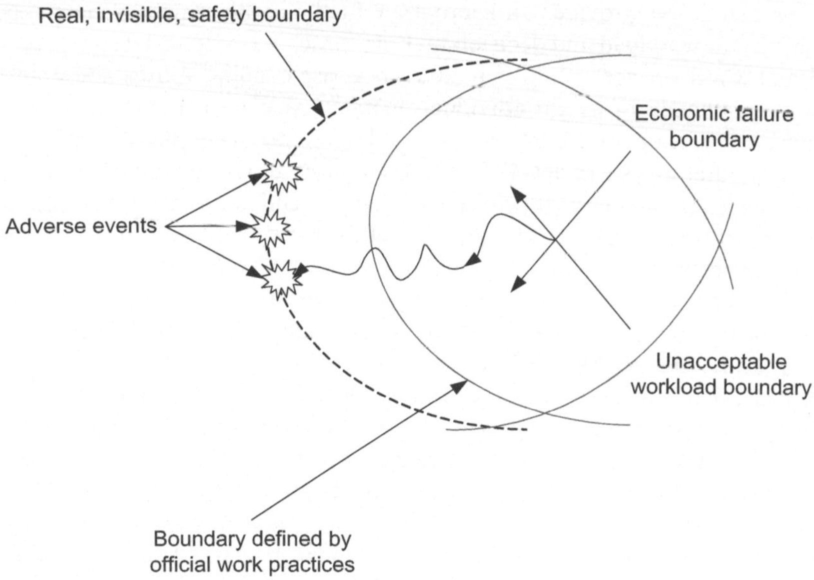

FIGURE 14.3 Rasmussen's dynamic safety space (Adapted from Rasmussen 1997).

behaviour can cause the system to cross safety boundaries and accidents to happen (Qureshi 2007; Rasmussen 1997). The key, then, is to detect in advance a) where those boundaries are and b) where the system is travelling in relation to them.

Mapping Migration

Unfortunately, there are few human factors and ergonomics methods that are capable of describing where boundaries are situated and whereabouts organisations may be in relation to them. Further, the ability to dynamically track an organisation's migration is not readily supported by current methods. Whilst information on so-called lagging indicators can provide an indication of proximity to a safety boundary, there is an absence of human factors and ergonomics methods that use leading indicators to track organisational performance and safety. Work currently being undertaken by the authors of this book is exploring the capacity for EAST to be used as a framework for a. identifying leading indicators, and b. tracking organisational behaviour and ultimately migration towards and away from safety boundaries.

SYSTEMS CONCEPTS

Salmon et al. (2017) argue that many human factors and ergonomics concepts could be better understood by adopting a systems thinking lens. Citing accident causation and distributed situation awareness as examples of where a systems thinking approach has proved successful, Salmon et al. (2017) argue that other concepts such as mental workload and decision making should be viewed through a systems lens. They also argue, however, that we do not possess the methods to support this. It is our view that EAST provides an appropriate method with which to view concepts such as mental workload and decision making through a systems lens.

Mental workload (see Young et al. 2015) is a case in point. Whilst it has predominantly been thought of as an individual operator concept, it is increasingly being examined at the team level (Helton et al. 2014), and there is no reason why it cannot be considered at a systems level. Just as Stanton et al. (2006) describe how situation awareness is an emergent property of systems and is distributed across operators, mental workload can be also thought of in this way. Moreover, similar to the transactions in awareness described by Stanton and colleagues, transactions in mental workload between operators are readily apparent in sociotechnical systems. An example of this is the workload shedding by air traffic controllers, dividing sectors up as the air traffic increases (Walker et al. 2010).

There are many methods available to support the assessment of individual operator workload, including the highly popular NASA TLX (Hart and Staveland 1988) and many similar subjective rating scales (see Stanton et al. 2013). In addition, there are other individual focused methods, such as psychophysiological measures. Further, methodologies that support the assessment of team mental workload are emerging (Helton et al. 2014), although these are not without their problems. Unfortunately, methods that can specifically consider mental workload at a systems level do not yet exist. Such a method, something akin to 'distributed mental workload' assessment, would need to consider the workload of multiple agents, how interactions between actors shapes each other's workload, how different levels of workload dynamically shift throughout task performance and, further, what wider systemic factors constrain or facilitate workload. In addition, under a systems view, workload across system levels and even non-human actors should be considered.

Again, it is our opinion that EAST can be used to provide such an analysis. Specifically, EAST considers multiple agents and interactions between them (via the social network), the tasks being undertaken by different agents (via the task network), and the information load associated with different tasks and agents (via the information network). Early applications of the EAST framework used the information network to provide an indirect assessment of operator workload by calculating the total number of information nodes being used by different operators during different scenario phases (e.g. Salmon et al., 2008). This provides a dynamic indication of workload as different tasks are undertaken. As such, EAST can be used to assess the level of workload imposed on different agents, the distribution of this workload across the system, and how this workload is exchanged across agents.

HUMAN FACTORS AND ERGONOMICS METHODS IN DESIGN

Human factors and ergonomics has a key role to play in the design of new technologies, interfaces, training programmes, procedures and even overall systems. Despite this, a criticism of many human factors and ergonomics methods is that they do not directly contribute to design – that is, they are not used by designers to design (Salmon et al., 2017).

The requirement to shift the emphasis on human factors and ergonomics to the front end of the design life cycle is well known (Norros 2014; Salmon et al., 2017). However, it requires human factors and ergonomics methods that can be used by designers and design teams to design or that can at least be integrated with existing design approaches. It also requires human factors and ergonomics researchers and practitioners to take the lead to facilitate design efforts, which would be a paradigm shift in the way complex systems are engineered.

Whilst a significant shift is required, it is notable that many of our methods are suited to directly and indirectly informing design. Rather than develop new methods per se, the pressing requirement seems to be the development of processes that bridge the gap between analysis and design. Specifically, processes are required to translate human factors and ergonomics analyses into specific design requirements. One such approach that has been developed recently for this purpose is the Cognitive Work Analysis design toolkit (CWA-DT) (Read et al. 2015; 2017).

The CWA-DT assists CWA users to identify extract insights from CWA analyses and to use these insights within a participatory design paradigm. It promotes the collaborative involvement of experts (i.e. human factors and ergonomics professionals, designers and engineers), stakeholders (i.e. company representatives, supervisors, unions) and end-users (i.e. workers or consumers) to solve design problems based on insights gained through CWA.

Underlying the CWA-DT is both the design philosophy of CWA (i.e. 'let the worker finish the design') and the related sociotechnical systems theory approach, which aims to design organisations and systems that have the capacity to adapt and respond to changes and disturbances in the environment (Trist and Bamforth 1951; Cherns 1976; Clegg 2000; Walker et al. 2009). Consequently, the CWA-DT provides design tools and methods that encourage consideration of the values underlying the sociotechnical systems approach (and indeed underpinning human factors ergonomics more generally). These include the notion of humans as assets or adaptive decision makers rather than error-prone liabilities; of technology being designed as a tool to assist humans to achieve their goals, rather than implemented because of assumed efficiency or cost savings; the consideration of the needs and capabilities of a diverse set of end-users; and of design focussing on promoting the quality of life or wellbeing for end-users. Further, the consideration of sociotechnical design principles, such as minimal critical specification, boundary management and joint design of social and technical elements, intends to achieve the design of systems that can operate within their safety and performance boundaries both on implementation and in an ongoing fashion through continual monitoring and re-design.

Whilst the approach was originally designed for use with CWA outputs, there is no reason why it cannot be used in conjunction with the outputs of other systems human factors and ergonomics methods such as EAST. Indeed, Salmon et al (2018) recently used the CWA-DT in conjunction with both CWA and EAST as part of an intersection design process. This involved using insights derived from the EAST analyses of road user behaviour presented in Chapter 7 as part of a CWA-DT participatory design process to design three new intersections. An emphasis was placed on creating designs that catered to the needs of all forms of end-user, rather than designing exclusively for drivers and motorised vehicles. The intention was to create

FIGURE 14.4 The EAST models sociotechnical systems by combining task, social and knowledge networks into composite networks that can be systematically degraded into all possible configurations.

intersection design concepts that, once implemented, would support safe interactions between drivers, cyclists, motorcyclists and pedestriansby ensuring that all road users are aware of each other and understand each others' likely behaviours. The design brief was to replace one of the intersections from the EAST study presented in Chapter 6 (See Figure 14.4).

When we evaluated the three designs with drivers, cyclists, motorcyclists and pedestrians, two of the designs performed best against key criteria: alignment with sociotechnical systems values, attainment of key intersection functions (such as to minimise collisions, maximise efficiency, maximise compliance, optimise flexibility), and user preferences.

The first design is known as the "turning team" design (see Figure 14.5). It works on the premise that different road users could work effectively as a team when proceeding through the intersection. To do this the design aims to make drivers explicitly aware of other forms of road user (to connect the team) and provides each with a clear and dedicated path through the intersection. The design aims to clear cyclists from the intersection before allowing motorised traffic to enter. Other features include a pedestrian crossing path wide enough to accommodate cyclists who are not comfortable with using the road, motorcyclist filtering lanes, and phasing of traffic lights based on road user type and direction of travel.

The second design is the "circular" concept. It explicitly separates motorised and non-motorised traffic. A circular pathway around the intersection is provided for pedestrians and cyclists to use. This pathway links with cycle lanes running down the centre of the road, separated by a kerb from the roadway. On the roadway, this design provides a separate bus lane and a motorcycle zone at the front of the intersection to encourage motorcyclists to filter to the front. Finally, the design incorporates signs warning motorists to be on the lookout for cyclists and for motorcyclists filtering through the traffic from behind (Figure 14.6).

FIGURE 14.5 Turning teams intersection design concept.

CONCLUSIONS

Following on from Salmon et al. (2017), this chapter examined some of the challenges posed for human factors and ergonomics methods by the contemporary paradigm of systems thinking and discussed how the EAST framework can respond to the challenges posed. Although human factors and ergonomics researchers and practitioners have a range of methods at their disposal, the popular paradigm of systems thinking requires analyses that go beyond their capabilities. The discussion presented in this chapter has suggested that

- accident analysis methods, though high on explanatory power, do not describe accident causation in a manner that is congruent with contemporary models. Based on previous accident analysis applications, EAST could be used to describe the role of normal performance in accident causation;
- despite there being a range of appropriate candidate methods, we currently do not have a method that supports the prediction of accidents. EAST is

beginning to show its capacity for predicting system behaviour and can be used to predict specific failure scenarios;

- the migration of systems towards and away from safety boundaries has not yet been dealt with by ergonomics methods. EAST could potentially be used to identify leading indicators and then track the migration of system behaviour towards and away from system boundaries;
- various ergonomics constructs may be suited to systems level analysis, but there is a lack of ergonomics-based systems methods to support this. EAST can be used to describe and assess concepts such as mental workload and decision making from a systems perspective; and
- despite its critical role in the design process, few ergonomics methods are actually used by designers to design. The recently developed CWA-DT (Read et al., 2015; 2017) can be used in conjunction with EAST to create novel designs that align with core sociotechnical systems theory values and principles.

FIGURE 14.6 Circular intersection design concept.

In closing, we would like to emphasise the important role that new and emerging systems thinking-based methods such as EAST have in ensuring that human factors and ergonomics remain relevant. In a world where complexity is apparently increasing, and problems are becoming more difficult to resolve, methods that have the capacity of describing systems in a manner that allows us to understand what components exist in the system and how they interact together are required. Only by taking the overall system as the unit of analysis, and by considering all components and how they interact together to shape behaviour can we begin to understand systems and their emergent properties. EAST allows human factors and ergonomics researchers and practitioners to do this. We hope that the reader has found this book useful, and that further applications of EAST are undertaken. In particular, we encourage applications in new domains and focussed on problems not previously examined such as terrorism, cybersecurity, and artificial intelligence.

REFERENCES

Cherns, A. (1976). The principles of sociotechnical design. *Human Relations*, 29, 783–792.

Clegg, C. W. (2000). Sociotechnical principles for system design. *Applied Ergonomics*, 31, 463–477.

Dekker, S. (2011). *Drift into Failure: From Hunting Broken Components to Understanding Complex Systems*. Ashgate: Aldershot, UK.

Dekker, S. (2014). *The Field Guide to Understanding Human Error, Third Edition*. Ashgate: Aldershot, UK.

Dekker, S. W. A., Hancock, P. A. and Wilkin, P. (2013). Ergonomics and sustainability: Towards an embrace of complexity and emergence. *Ergonomics*, 56(3), 357–364.

Deublein, M., Schubert, M., Adey, B. T., Köhler, J. and Faber, M. H. (2013). Prediction of road accidents: A Bayesian hierarchical approach. *Accident Analysis & Prevention*, 51, 274–291.

Goode, N. Salmon, P. M., Lenné, M. G. and Hillard, P. (2014). Systems thinking applied to freight handling operations. *Accident Analysis and Prevention*, 68, 181–191.

Griffin, T. G. C., Young, M. S. and Stanton, N. A. (2008). Investigating accident causation through information network modelling. *Ergonomics*, 53(2), 198–210.

Griffin, T. G. C., Young, M. S. and Stanton, N. A. (2010). Investigating accident causation through information network modelling. *Ergonomics*, 53(2), 198–210.

Grote, G., Weyer, J. and Stanton, N. A. (2013). Beyond human-centred automation – concepts for human–machine interaction in multi-layered networks. *Ergonomics*, 57(3), 289–294.

Helton, W. S, Funke, G. J, and Knott, B. A. (2014). Measuring workload in collaborative contexts: Trait versus state perspectives. *Human Factors*, 56(2), 322–332.

Hollnagel, E. (2012). *FRAM: The Functional Resonance Analysis Method: Modelling Complex Socio-technical Systems*. Ashgate: Aldershot, UK.

Hollnagel, E., Leonhardt, J., Licu, T. and Shorrock, S. (2013). From Safety-I to Safety-II: A white paper. Eurocontrol. Available at: https://www.eurocontrol.int/sites/default/files/content/documents/nm/safety/safety_whitepaper_sept_2013-web.pdf (Accessed 21 July 2015).

Hutchins, E. (1995). *Cognition in the Wild*. MIT Press: Cambridge, MA.

Leveson, N. G. (2004). A new accident model for engineering safer systems. *Safety Science*, 42(4), 237–270.

Leveson, N. G. (2011). Applying systems thinking to analyze and learn from events. *Safety Science*, 49, 55–64.

Moray, N. (2008). The good, the bad, and the future: On the archaeology of ergonomics. *Human Factors*, 50(3), 411–417.

Norros, L. (2014). Developing human factors/ergonomics as a design discipline. *Applied Ergonomics*, 45(1), 61–71.

Perrow, C. (1984). *Normal Accidents: Living with High-Risk Technologies.* Basic Books: New York, NY.

Qureshi, Z. H. (2007). A review of accident modelling approaches for complex critical socio-technical systems. *Conferences in Research and Practice in Information Technology Series*, Vol. 336.

Rafferty, L. A., Stanton, N. A., (2012). *The Human Factors of Fratricide.* CRC Press, Boca Raton.

Rasmussen, J. (1997). Risk management in a dynamic society: A modelling problem. *Safety Science*, 27(2/3), 183–213.

Read, G., Salmon, P. M., Lenne, M. G., Jenkins, D. P., (2015). Designing a ticket to ride with the Cognitive Work Analysis Design Toolkit. *Ergonomics.* 58:8, 1266–86. doi: 10.1080/00140139.2015.1013576.

Read, G. J. M., Beanland, V., Lenne, M. G., Stanton, N. A., Salmon, P. M., (2017). *Integrating Human Factors methods and systems thinking for transport analysis and design.* CRC Press, Boca Raton, FL.

Reason, J. (1990). *Human Error.* Cambridge University Press: New York, NY.

Salmon, P. M., Stanton, N. A., Lenné, M., Jenkins, D. P., Rafferty, L. A. and Walker, G.H. (2011). *Human Factors Methods and Accident Analysis.* Ashgate: Aldershot.

Salmon, P. M., Lenné, M. G., Read, G., Walker, G. H. and Stanton, N. A. (2014a). Pathways to failure? Using Work Domain Analysis to predict accidents in complex systems. In: T. Ahram, W. Karwowski and T. Marek (Eds), *Proceedings of the 5th International Conference on Applied Human Factors and Ergonomics AHFE 2014*, Kraków, Poland, 19–23 July 2014.

Salmon, P. M., Goode, N., Lenné, M. G., Cassell, E. and Finch, C. (2014c). Injury causation in the great outdoors: A systems analysis of led outdoor activity injury incidents. *Accident Analysis and Prevention*, 63, 111–120.

Salmon, P. M., Walker, G. H., Stanton, N. A. (2016). Pilot error versus sociotechnical systems failure? A distributed situation awareness analysis of Air France 447. *Theoretical Issues in Ergonomics Science*, 17(1), 64–79.

Salmon, P. M., Walker, G. H., Read, G. J. M., Goode, N. and Stanton, N. A. (2017). Fitting methods to paradigms: are ergonomics methods fit for systems thinking?. *Ergonomics*, 60(2), 194–205.

Salmon, P. M., Read, G. J. M., Walker, G. H., Lenne, M. G. and Stanton, N. A. (2018). *Distributed situation awareness in road transport: theory, measurement, and application to intersection design.* CRC Press, Boca Raton, FL.

Stanton, N. A. and Harvey, C. (2017). Beyond human error taxonomies in assessment of risk in sociotechnical systems: A new paradigm with the EAST 'broken-links' approach. *Ergonomics*, 60(2), 221–233.

Stanton, N. A. and Stammers, R. B. (2008). Bartlett and the future of ergonomics. *Ergonomics*, 51(1), 1–13.

Stanton, N. A., Salmon, P. M., Rafferty, L., Baber, C., Walker, G. H. and Jenkins, D. P. (2013). *Human Factors Methods: A Practical Guide for Engineering and Design, Second Edition.* Ashgate: Aldershot, UK.

Stanton, N. A., Stewart, R., Harris, D., Houghton, R. J., Baber, C., McMaster, R., Salmon, P. M., Hoyle, G., Walker, G. H., Young, M. S., Linsell, M., Dymott, R. and Green, D. (2006). Distributed situation awareness in dynamic systems: Theoretical development and application of an ergonomics methodology. *Ergonomics*, 49, 1288–1311.

Svedung, I. and Rasmussen, J. (2002). Graphic representation of accident scenarios: Mapping system structure and the causation of accidents. *Safety Science*, 40, 397–417.

Trist, E. L. and Bamforth, K. W. (1951). Some social and psychological consequences of the longwall method of coal-getting: An examination of the psychological situation and defences of a work group in relation to the social structure and technological content of the work system. *Human Relations*, 4, 3–38.

Underwood, P. and Waterson, P. (2013). Examining the gap between research and practice. *Accident Analysis & Prevention*, 55, 154–164.

Underwood, P. and Waterson, P. (2014). Systems thinking, the Swiss Cheese Model and accident analysis: A comparative systemic analysis of the Grayrigg train derailment using the ATSB, AcciMap and STAMP models. *Accident Analysis & Prevention*, 68, 75–94.

Walker, G. H., Stanton, N. A., Salmon, P. S. and Jenkins, D. P. (2009). *Command and Control: The Sociotechnical Perspective*. Ashgate: Aldershot, UK.

Walker, G. H., Gibson, H., Stanton, N. A., Baber, C., Salmon, P. M. and Green, D. (2006). Event analysis of systemic teamwork (EAST): A novel integration of ergonomics methods to analyse C4i activity. *Ergonomics*, 49, 1345–1369.

Walker, G. H., Stanton, N. A., Ba-ber, C., Wells, L., Gibson, H., Salmon, P. M. and Jenkins, D. P. (2010). From ethnography to the EAST method: A tractable approach for representing distributed cognition in air traffic control. *Ergonomics*, 53 (2), 184–197.

Walker, G. H., Salmon, P. M., Bedinger, M., Cornelissen, M. and Stanton, N. A. (2017). Quantum ergonomics: Shifting the paradigm of the systems agenda. *Ergonomics*, 60(2), 157–166.

Woods, D. D. and Dekker, S. (2000). Anticipating the effects of technological change: A new era of dynamics for human factors. *Theoretical Issues in Ergonomics Science*, 1(3), 272–282.

Young, M. S., Brookhuis, K. A., Wickens, C. D. and Hancock, P. A. (2015). State of science: Mental workload in ergonomics. *Ergonomics*, 58(1), 1–17.

Index